SETI pioneers

$35.00

DATE			

SETI PIONEERS

SETI PIONEERS

Scientists Talk About Their Search for Extraterrestrial Intelligence

DAVID W. SWIFT

THE UNIVERSITY OF ARIZONA PRESS

TUCSON

The University of Arizona Press

Copyright © 1990

All Rights Reserved

This book was set in Baskerville and Gill Sans

Manufactured in the United States of America

This book is printed on acid-free, archival-quality paper.

Designed by Laury A. Egan

93   92   91   90   89   5   4   3   2   1

Library of Congress Cataloging-in-Publication Data

SETI pioneers : scientists talk about their search for
   extraterrestrial intelligence / David W. Swift.
                    p.   cm.
        Includes bibliographical references.
           ISBN 0-8165-1119-5 (alk. paper)
1. Life on other planets.   2. Interstellar communication.
      3. Scientists—Interviews.   I. Swift, David W.
   QB54.S44   1990   574.999—dc20   89-20214

To Catherine

At age one, she pointed toward
the full Hawaiian moon and
exclaimed, "Light!"

Contents

Preface

The moment when scientists started to search for extraterrestrial intelligence (SETI) was a significant point in Earth's history. For the first time, a group of the world's most qualified researchers sought evidence that we are not alone in the universe. To learn about their search, to preserve historic information not otherwise available, and to discover more about the origins of new scientific fields, I interviewed the first scientists engaged in this fascinating quest.

In the responses presented here, these sixteen pioneers tell about their involvement in the search. Their family backgrounds, their experiences in school, the development of their ideas, the reactions of friends and colleagues, and their speculations about the nature of extraterrestrial life are among the many topics covered in this book.

Since the search for extraterrestrial intelligence began in earnest in the early 1960s and most of the first generation of scientists are still alive, I was concerned that their experiences and perceptions be preserved. This book is the result of that concern. Though the search for alien beings may continue for thousands of years, this first generation and the period are unique.

Essentially all of the principal SETI pioneers were interviewed—not just a sample, as is the more typical procedure. Additional perspective is provided in an epilogue by a leading SETI scientist of the second generation. The interviews were conducted in the early 1980s, some in person, others by phone, and three by another SETI pioneer. The study is worldwide. While most of the scientists interviewed resided in the United States, others were living in Japan, Switzerland and the Soviet Union.

Few other studies have documented the emergence of a new field of science while it was actually happening. Fewer still have presented this emergence from the participant's perspective, rather than as interpreted by someone else.

The list of questions was standardized, but the scientists were encouraged to bring up other points as well. Thus we have both uniform, comparable data and additional observations.

The scientists' own words provide a sense of individuality often missing in histories, biographies and sociological studies. This is *not* a digest of their ideas; it is a verbal portrait, depicting each scientist's style of thinking and talking. The words they used, the length of their

SETI Pioneers. Seated, *from right to left*, John Billingham, Bernard Oliver, Frank Drake, Jill Tarter, and Charles Seeger, with Vera Buescher, administrative assistant, *far left*, at the NASA Ames Research Center, California, 1989.

SETI Pioneers. Iosef Shklovskii, center, and Nikolai Kardashev, right, in 1965 with Russian colleague Gennady Sholomitskii in Moscow after the announcement of strange radio signals from what turned out to be one of the first quasars detected. *Wide World Photos*

remarks, the manner in which they answered questions—and even the questions they chose not to answer—all contribute to an understanding of their personalities. Taken together, the interviews, ranging from concise to discursive, from a few pages to dozens, suggest the variety within this unique group.

I interviewed the scientists about many aspects of their lives, and present here their responses almost verbatim to preserve the character of each person's speech, edited to eliminate repetitions or to smooth the occasional awkward grammar of spontaneous conversation.

This book is not intended to be read straight through at one sitting. It is a series of conversations with SETI leaders, a source book for researchers as well as for general readers who want information about the pioneers as people. It provides vicarious contact with some of the most imaginative scientists of the era.

Acknowledgments

The SETI Pioneers were cooperative, not only in sharing with me their own thoughts but also in helping to contact other key people, clarifying details and reviewing the manuscript.

In addition to the scientists themselves, valuable assistance was provided by a number of other intelligent terrestrials. Vera Buescher tracked down elusive information and informants at NASA Ames Research Center. William Carver helped greatly in refining the manuscript and finding a suitable publisher. Maurice Richter, Fritz Rehbock and Ron Westrum supplied stimulating observations about science as a social process. Jon Lomberg offered insights into the thinking of one of the most controversial pioneers. Dale Cruikshank, himself a pioneer in the study of the outer planets, gave welcome assurances that this was a worthwhile project. Eric Johnson's computer competence and his patience in processing the files are greatly appreciated.

Support of many kinds was provided by Bob Arnold, Ken Burtness, Roger Clark, Mary Connors, Grania Davis, Ben Finney, Louis Friedman, Michael Klimenko, Paul McCarthy, Carol McCord, Thomas McDonough, Patricia Polansky, Robert Retherford, Harlan Smith, Glen Southworth, Peter Sturrock, Paula Szilard and Jurrie Van der Woude.

Lois Swift helped in many ways, including editing, word processing, encouragement and constructive criticism.

The universe is infinitely wide.
Its vastness holds innumerable atoms,
Beyond all count, beyond all possibility of number
Flying along their everlasting ways.
So it must be unthinkable that
Our sky and our round world are precious and unique . . .
Out beyond our world there are, elsewhere,
Other assemblages of matter making other worlds.
Ours is not the only one in air's embrace . . .
You'll never find one single thing,
Completely different from all the rest
Alone, apart, unique,
Sole product, single specimen of its kind . . . .
There are other worlds, more than one race of men,
And many kinds of animals.

Lucretius
70 B.C.

Our civilization is within reach of one of the greatest steps in its evolution: knowledge of the existence, nature and activities of independent civilizations in space. At this instant, through this very document, are perhaps passing radio waves bearing the conversations of distant creatures—conversations that we could record if we but pointed a telescope in the right direction and turned to the proper frequency.

Astronomy Survey Committee
National Academy of Sciences
1972 A.D.

From Fringe to Frontier: An Introduction

Are we alone in the universe, or are other beings out there, somewhere among the stars?

This is a fundamental question, with implications for a vast array of human concerns. Science, religion, politics, philosophy—few topics would not be affected by the answer. It touches such basic issues as the nature of life and our place in the cosmos. Are we unique? Is it possible for a technological civilization to avoid destroying itself? Do we have a future?

The question of extraterrestrial life is not new. It was pondered thousands of years ago by poets and philosophers and has long been a topic for novelists. In our time it is a familiar theme of movies, science fiction and television. For the general public, it is no longer an issue. Surveys indicate that the majority of the population believes that extraterrestrial intelligence does exist.

Now scientists, too, have become involved. Their participation suggests that the possibility of alien life must be taken seriously. It can no longer be dismissed as fantasy, without theoretical basis.

Scientific interest essentially began in 1959, when physicists Philip Morrison and Giuseppe Cocconi published a report suggesting that communication with extraterrestrials was possible, and astronomer Frank Drake used a radio telescope to look for signals. Such activities came to be known as the "search for extraterrestrial intelligence," or SETI.

Only a handful of scientists have engaged in SETI, but its emergence was nevertheless significant. It heralded a new stage in thinking about extraterrestrial life: the transition from an individual concern to a group concern. For the first time, a group of our planet's best informed thinkers gave serious attention to determining whether we are alone in the universe. Through the ages, isolated individuals had wondered about it, but now a number of scientists were working together.

Their entry lifted the ancient question to a new level, giving it respectability, vitality and momentum that one person alone could not have sustained. They discussed ideas, devised search strategies, adopted a title, obtained government funding, developed equipment, and attempted to mobilize support among other scientists and the public. Their efforts eventually led to the establishment of SETI as a research and development section within NASA, thereby gaining not only funding but legitimacy as well.

To fully appreciate these achievements we must be aware of the obstacles confronting the SETI pioneers when they began. Major changes have occurred since that time, obscuring earlier difficulties. By the 1980s the scientific community viewed extraterrestrial intelligence as a possibility worth considering, albeit at a low priority, and space travel, at least to nearby planets, was taken for granted.

When SETI started, however, in the late fifties, the prevailing view of the cosmos was much more conservative, more earth-centered. Most scientists assumed that we were alone in the universe and that space flight was highly improbable. In 1956, for example, the Astronomer Royal of Great Britain declared that the idea of space travel was "utter bilge." Few astronomers were studying the planets in our own solar system; fewer still suggested that other stars might have planets, too. Extraterrestrial life, in any form, was rarely discussed, and intelligent life was dismissed as science fiction.

Why then did SETI start at that particular point in Earth's long history? Technology is an obvious answer. Although people have wondered about extraterrestrial life for thousands of years, we had finally developed the means for actually conducting a search. These included not only equipment—such as spacecraft, radio telescopes and computers—but also knowledge. We had learned enough about the structure of the universe, the laws of nature and the origins of life, to attempt things previously impossible.

But the existence of appropriate technology is not the whole story. Although it tells why the search began at that time, it does not explain why this particular handful of scientists used the technology while many of their colleagues did not.

We must therefore look at these scientists as individuals. What kind of people were they? How did they get involved in such an unorthodox topic, beyond the boundaries of respectable science? Was it because they themselves were fringe people, out of touch with mainstream science, with nothing better to do?

No. The leaders of the search for extraterrestrial intelligence were successful, internationally respected scientists. They had doctorates or the equivalent in the physical or biological sciences and were affiliated with leading research organizations.

Consequently, they had many other options from which to choose, and most of these options were more rewarding than SETI in terms of prestige, research funding and professional power. Furthermore, there was no assurance of obtaining definitive results within a reasonable time, or even a lifetime. The probability of quick contact was assumed to be low, even by SETI advocates. Thus, by usual criteria, searching for extraterrestrial intelligence was one of the *least* rewarding activities open to them. Why, therefore, did they select it?

To find out, the interviews examine their experiences, background and thoughts in order to discover not only *what* they thought about SETI, but also *why* they thought about it at all.

The relevance of these interviews extends beyond SETI to science in general. Modern scientific documents minimize the personal, human side of science, presenting facts about the research itself, but not about the people who conducted it. Consequently, the vast repositories of technical reports reveal few clues about scientists as human beings, or about the actual processes of science.

During the Voyager 2 spacecraft's 1981 encounter with Saturn, geologist Gene Shoemaker observed:

It's not immediately evident, at the surface, how science really works. The classical description in the textbooks is generally wrong. It describes an ideal case that never exists. You must actually get in and see the human and sociological interactions involved in this process. It is quite different from what is usually described in textbooks.

When you're confronted with some new data, or thinking about old problems, it would be interesting to know what the alternatives were that were considered by earlier investigators and the process by which they arrived at the conclusions they did. That doesn't get recorded in the normal scientific literature.

Astronomer Robert Strom expressed similar feelings about earlier scientists:

"I'd like to know them personally, and actually sit down and talk to them, and see what their personalities were like, and what type of questions they asked, because in science the main thing is to ask the right questions. Where their interests lie. Something about their private life, too. How they really viewed what they were doing, how they viewed their level of science at that point, and what they hoped they would gain by what they were doing. These are the things I'd really like to know."

And these are the kinds of issues raised in the interviews, along with the pioneers' thoughts on SETI. But first, a brief review of SETI's background will set the stage for the more personal accounts of the individual pioneers.

Historical Background

Because manned flight beyond our solar system is not yet feasible, the search for alien intelligence has focused on signals. During the nineteenth century various methods were proposed for sending messages into space, with the hope that someone—or something—would reply.

For example, mathematician Karl Friedrich Gauss suggested planting broad bands of forests in Siberia in the shape of a right-angled triangle. Inside the triangle wheat would be planted to provide a uniform color. An elaboration of this basic scheme would have included squares on each side of the triangle, to form the classic illustration of the Pythagorean theorem.

Charles Gros urged the French government to build a gigantic mirror that would reflect sunlight towards Mars. Joseph von Littrow, a Viennese astronomer, is said to have suggested that canals be dug in the Sahara Desert to form geometric figures twenty miles on a side. At night kerosene would be spread on the water and set afire.

By the century's end, however, effort was shifting from sending signals to searching for them. The development of radio provided a new method for interplanetary communication. One of the first to respond was Nikolas Tesla, who made many contributions to the young science of electricity. In 1899 he built a two-hundred-foot transmission tower with which he claimed to have observed "electrical actions" that appeared to be signals. Ruling out explanations such as the Sun, auroras, and earth currents, he reported: "The thought flashed through my mind that the disturbances I had observed might be due to intelligent control. . . . The feeling is constantly growing on me that I had been the first to hear the greeting of one planet to another" (Sullivan 1966:179).

In 1920 another radio pioneer, Gugliemo Marconi, reported that the stations of his company, which were transmitting messages across the Atlantic, had been hearing strange signals that he thought might be coming from another planet. And during the close approach of Mars in 1924, the U.S. Army and Navy ordered their radio stations to avoid unnecessary transmissions and to listen for unusual signals; the army's chief of the code section was standing by to decipher any messages received.

By the early 1930s the suggested use of radio as a means of communication with extraterrestrials extended beyond the solar system to other stars. Articles by E. W. Barnes suggesting this very long-distance application appeared in *Nature* in 1931, and the next year Hiram Percy Maxim wrote in *Scientific American*:

John Kraus adjusting the three-band rotary beam antenna of his station W8JK from the elevated walkway in the 1930s. *By courtesy of John Kraus*

Radio waves represent our first tool with which it may prove possible to carry a signal across the great reaches of astronomical space. . . . If life does exist somewhere else, and it is reasonable to expect that it does, then someday someone is likely to encounter, by means of radio, an extra-terrestrial intelligence. [1932: 201]

These statements, though appearing in respectable journals, did not catch the attention of the scientific community, but during those same years two dramatic new developments in radio began which, in later decades, would make such proposals more credible: radar and radio astronomy.

The British government, concerned that Germany was rearming, perceived that microwaves—radio waves of very short wavelength— could detect the approach of hostile aircraft despite clouds or darkness, so in 1935 the first of a series of secret radio detecting and ranging stations were constructed along the English Channel. During World War II, radar was so effective that tremendous effort was given to its further development. By 1946 signals were being bounced off the moon, and within the next fifteen years radar was used to examine the surface features of Venus, always blocked to visible observation by perpetual cloud cover.

Radio astronomy also developed rapidly during this period. It was born accidentally in 1931 when Karl Jansky of Bell Telephone Laboratory, trying to trace the source of interference with transatlantic radio, discovered radio emissions from the center of the galaxy. Then

Grote Reber, a radio amateur, built a 31-foot antenna in his yard and found the entire Milky Way to be a source of radio emissions.

It was subsequently demonstrated that these emissions were from clouds of hydrogen, shedding energy at a wavelength of 21 centimeters. Fortunately, this wavelength travels freely through space and the Earth's atmosphere, enabling astronomers to examine phenomena such as the spiral arms of our galaxy, which are blocked to visible observation by dust clouds and other barriers.

This stimulated construction of radio telescopes in several nations and set the stage for the landmark report by Cornell physicists Giuseppe Cocconi and Philip Morrison. Writing in *Nature* in 1959, they stated that the 21-cm line provided a ready-made beacon that would be known to all advanced technological civilizations: "the presence of interstellar signals is entirely consistent with all we now know, and . . . if signals are present the means of detecting them is now at hand" (p. 846).

In view of the profound importance the detection of interstellar messages would have, Cocconi and Morrison concluded that "a discriminating search for signals deserves a considerable effort. The probability of success is difficult to estimate, but if we never search, the chance of success is zero" (1959:846).

At the same time, mutually unaware of the others' work, Frank Drake was preparing for just such a search, using the 85-foot radio telescope at the National Radio Astronomy Observatory at Green Bank, West Virginia. Beginning in the spring of 1960, his Project Ozma devoted about four hundred hours to observing the two closest northern hemisphere stars resembling our sun: Tau Ceti and Epsilon Eridani.

The third major event launching the serious consideration of SETI was a meeting held at Green Bank in 1961. Organized by the National Academy of Sciences to assess the possibility of communicating with other worlds, it was attended by eleven distinguished scientists, including Cocconi, Morrison, Drake, Bernard Oliver, Melvin Calvin and Carl Sagan.

The gathering of these eminent scientists would have been important enough in itself. Additionally significant was the role of the National Academy of Sciences. The fact that the NAS not only approved a discussion of ETI but actively organized it indicates that the topic was taken seriously by the officials of this prestigious scientific organization. This did not mean that SETI's struggles were over— research funds would not be obtained for years, and only a handful of scientists devoted any time to SETI—but at least the topic was no longer taboo at the highest level of American science.

The 85-foot radio telescope at Green Bank, as it looked at the time of Frank Drake's Project Ozma. *By courtesy NRAO/AUI*

A second noteworthy aspect of the Green Bank meeting was the formation of an informal group studying extraterrestrial intelligence. Several of the participants had previously communicated with each other about ETI but usually along with other topics. Now for the first time they were convening for the express purpose of examining this specific subject.

A third result of the Green Bank meeting was the Drake Equation for estimating the number of civilizations in the Galaxy capable of communicating with one another. The equation incorporates estimates from many disciplines, including psychology, history, astronomy and biology. Depending on the values selected, the most common result is a million communicating civilizations. While the estimates vary enormously, the Drake Equation continues to stimulate debate.

Shortly after the conference Drake sent the other participants a

One of the world's first parametric amplifiers, used by Frank Drake in Project Ozma. *By courtesy NRAO/AUI*

message of the kind that extraterrestrials might send to us. It was series of pulses and gaps—a binary code, the simplest of all communication systems because it uses only two symbols. Bernard Oliver added refinements to the message and Morrison devised a scheme for coaching extraterrestrials in establishing a television link with us.

These three events—the Cocconi-Morrison report, Drake's Ozma search, and the Green Bank meeting—heralded the birth of SETI as a serious scientific endeavor. Each served a particular function: Cocconi and Morrison provided a theoretical framework and a challenge for testing it. Drake demonstrated that empirical approaches could be made with existing technology. Green Bank signaled the emergence of SETI as a group concern sponsored by the nation's foremost scientific organization.

Not all scientists accepted radio as the most likely mode of interstellar communication. Ronald Bracewell, an electrical engineer at Stanford, suggested that other beings, instead of building beacons, may have launched sophisticated interstellar probes to explore alien solar systems, to monitor events, and to report back any noteworthy findings.

Freeman Dyson of the Institute for Advanced Study at Princeton suggested that advanced civilizations might construct a tremendous sphere around a star to conserve its energy. We might detect such spheres by their infrared radiation. Dyson added that we should also watch for radiating "skid marks" made by alien spaceships as they maneuver in interstellar space.

In 1963 Soviet astronomer Nikolai Kardashev reported unusual radio signals from the astronomical object CTA-102, and suggested that they were of artifical origin. The source was subsequently identified as a quasar.

Kardashev is also known for his classification of civilizations according to their energy outputs. A Type I civilization is capable of using all of the energy falling on its planet from its sun. A Type II civilization could use the total energy output of its sun, and a Type III civilization could use the total energy output of the entire galaxy.

Also in the Soviet Union, Iosef Shklovskii was writing *Intelligent Life in the Universe*. Sagan revised it after it was translated into English. The book, published in 1966, stimulated many scientists to think seriously, for the first time, about SETI, while Walter Sullivan's *We Are Not Alone* was reaching a large general audience.

In 1967 a flurry of interest over the possible detection of intelligent signals was later resolved when it was realized that they were of natural origin. These objects soon came to be known as pulsars.

The following year Vasevelod Troitskii examined twelve stars and in 1970 inaugurated an all-sky search. It used a network of coordinated sites reaching eight thousand kilometers across Asia. The most ambitious project up to that time, it logged over seven hundred hours.

In 1971 John Billingham, at NASA's Ames Research Center in California, organized the first serious engineering study to design a SETI system. The ten-week project, led by Bernard Oliver, envisioned an array of antennas called Cyclops. The system would start with one antenna, then add more as needed, to do a wideband search.

In December 1973 John Kraus and Bob Dixon began a SETI search on the Ohio State University radio telescope. The search was still continuing in the late 1980s. It examined many channels and turned up a tantalizing signal that remains unexplained.

The following year in California, Charles Seeger became the first scientist to devote full-time to SETI. A veteran radio astronomer, he joined Billingham's small team at Ames Research Center. A couple of

The Ohio State University radio telescope "Big Ear," designed and constructed by John Kraus and his students. An office building stands just right of center in the photograph. By courtesy of John Kraus

Computer printout of a still unidentified signal, known as the Wow! Signal, obtained in 1977 by the Ohio State University radio telescope.

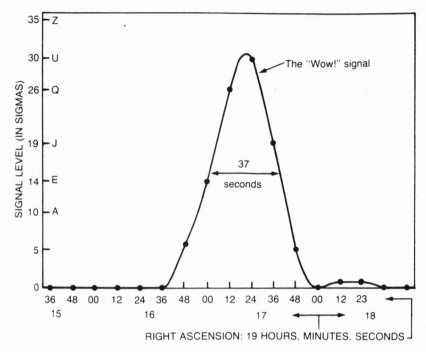

The unidentified signal detected by the Kraus Ohio State University team was thirty times stronger than the background.

years later the team was strengthened further by the addition of Jill Tarter, who had just completed her Ph.D. at the University of California, Berkeley. Tarter devoted 80 percent of her time to SETI, far more than anyone else except Seeger.

The first deliberate message from Earth was sent in 1974 by the staff of the Arecibo Observatory, under the guidance of Drake and Sagan. The message consisted of ones and zeroes in the form of pulses. There were 1679 of these pulses, the product of two prime numbers. The message, containing information about us, our planet, and our place in the solar system, was aimed at the Great Cluster in Hercules, a group of 300,000 stars, 24,000 light years away.

Similar information was also being prepared in visual form, to be carried into space by our unmanned spacecraft. One message devised by Sagan was mounted on a plaque attached to the Pioneer 10 and 11 spacecraft. The later Voyagers, launched in 1977, carried recordings of pictures as well as sounds of Earth.

These messages rekindled controversy as to whether we should proclaim our presence to other, potentially hostile civilizations far

The Arecibo Observatory with its 305-meter radio telescope, the world's largest, at Arecibo, Puerto Rico. The Arecibo Observatory is part of the National Astronomy and Ionosphere Center, which is operated by Cornell University under contract with the National Science Foundation. *NASA Ames Research Center*

more advanced than ourselves. "No," said one school of thought. "It is prudent to remain silent in a jungle." Nevertheless, our commercial radio and television, along with still more powerful radars, have been unintentionally announcing our existence for decades.

By the end of the 1980s there had been at least forty-eight searches for ETI. Only two of these had systematically examined more than a few stars in a few channels. One was Kraus's Ohio State survey.

The other was the work of Harvard physicist Paul Horowitz. In 1981 he designed portable "Suitcase SETI" equipment. From Arecibo he observed two hundred sunlike stars. Moving to Harvard's abandoned 84-foot radio telescope, the project was renamed Sentinel. In its first week of operation, during March 1985, it completed as much searching as all previous efforts, of all nations combined. Expanded to 8.4 million channels in 1985, it became META—Megachannel Extraterrestrial Assay—a full-time, full-sky survey.

After the Green Bank meeting, organized support for SETI in the United States lapsed for at least a decade. The Cyclops report of 1971, for all its grand vision, was merely a summer study sponsored jointly by Stanford University and Ames. Then in 1974 NASA established an Interstellar Communication Study Group under the leadership of John Billingham. The group arranged a series of SETI workshops chaired by Philip Morrison. The report of these workshops (SP 419 SETI) reached a wide audience and was published commercially as a Dover paperback.

Nevertheless, funding for SETI remained minimal until the early 1980s, when two prominent scientific committees endorsed the search. Then the government allocated $1.5 million a year to start a SETI program. While this funding was encouraging, SETI advocates noted that it amounted to less than a penny a year for each American.

By contrast, Soviet scientists and their government supported SETI for over twenty years. Though hampered by less sophisticated equipment, they formed a permanent section on the Search for Cosmic Signals of Artificial Origin within the Soviet Council on Radio Astronomy, and organized national meetings on SETI in 1964 and 1975 and international conferences in 1971 and 1981.

Few other nations paid much attention to SETI. In Japan, for example, physicist Kunitomo Sakurai worked virtually alone, translating into Japanese James Gunn's *The Listeners* and writing his own book on SETI, the first book on the subject to be written originally in Japanese.

At the end of three decades, SETI was at a crossroad. It had the approval of the scientific community. Systematic search plans had been formulated. Improved technology was being developed. Funding, even though minuscule, was forthcoming from government and private sources.

However, other factors worked against these favorable currents. Manmade signals generated for communication and defense were steadily encroaching on the same frequencies most suited for SETI. In a few years these could drown out possible alien signals. Time was running out in the most easily examined part of the radio spectrum. Consequently, the search for extraterrestrial intelligence was threatened by a problem not present in earlier days.

Although no one knows how it will end, the SETI pioneers can tell us how it began.

Philip Morrison

ASTROPHYSICIST

Born November 7, 1915

Somerville, New Jersey

Institute Professor at Massachusetts Institute of Technology, Philip Morrison seeks to explain supernovae, cosmic rays, cosmic x-rays, active galaxies, quasars and cosmology. In 1959 he and Giuseppe Cocconi were the first scientists to call upon the professional community for a coordinated search for extraterrestrial signals. Their influential report in the British journal *Nature* marked the start of the SETI movement.

Morrison has continued to develop the interest of scientists in the search. He contributes frequently to the literature on communication beyond the earth, and has been a leader in conferences on the subject, including Byurakan, USSR 1971; Boston University 1973; numerous NASA symposia; and the IAU in 1980 and 1984.

He received a B.S. from Carnegie Institute of Technology in 1936 and a Ph.D. in theoretical physics in 1940 from the University of California, Berkeley, under the supervision of J. R. Oppenheimer. For the next two years he taught at San Francisco State College and the University of Illinois at Urbana, and from 1942 to 1946 worked on the Manhattan Project. Then he joined the physics faculty at Cornell. Since 1965 he has been at MIT.

A group leader in the design and testing of the first atomic bomb, in 1945 he rode in the back seat of the automobile carrying the bomb's plutonium core from Los Alamos to the desert site of the first test.

After the war he took no part in arms development, and became an active advocate of nuclear arms control. He opposed the development of the hydrogen bomb, and in 1948 was a founder of the Federation of American Scientists, serving as its chairman from 1973 to 1976. He has worked on or contributed to several books on disarmament, including *The Price of Defence*, the first volume to propose a detailed alternative defense posture for the United States.

He also interprets science and technology for the public. He appears frequently on television, has written a physics textbook for non-science students and more than seventy articles for the *New Yorker*, *Saturday Evening Post*, *Newsweek* and other popular magazines. Since 1965 he has been the book editor for *Scientific American*.

He is the author or coauthor of more than eighty scientific articles, including "Are Quasars Giant Pulsars?," *Astrophysical Journal of Letters*, (1969); and "The Radio Rings of Her A," *Nature*, (1988).

Among his awards are the Presidential Award, New York Academy of Sciences, 1980, and the AAAS-Westinghouse Award for the Public Understanding of Science, 1988.

His professional memberships include the American Physical Society, the American Astronomical Society, the International Astrophysical Union, the National Academy of Science, the American Association of Physics Teachers and the American Philosophical Society.

Interviewed June 1981 at Cambridge, Massachusetts

Where did you live during childhood and when you were growing up?

In a residential suburb of Pittsburgh, Pennsylvania.

Tell me about your earliest activities in science.

I was one of those kids who had a lot of interest in scientific things. I remember very early getting a gift of a microscope that didn't work well from a relative who was a physician. And I also had a chemistry set. I was always fooling around with electrical, chemical and mechanical things.

When I was five years old, radio station KDKA announced that it would broadcast the returns of the November 1920 national elections. This was the first time public broadcasting had been done, and there was in Pittsburgh an immediate run on radios. My father bought me a crystal set a week before the broadcast. I was very excited and spent all my time listening to the damned thing.

What about during adolescence? What activities do you recall from those years?

I was caught up in radio from about ten on. By that age I was very serious about it. I had my amateur's license and was building radios myself. Later, when I entered Carnegie Institute of Technology I helped to establish the amateur radio station there, W8NKI. It's still on the air.

In 1936 we radio club members became transient local celebrities. A flood disabled all communications in Pittsburgh, but the college radio station had its own emergency power source, so we stayed on the air and helped coordinate relief operations. The National Guard kept a car downstairs and drove us undergraduates back and forth. We were for a few days one of the main links in emergency relief.

How was your elementary schooling?

I began school late because I had a case of polio, which made it impossible for me to go to school until I was in the second or third grade. The first three or four grades were very good. I went to a small private school, very challenging and interesting. I still remember it very well. When I transferred to sixth grade in public school I didn't think very much of the school, but I was not antagonistic. I enjoyed the other kids and reading and working. I didn't complain a lot.

Did it give you helpful preparation for a scientific career?

Overall it did, yes.

What about secondary school?

It was a very good secondary school generally, but the science there was not good. We kids who were interested in science tended to think we knew more than our teachers did about physics and chemistry, and we were probably right. Yet I thought high school was very good; not because of learning science but because of literature and meeting many new people, and growing up among them.

When did you first think about majoring in science?

It was before I went to grade school. As I said, I had always been interested in science, and I had hoped to become a radio engineer when I went to college. But once I was there I realized I really preferred physics. Essentially I preferred the people I met in physics and the kinds of problems they were involved in.

What was it about the people?

The physicists seemed more imaginative. They asked questions, didn't depend on stereotyping, and they had more attractive goals than engineers.

So it was partly a matter of their goals and interests?

Yes. The engineers were concerned with conventional designs which they could not change very much, only a little bit, whereas the physicists were interested in novelties and in deep understanding. They weren't so constrained by economics and production considerations.

I was more interested in that than in becoming any sort of designer, so I majored in physics.

It was not a difficult decision because there were no jobs anyway; it was in the 1930s, in the middle of the Depression.

When did you first think about the possibility that extraterrestrial intelligence might exist?

It is, of course, a famous idea in science fiction, and I read a lot of science fiction at an early age. I don't remember anything that was particularly salient. I think the best works I read were the novels by H. G. Wells.

So it was reading, primarily H. G. Wells, that suggested the idea of ETI to you?

No, I wouldn't put it that way myself. I think what we did, if we did anything at all in the SETI business, was not to raise any question of possibilities, but to raise a question of empirical tests, which is almost the reverse. That is a more mature way to think about it. It's perfectly true that the imagination of science fiction, or the Bible or anything you want, suggested extraterrestrial life. It's commonplace in the culture.

The little thing we thought of, which came as something novel, was that there might be a way to find out if that was so, rather than passively waiting for strange creatures to appear, like Wells' *War of the Worlds*. What I'm pointing out is that the idea, the concept, of extraterrestrial life is ubiquitous.

Did you discuss this possibility with other people?

Not especially.

During your college years, were there any ideas, books, teachers, classes, other students or events which stimulated you to think about ETI and the possibility of searching or communicating with it?

No, I never thought of it except in the general way that it's in the culture.

What about communicating with, or at least searching for ETI? When did you first think about this possibility?

Well, I was always interested in radio. Maybe that's the connection. But I certainly didn't formulate the idea of radio search for ETI until 1959, a few months before we published our paper in *Nature*. I hadn't done much with radio for ten or fifteen years before that.

Who or what suggested this to you?

During a chamber music performance in the Cornell Student Center I first began to think of the possibilities of gamma-ray astronomy. By the end of 1958 I had published a summary of what might be learned from gamma-ray astronomy. I badly underestimated the experimental difficulties—a common tendency with most theorists—and it was not until fifteen years later that innovative experiments produced results.

However, the paper was challenging, and a few months later, in the spring of 1959, Giuseppe Cocconi came into my office. We were thinking about gamma rays of natural origin when we realized that we knew how to make them, too. We were making lots of them downstairs at the Cornell synchrotron. So Cocconi asked whether they could be used for communication between the stars.

It was plain that they would work, but they weren't very easy to use. My reply was enthusiastic yet cautious. Shouldn't we look through the whole electromagnetic spectrum to find the best wavelength for any such communication? That was the germ of the idea.

So at the time you wrote the earlier report, on gamma rays, you weren't particularly thinking of SETI?

True. It did seem maybe it could be used, in passing, without serious analysis. My first serious thoughts came when Cocconi walked into my office. I had had vague ideas about it just in those few months because of the gamma ray paper, and he brought them to a head.

You, Drake, Sagan and Cocconi were all in Ithaca, at Cornell University. Was this just sheer coincidence that four leaders in this field would be thinking about SETI in the same place?

It's a chronological question to some extent. Drake had already left when we first thought of this thing. He had been a graduate student, but he had already left a year or two earlier. Later he came back to Cornell. Sagan was not to come for years. So we had no special connection with Drake or Sagan. It is quite probable that the interest in the big Arecibo dish was a unifying element behind all of this—the Center for Radio Physics and Space Research. But there was no personal connection.

Starting with the conversation you had with Cocconi, what was your purpose for writing the Nature paper? Whom were you trying to reach?

Well, after we talked about it a bit, we realized that it was rather a good idea. We felt that it should be called to the attention of the community of scientists who could criticize it, and also to the radio astronomers who would have to carry it out in some way. We were not that confident that we knew all about it. We were very amateurish in radio astronomy. So we thought it should be published and criticized and elaborated on by experienced radio astronomers.

It was obviously not an article that could be easily put into an astrophysical journal or something like that because we didn't have results. It was a speculative thing. Therefore it fitted *Nature*, which often prints short, speculative papers.

What sort of response did the radio astronomers give you? Were helpful suggestions forthcoming?

No, they were mostly very antagonistic. Lovell [Sir Bernard, founder of the Jodrell Bank Radio Observatory] was publishing, too. Cocconi wrote to Lovell telling him about it in advance, because he had the biggest dish. I'm not sure Lovell even answered. He had a very negative reaction to the whole thing. But he later said that he was wrong, in his autobiography.

Aside from radio astronomers, what other responses, one way or the other, did the article elicit?

It got a huge newspaper and popular media coverage, which we didn't anticipate. Just at that point, both of us left Cornell. I went away on leave and Cocconi went away for good, essentially. He left the employment of Cornell, and did not return. I didn't get back to Cornell for a whole year.

I met him in Geneva, in July or August, to finish the paper. We sent the paper in from Geneva.

The media kept chasing me because I was going around the world. In every city I visited there would be messages from reporters wanting to talk to me, for six months and more after the publication.

The radio astronomers' response was negative and the public's was enthusiastic and curious?

Yes.

Around the time of the Nature article, what were the responses of your colleagues, not necessarily to the article specifically, but to your interest in SETI?

I think they were sharply divided. Most felt it was not a good idea, probably rather foolish, certainly completely speculative, and hardly worth discussing. On the other hand, a number of people were much intrigued by it. They immediately became sympathizers. It was very popularly regarded by the physics-astronomy community, in the sense that while people were not convinced, a great many people wanted to hear about it and talk about it. I made many talks at many physics departments and various societies. There were huge numbers of invitations, of which I accepted many.

This went on for several years. It continues to this very day, in spite of the fact that the topic has been beaten to death. I still get invited every year to give three or four such talks. I don't accept them all, though I give one or two.

How do your colleagues look upon that now?

You should understand it's not something I spend much time on. It's a small activity. I write some paper every couple of years and I went to ten or fifteen meetings for NASA and spent weeks writing it up for them; but we are talking about twenty years of activity, during which, at the most, SETI would use a few months of my time, if I added it all up. It's not a major part of what I do.

Maybe five percent?

Or even less; maybe two percent. I think nobody in the department would grudge me a few percent of my time. Many of them are not interested in it themselves. We can always get a conversation started about it, even though they don't take it very seriously. About a year ago Frank Drake came and talked about it to the MIT physics colloquium, which I was surprised and pleased to see.

But in general, my colleagues don't think much about it at all. Occasionally they remember when they see or hear some public comment, and they might mention it to me.

What about your family? What do they think about your ETI interests?

Well, they know about it. They know it's a cause of notoriety. They're somewhat amused, I think, and pleased, but it's not a major question for them.

What about your friends?

I suppose the nonscientific ones regard it as a very interesting and challenging question, much more than do the scientists.

What about your superiors, bosses, etc?

I don't know. I don't suppose they think about it much. Once you get to be an elderly professor nobody pays much attention or cares what you do anymore.

What about your chairman at the time of your Nature *article?*

I was an independent worker. Nobody paid us the slightest attention. I was a theorist and I published a paper. As long as I published papers of some consequence, my colleagues couldn't very well complain.

Were you a member of the physics department?

Yes. The Cornell physics department didn't care. In fact, at every university we have professors who do many crazier things than that, and nobody says a word to them. I'm not saying those acts don't affect the attitude of their colleagues. Probably they do, but this was a very minor peccadillo from the point of view of my colleagues. Perhaps I was wasting a little time but it was popular, and once it became popular it made the newspapers.

Take someone like Carl Sagan, who is quite an able scientist and a very capable person. I know—I don't have to say 'I think'—I *know* he has acquired many enemies by his high public visibility. It's got nothing special to do with SETI; it has to do with being prominent in the newspapers.

Is it because other scientists resent his prominence?

Well, you can say that boldly, or you can say more judiciously that some scientists regard what he has done as a superficial kind of bid for fame and fortune. There's no question that quite a few people are antagonistic to him, mainly because of his prominence.

So if he were not prominent, they would not be so irritated by him?

That's right. If he was just doing some little odd thing that he published in a geophysical journal they wouldn't mind at all, but when he's on TV and writes books they find all sorts of fault with him. It's perfectly true. I know; I've sat in meetings where people were deciding to make awards, and their opinion of Sagan was clearly very much colored by this attitude.

Have you yourself encountered any really negative reactions because of your interest in SETI?

It's hard to say. I'm not directly aware of any, but I'm sure that there were quite a few people who didn't approve, especially in my earlier years. It injured my reputation in certain circles but not with any very serious outcome. A younger person would probably have had a lot harder time doing it; we were around forty-five at the time.

It really depends where you work. If someone works beside Frank Drake and comes up with a proposal to do something like this, it's clearly quite all right. But if he goes to some other radio astronomer, the director of another observatory, he might be severely rebuffed, but such a person is probably going to understand that before he makes any approach. The risk exists but I don't think it's apt to be a special problem.

What effect, if any, has your interest in ETI had upon your career as a scientist?

I don't know—not much. It brought public notoriety, but that isn't very important in the scientific world.

Before going on to discuss your ideas about SETI, let's get a little more background information. What was your father's occupation?

My father was a commercial businessman, a small enterpriser. I don't think he ever went to high school; he might have, but I don't think so. I'm sure he didn't graduate from high school.

What about your mother? What formal education did she have?

She graduated from high school and perhaps went to a year or two of college, or something like that. She stayed at home, a rather conventional family pattern in the twenties and thirties.

Did she have occupational skills or work experience other than that of the typical wife and mother of that time?

No, she did not.

How would you describe them?

Friendly, outgoing to a degree. They weren't party-oriented, but they were certainly friendly and warm, and created a nice social atmosphere.

How did they influence your eventual choice of science as a career?

I decided on my own to study science. I think my parents were a little bit disturbed by my decision, but it was such a peculiar historical time, in the Depression, that they didn't feel they could put any great pressure on me to do something else. They hoped that I would go into a profession that seemed more practical, like medicine, law—the professions they knew about. Like many people of their social milieu, they admired professional people, educated people. They had no particular knowledge of or interest in science; it was remote from their point of view.

And what about ETI? How did they influence you in this respect?

They had no idea of it until the papers carried the story. Then they grasped the idea readily.

Were there other people we haven't mentioned who influenced your ideas about science and ETI?

No, but something of a coincidence, which I can't fully explain, is that volume three of Needham's famous *History of Science in China* came out in 1959, and I bought a copy then and read it with great interest. It does allude to this general idea, an idea very old in Buddhism. So it's conceivable that you could trace some connection to that, but I didn't think of that at the time at all; I thought of it only in retrospect.

What religious beliefs did your parents have? I'm asking because of the possibility that religion might play a part in attitudes toward extraterrestrial life.

Well, it might, but I think that it's just one of the permissive routes; it isn't an essential factor. My parents were Jewish. Their beliefs were conventional but not very deep. They belonged to the Jewish community; they went to services infrequently, on special occasions—funerals and high holidays. I went to Sunday school, at first very seriously, then less and less engaged.

Did you have brothers or sisters?

I had a sister, younger than myself. She died rather young.

What about your friends when you were growing up? Did they share your interests in science and ETI?

Yes, in science, but not ETI. ETI's not a category that I can remember being very much interested in when I was at that age. I know if you had asked me, I would've said, "Of, course." That was solely because of science fiction; we knew nothing about it in science.

What were you like during childhood and adolescence?

I had many friends and clubs and so on.

What other professional interests do you have, apart from SETI?

I'm a theoretical physicist and I teach courses in physics and astronomy. I do research with graduate students on a variety of things. Overall I would describe myself as a high energy astrophysical theorist.

What recreations do you enjoy?

Well, I don't really have any highly organized ones, but I would very much like to travel. I enjoy canoeing, and I like music and cooking and film.

What was the most memorable moment or event in your life?

That's a hard question. I suppose the most striking one, that is not very personal, was the sight of the explosion of the first test atomic bomb. It was a tremendous experience.

What has been your greatest satisfaction in life?

It might be the same event. That seems very strange now, but at that time it surely was. These things always change once you reevaluate them.

Let's think back now to some of your SETI related activities; the 1961 meeting at Green Bank, for example. Looking back on that meeting, what stands out in your mind?

Since I've kept in touch with many of those people ever since, the meeting itself is rather blurred. I think the most disappointing part was the experience of the dolphins. At that time we were quite enthusiastic about it because John Lilly came and told us about communication with dolphins. Within a few years the subject had pretty much dissipated, and Lilly's work was not found to be reliable.

In a different vein, that Green Bank meeting gave rise to the Order of the Dolphin. It wasn't an organization in the regular sense; it was only commemorative. Notice came to us at this small meeting in Green Bank that one of the dozen people there, Melvin Calvin, had just been awarded the Nobel Prize. That was very exciting.

He felt happy about that, so he caused to be made these little pins which had silver dolphins on them, which he sent to all of us. It wasn't that we ever had meetings or chose officers of the Order of Dolphin. It was just a souvenir of the particular time we were together. But the members provided a very useful list of a dozen or fifteen other people interested in extraterrestrial intelligence. They had been co-authors of papers or had been on symposiums.

The prime mover for setting up that meeting was J.P.T. Pearman?

Yes.

It was his initiative in calling the meeting?

I can't say it was done at his formal initiative; somebody else may have suggested it to him. There was no question that a meeting was called for, and many people were interested in getting such a meeting going. I suppose the proposal came from the Space Science Board, of which he was a staff person. I don't know which members of the Space Science Board supported the project, but I would guess that their minutes would show that.

You were chairman of a NASA SETI advisory committee. Could you comment on this?

The people who tried to organize SETI within NASA wanted a study made, because in those days NASA was seeking another set of

goals for what to do after the Moon. It still is, in fact; but in those
days they made quite a number of exhaustive investigations for the
long run—colonies in space, shuttles, and other things. One study was
at Ames, because of the enthusiasm there around Barney Oliver and
John Billingham. They had conducted the famous Ames summer
study which gave rise to Cyclops.

I went out there and talked to those people for the first time around
1969, when NASA still had all kinds of money. I worked with them to
seek a decision in favor of making a study of this kind. Then they,
John Billingham and his associates at Ames, went back and decided
how to set it up. They compiled a small list of people who might be
chairman. They had me on the list and called me. I accepted.

What do you recall about your chair duties or activities?

Primarily it was an educational activity, to make an official study so
nobody could ignore all the possibilities we signaled. One important
issue was to try to make sure any opposition was fairly confronted and
incorporated in the right way. Secondly, we hoped very much we
would be able to leave a residue of small continuing practical activity
within the NASA structure. So we organized meetings and got all the
people together and listened to what experts had to say.

It was quite interesting, going into the details and discussing them
and finally writing a document and getting everyone to agree to it. I
was rather pleased because many of these people were, to begin with,
unenthusiastic, to say the least. Many were actually rather opposed
to the idea but had enough open-mindedness to say, "Let's listen
to it," to see if the idea had some merit. That's probably why most
were there.

Is this spelled out in the NASA volume on SETI?

I don't think it talks about meeting opposition very much. That's
the sort of thing you don't say or put in quotes. From the names of
the people attending, you could see that before this time those people
had been not at all involved, or even that they had been antagonistic;
now we had a small new wave of support.

Would you care to mention who was antagonistic?

Jesse Greenstein has said so. He's the most interesting. He's a senior
optical astronomer at Cal Tech, Palomar Observatory. Radio astronomer Bernie Burke is a less clear-cut case.

Originally opposed or skeptical?

Yes. Now he is somewhat less skeptical; at least he is tolerant. But
now, since publication of the NASA document three or four years

ago, there has arisen a very powerful public opposition which I am somewhat unhappy about, but there's not much to do about it.

The most interesting name is Michael Hart. He in particular, and quite a few other people, present a curious argument, which I call neoMalthusian. It claims that human society—or other intelligent life—must expand indefinitely, filling the galaxy within a geological epoch or less.

But since there are no signs that this has happened, the neo-Malthusians say there isn't anyone else out there at all. Since we don't see people visiting us all the time, we must be the very first to arise because, they assume, it must in time become very easy to travel in space; it must be easy to modify the universe, to turn it into a park.

These SETI opponents don't like to think that it might be difficult to do those things. So, if you take that point of view, and note that the universe has not obviously been colonized or modified, the only explanation is that we are the first, the only ones. There is nobody, or hardly anyone else out there; therefore it is not worth searching for them.

How would you assess the impact of this view on SETI?

Oh, it's powerful. I think it contributed to discouraging congressional support of SETI, but I don't think this was intentional. I think that most of that group, if asked, would say we should still try, but the effect of their skepticism, which comes strangely enough from an even more optimistic view of history . . .

You're saying that they may not intend to block SETI?

No, I don't think they intend to block it.

Their own honest skepticism is being picked up by the legislators?

I don't think it's *skepticism*. I think theirs is rather an honest *enthusiasm*, really not skepticism at all. They're on the opposite side from the skeptical congressmen, but they reach the same conclusions because they feel "Yes, of course there'll be all kinds of space travel and colonies everywhere. It's just a question of time; 200 years or 500 years or 100,000 years, before occupying the stars." Since that's *not* the case, the only reason must be that we are alone.

Curious paradoxical reasoning. My position is that it's not that easy. They overestimate the ease of space travel, and the danger of pollution of the universe with probes. In my opinion it's exactly neo-Malthusianism: it's exactly a parallel argument.

They believe the opposite from the congressmen?

Yes. The congressmen are rather Aristotelean while these people are neoMalthusian. Aristoteleans assume that we are truly the center of the universe. Earth is God's little blue footstool, and no other place can harbor our intelligence. So let's clear up the Earth before we find out anything about other worlds.

This is opposite to the belief of the Malthusians, who assume there *could* be civilizations elsewhere, much more powerful than we are, yet we see no signs of these very powerful beings. Therefore it must be because they don't yet exist at all.

There's a rather well done book on the subject called *Are We Alone?*, by two people at Maryland. I don't know Hart very well but I think that, reading between the lines, he is perhaps moved also in part by a religious concern for demonstrating that we are the only self-conscious beings created.

There was another piece from this perspective, by Freeman Dyson, some years ago. He was the first one to publish it, though it's an old idea. He spoke about looking for a park. If people are out there, the Galaxy would slowly turn into a park. But, it's not a park; it's still wild. Since it isn't a park, it means there's nobody yet powerful enough to turn it into a park, to control it, to design it. Since that must happen in a rather short time, he argues, we must be the very first.

A third view is the one I hold, the Copernican view that we are typical in the universe. Why should this be the only place where there is life? There may be many worlds broadly like ours, but not exactly like it. We are finite beings with finite capabilities, growing but always limited, by no means sure either of becoming universal or of remaining alone. Perhaps we can find the others, and that is what we ought to do.

You have suggested that one of the deterrents to SETI is that it might take a long time, so that any particular person involved in it would have little chance of success?

Yes, that's a major problem in organizing a long systematic search. We believe we have solved that problem as far as it can be solved, by a straightforward technique, also well known: we would not ask any single scientist to spend the bulk of his time doing the search, because no single person is likely to succeed. If we recruited such a person it would be an imposition to require a lifelong duty.

What we must do instead is ask many people to each give a small part of their time. Then everybody is doing some service for the general welfare, for this long-range exploration. Yet nobody suffers from it, because the opportunity lost by not doing something else is only small.

We'd like to compensate even that small lost-opportunity cost by making the equipment useful for this purpose available to such people for other purposes during the time when it is not being used for SETI. That would be very advantageous.

That's the way we would like to organize a SETI search: not to build a dedicated dish or anything of that sort for the moment, but to lend the equipment or give it to a number of people who have dishes, in exchange for a promise to devote some of their time to SETI.

Depending on the enthusiasm, we could gradually build this up until finally we could even imagine building a number of dishes. But even then I would not dedicate any big dish solely to this purpose. I would just staff that dish with people who would do many other things, but would use some of their time for this. Nobody suffers, and the cost is not prohibitive.

We have to do a good, thorough search. Everyone agrees we must have an extensive survey over many directions, many frequencies, many intensities, many times, and so on. You can't expect to guess the right answer right away. Yes, there are certain criteria that would improve your chances. But even so, everyone agrees it must be done right, looking through various parameters to make the search.

Imagine what it would be like if all you were doing was adding one negative result after another, day after day, month after month, year after year; perhaps decade after decade. Perhaps one great day you succeed. That's wonderful, but what if you do not succeed? You can't guarantee that you'll have any success at all. So you've spent a whole lifetime just working hard at the radio telescopes. You've made no contributions to science, found nothing, nobody knows your name or your work, and you find that you've done nothing except carry out the overall plan; it's a great loss of a unique lifetime. It won't come back to you!

I'm afraid the only people who would do such things are people who don't care very much for achievement, who are willing to immerse themselves routinely in this long-range goal. There are very few of those people, and I'm not sure they're the best people to carry out the search. Their work would become stereotyped, routine. They would not be alert to novelties and changes and nuances as would a person who is very lively and hasn't done very much of it. He is fresher, more likely to be aware of novelties and nuances. Every year he or she has done something on SETI but remains untired. Such an astronomer has made all kinds of other results and discoveries, used new techniques and watched the science grow.

One science-fiction writer imagined a station in New Mexico which dedicated year after year, decade after decade, a hundred years, to this one activity. It's very hard to imagine that being done, because what will those people do after twenty-six years? They'll be more interested in the bowling club, in early retirement and pensions. They won't be interested anymore in the search. It's very hard to keep that going forever.

But if there were dozens of places, each one spending only one week a year, with ten able people who are doing something else fifty weeks of the year, the search would go on very, very well. They would argue with each other, compete with each other. The whole thing would be very much better under that scheme, yet it would have the same overall size and cost.

They should have a plan. They should meet every year or three to steer their plan: "I'm going to try this new technique. You try that one." They might exchange places, and so on. In five years' time they might say, "that was not a good idea; we can do it better this way." The search would become a living science; not simply a predetermined activity which is just looking through a haystack for a needle.

It *is* a needle in a haystack. We can't condone people spending all their lives searching for it. Once the needle is found it would be a great success, but no one person has much chance of finding that needle. The day he retires they hire a young woman who finds the needle the next day! That's a sad state of affairs!

Thinking that through—not the first time in astronomy that this has occurred, by the way—we feel that the right way to do it is not to place the burden on a small set of persons but to try to spread it around.

The situation is similar with respect to supernova; not quite the same, but similar. A supernova will occur in our galaxy on the average (we don't know the numbers terribly well) about once in a hundred years; it might be 30 or 40 years, it might be 200 years—we don't know. I won't mislead you if I say about a hundred years. When that supernova occurs in our galaxy it will immediately answer many riddles about supernovas which we can't answer when they occur only distantly, because we don't have enough sensitivity.

On the other hand, to answer those questions very well, it would be admirable if we were ready for it, so the first hour that it comes we would be able to measure it. At least on the first day! Absolutely the first week, because the most important signals would come at the very beginning. But we *won't* in fact be ready for it because nobody is prepared.*

The right thing to do would be get people to prepare. It doesn't

*Six years later, Supernova 1987A occurred.

cost very much. Make sure the apparatus is ready to go at a moment's notice. Hit the alarm, and up the rocket goes.

That would be wonderful, but nobody would do it because it's too uncertain. Nobody can plan for a thing like that which isn't going to happen for 30 years or 50 years or 100 years. Who would do it? It's very hard to get anyone to pay for it or to invest the time in it when you won't see the fruit of it at all.

It's just not the time in the history of the world when we have the kind of social consciousness which would enable people to spend time on something they cannot see maturing. Planting an oak tree is much the same thing, but at least people have planted oak trees in the past, so we have some feeling of obligation.

People say, "You should turn it over to some organization." I think that's probably true. The best organizations are voluntary ones, like a religious community. A monastery can do this very well, dedicated for centuries to specific tasks. Government organizations, like the ones that measure the weather or the magnetic field, do just that. But they need other, more changing activities along with it or their systems become stagnant. It's a serious problem.

It's hard to carry out long-range plans, especially in science, because the ideas and apparatus change very much during a hundred years. Whatever you prepare now is not going to look useful a century from now. That's why you have to plan in a rational way.

In order to make it work it has to become a social effort. It can't be one individual person. We've done about all that individuals can do. Paul Horowitz and Frank Drake twenty years apart each did all that a single person could. Now it takes money, at least to build equipment, and money involves a number of people supporting the project. So it isn't just one quixotic person being denied. Unless you can convince the people to dispose of some resources, you're not going to get resources to spend. We are just beginning to get some. Frank Drake carried out the most unlikely and important search of all, the very first one, with very limited means. His boss agreed, of course; otherwise he couldn't have done it.

Your Nature *article is such a landmark that we ought to preserve as much information as we can about the factors associated with it.*

In this paper we asked ourselves a question: what is the best way to communicate across space? This led us to a search proposition we then considered. Could we travel? No, it seems too slow, too expensive. Could we use neutrinos? That was too hard to do. What about electromagnetic radiation? Well, that is possible. Which are the possibilities? Visible light, for example, is too common in the universe, so artificial signals could not easily be distinguished from natural star-light. Even if we set off a thousand hydrogen bombs, the explosion would not be noticed from nearby stars a hundred light years away, because our sun's light and natural flare variations are too great.

Nature does not know very well how to make microwaves in the galaxy, so the Sun's radio signals are nothing compared to even little Arecibo in one narrow concentrated beam and band.

We went systematically through all the possibilities: after several weeks of reading, discussion, and calculations we came to the conclusion that microwave radio was the best, particularly at the 21 centimeter wavelength.

After you completed the article was there any difficulty getting it published?

No. I sent it to *Nature* in London by way of Professor Blackett, an imaginative and influential physicist who was then at Imperial College. I knew him personally, and requested his good offices in having the article accepted. He acted promptly and it soon appeared.

Was there anything else associated with the article that we should note?

Well, I've mentioned the response from the media as I traveled around the world in the following months. I went to Moscow, London, Rome, Israel, India, and Japan, and in each place reporters asked me the same questions over again. The world became caught up in SETI. That was the most conspicuous thing I remember from 1959 and 1960.

The next thing, in 1961, was our contact with the dolphin people. It took a year or two to realize that there was less there than met the eye.

Next, and quite exciting, was the possibility of life on Mars. I'm sure that the two were related notions. I was very enthusiastic about Viking. I think it was given a good deal of push by the very same idea that we got started. Then there was the actual landing on Mars, which pretty well demonstrated that no advanced life was there. It was a disappointment, a setback.

What was your connection with the Viking program?

I was on the Space Science Board for a few years. Occasionally, we discussed various Viking issues, but I didn't have any direct part to play. In the very beginning I was involved in some of the preliminary planning sessions. Later on I went down to the Viking launch and talked to the people there, and I was invited out to California when Viking landed. I'm a friend of Viking, but not a direct participant.

What about chairing the SETI meetings?

Well, that had some air of drama and conflict about it. I learned a good deal that I didn't know, especially about radio. I was pleased by the way some of the conventional astronomers came around to the idea. I became persuaded that detection of external solar systems was right around the corner. I still think it is!

I'm a great supporter of plans to get the right equipment. That was mostly because of George Gatewood. His stuff and Paul Horowitz's and Despain's and the other Berkeley-Stanford computer science people have brought the main novelty to this whole business, because of the great power of digital multi-channel analysis today.

A very useful service was performed by a student of Frank Drake's. His name is George Helou. He made a very nice calculation, which Frank stimulated, looking at the narrowest bandwidth which you can use in the Galaxy. He showed that there is a lower limit, which was about what one expected, but he did a better job than anybody else had done—pretty persuasive.

Squeezing down the bandwidth to its narrowest limit is probably the best idea for the search. I still favor a very large number of channels for the frequency search, which is exactly what Horowitz did. That's why I was encouraged by Horowitz. We never talked over the matter, but he read the papers and thought about it and came up with the same ideas, which seem very close to the idea we had in 1959, much improved by the power of new technique.

Another interesting activity you were associated with, involving extraterrestrial intelligence, was the Voyager recording.

The initiative was entirely taken by Carl and Frank, as far as I know. I took no initiative. They simply called me up and said they're going to put this record on the spacecraft and asked me if it was a good idea. I said, "Yes, it is." They asked, "What should we put on it?" So we had a discussion, mostly over the phone. I don't remember having an actual meeting about it; we might have, but I don't think so.

I just proposed a few images and the reasons for them. I urged that they get all the continents in and have various kinds of people represented. They agreed. It wasn't anything novel.

I remember one specific suggestion, which is the drawing from Newton's book, *The Principia*, showing a satellite in orbit. That was the first sign of the idea of putting things into gravitational orbits, which goes back to the very first substantial book on gravitational theory. It did appear on the record.

I think there were three or four other ideas, some of which were not included on the record—they couldn't put everything on it. They got several other friends to give them ideas. My role was quite peripheral.

It's a pity that the record can't be circulated. I knew it was going to be difficult, but I didn't know the problem was going to be so irrevocable. It's caught between the record industry, which is the marketplace apotheosized, and the government, which can't deal with the market. Between these two it's impossible to get that record circulated.

You mean a commercial recording which people could buy?

Yes. We originally thought it might sell a million copies, bringing millions of dollars either for the government or for some foundation to do research. It would be a very good thing. It would interest and excite a lot of people. Many children would hear it. But it can't be done.

Why not?

No record company will circulate it without exclusive rights, and the government can't give exclusive rights. At least, that's the way it appears to the two parties. I think wiser heads could intervene but no one would do it. So after several years they gave up.

A generous and wealthy man in Texas bought 10,000 copies of the record and had one sent to each high school in the USA. That was the closest to mass distribution that's ever occurred; he paid for it himself. The government didn't object to giving them away free. There was no record company to complain, because it wasn't on the market. That was the only way it could be done.

What stages has your thinking about ETI and SETI gone through? Have your ideas changed much since you first began thinking about it or have they remained pretty much the same?

Relatively the same. It's a curious case. So little has been done that I don't think much has changed. That's somewhat disappointing but I think it's true. There have been two developments in other, related areas. One is the development of electronic digital computers. It has transformed the whole search situation, making possible much more action than was possible before. The other change is the maturing of our astronomical knowledge, which hasn't really affected the basic ideas very much.

But what is remarkable is that there has been so little change in SETI itself since 1960. The Arecibo dish, which wasn't quite built then, was going to be the biggest dish; it still is the biggest dish. The twenty-one centimeter line still seems like a good bet, perhaps not so uniquely good as I originally thought. I still think it's the best, but I'm willing to admit there are other candidates.

We could have done more, but essentially the space business was cut off in 1972 or thereabouts, and for the last ten years we've gone less and less vigorously into space. Were NASA still as rich as it was in '70, we might have accomplished more.

Nevertheless, I do think I was right—I've always taken this view— that I would not undertake a crash campaign to convince people they should buttonhole everybody in Congress and say how important SETI is: to save the world, and that sort of thing. Maybe doing more of that might have started SETI going more easily, but in the end it would not have survived very long. My judgment is that it is better to work for the long term than for the short term.

I'm not trying to convince congressmen to do something they don't want to do. If they don't feel that SETI is worth funding, then please don't do it. They have to become convinced that it has worthwhile, important consequences. Once they are convinced, they'll do it. Until then I'll just point this out to them. I think that is the better approach.

What alternatives, what other possibilities about ETI have you considered?

I've tried systematically to consider all possible alternatives, but I always come back to the same conclusion: that the right thing is to look in the microwave region, and probably near the frequencies we first said. They still seem advantageous.

What would you do differently if you were starting over again to study ETI?

I don't know. I think I would have been wiser to anticipate the public reaction, but I'm not sure it would have made much difference.

How would you characterize the public reaction?

I think the public reaction, which varies a lot from what the newspapers are saying, is still relatively unsophisticated. It has to do with the works of imagination, like film, such as *E.T.*, and any number of things in the past. Most people in the press and so on have not grasped the central new idea, which is that there is an empirical road to gaining that knowledge. It doesn't matter what your view is on how extraterrestrials look or where they live—that is all more or less irrelevant.

That degree of abstraction is very hard for the public and the press to gain. If I were doing it again, I would've written a long piece on that to begin with. I don't know whether it would have made much difference but it seems a sensible thing to do.

Why is ETI important to you? Of all the thousands of topics a scientist might investigate, why did you choose ETI?

If you formulate the question that way, it's very hard for me to answer. I don't think that's a fair description at all of what the situation was like. It wasn't that I *chose* it; it wasn't that I spent five years going to school studying ETI and then I said, "Now I'll write a little paper on it."

It was an idea that came up in the course of a five minute discussion. We thought up the idea, which then raised a question in our minds, and we said, "Let's find out the answer to that question." And that is the answer we proposed. Since then most of the work has been refining the answer, in a practical, real, concrete, and not a theoretical way.

So it didn't come from studying ETI or any large-scale question like that. It came out of logical analysis of the electromagnetic spectrum, and the simple statement that radio waves offer a means of communicating across space. Once we realized that, we said, "Look, this is very interesting because this ancient subject of the imagination has now been brought into contact with experimental science." That's the way it happened.

What does SETI mean to you today?

I have a long connection with it, so I have certain obligations to try to have it prosper as much as possible. I try to keep alert and to read the literature. I get volunteer manuscripts all the time—people thinking about it; sometimes good, sometimes bad.

At the moment my main feeling is that I am somewhat disturbed by the tendency to say it can't be done, but I don't feel that there is an easy way to intervene. It began very long ago: the idea that if people haven't come here, it must be because we are the first. But I think that's very naive. I don't know if Freeman Dyson agrees with me, but I hope he's modified his ideas to some extent.

What has been the most difficult aspect of your ETI work?

For me there really hasn't been anything difficult because I have not spent a great deal of time or labor on it. I've only had good results, meeting interesting people and going to fine meetings and writing. It has been a pleasure and an opportunity rather than anything else. The way it's difficult is that after twenty years, only within the last half year has any systematic program been undertaken. So it's been a disappointment rather than a difficulty. Perhaps they are related.

How do you think most scientists today view SETI?

I think most would say it's legitimate, but not important.

What about public attitudes toward SETI?

I think much of the public actually believes that the chief purpose of radio astronomy is SETI!

What is your guess about the probability that ETI exists?

In this very room I must have made a dozen television programs about it. I said last week to CBS, Walter Cronkite's people, that it isn't the right question to ask me or anybody else, "What is the probability of something being out there?"

That turns the idea around and distorts it. The question is really, "Should we do something to find out?" If you ask wise men or scholars from now til doomsday, you'll never find out. To find out you must do something empirical.

So we are trying to say, "Here is our method which still, after twenty years of criticism, looks as though it is a viable method of communication across space. Should we not take that hint and do something about it?" Just sitting and talking is not doing anything, so I don't think they should ask how probable is it, or why does Hart claim there isn't anybody else out there and I believe there is?

I think Hart would say—any judicious person would say—that his case is not proved any more than my case is proved. Both of us are trying to reason out plausible cases. But we can go beyond making plausible cases; we can try an empirical technique to find out.

Realizing that it is only a guess, when do you think contact with ETI will be made?

I don't really know, of course, but I've always said it surely will take decades of systematic searching, and maybe a century. It depends on how hard you search.

What might ETI be like? What form might it take?

I always decline that question because that's usually what people want to hear but it's exactly what we *don't* know, and that really disconnects it from science. It's in the domain of open speculation.

What scientists or other people in the past do you admire?

I've been lucky enough to know quite a few of the recent physicists of the highest sophistication. I admired them very much, knowing them as individuals: Niels Bohr and Robert Oppenheimer, to name two of the well known. I admire Einstein intensely.

Were there any in past centuries you admire?

Certainly, but there's nothing very novel about it: Rutherford, Maxwell, Faraday, Newton—all the great names of physics we deal with every day. I'm also very fond of one less usual person, just because knowledge of him came to me when I was a small school child. That is Francesco Redi of Florence, maybe around 1700. He did the simple, wonderful experiments that established the fact that life did not appear spontaneously in rotting materials.

Who are the outstanding people in SETI today?

Frank Drake, Otto Struve, and Cocconi and I initiated the modern interest in SETI. Our two groups—the experimenters on one side and the theorists on the other—didn't know about each other, but both were doing the same thing at the same time. I think our *Nature* paper was the initial modern publication. Of course, the idea had been talked about before, but it had been dropped from the literature for decades. Then we—both groups, both pairs—saw the connection with radio astronomy, and called people's attention to the rationality of working in that direction.

What role have these people played in SETI?

Giuseppe and I published a theoretical paper which called wide attention to SETI, and it's useful because it does exactly that; it plants the seed. It alerts the profession to it. But Frank's contribution is more impressive because he really did something. He did it himself, instead of only publishing an idea. Theorists don't have to do any experiments; they can publish mere ideas in advance. The two modes are complementary; they happened at the same time.

How about Sagan? How would you characterize his role in SETI?

I think it's very effective, especially in the public domain, the public eye. I don't think he's actually done very much directly bearing on the technical problems. In our NASA committee, for example, he often could not come because his time was so devoted to these other matters. There's no doubt that he's had an impressive impact on the public about the whole question. Some people fault him for his enthusiasm, which I don't do because he is absolutely sincere. He really believes what he says. Some people feel that he's too enthusiastic about it.

I must say I think Frank Drake is an enthusiast, too. I would not have written that article about eternal beings that you might contact by SETI in the way Frank did. It's possible but it's very tendentious; in a way it appears to be special pleading. I think it was a small mistake to do it that way, just as I think Cyclops was a little bit of a mistake.

Why do you think Cyclops was a mistake?

In its time, it was fine, but one might have looked ahead. It gave everyone the feeling that *any* SETI search was a ten-billion dollar search, a feeling very hard to dispel. Most people not acquainted with it still think that is the case.

What about Calvin's role?

It was very helpful in the beginning, especially because of his obvious sympathy with the problem. He is much concerned with the origin of life, so it was easy to transfer that. Since he is such an admired and well-connected scientist it was a great help to have him associated with SETI.

Do you feel that he gave some credibility to SETI?

He certainly did. One nice feature of SETI is that intellectually it involves every branch of science. As people have often said, the famous Drake Equation consists of factors taken from half-a-dozen different branches of human knowledge. Its interdisciplinary nature was very well brought out by Calvin's early sympathy and association with us. The fact that he was so prominent and won the Nobel Prize the very year he did this was certainly a good thing for getting us started.

What about Bernard Oliver? How would you characterize him?

Barney Oliver is one of the most devoted and energetic persons ever connected with the field—and again, a very important person. I think he had the whole idea independently. He was enthused about the fact that Drake was doing something. He soon came to visit Drake. He's kept Ames interested and going all this time. If he were not out there, interest might have died out at Ames.

Su Shu Huang [astrophysicist at UC Berkeley and NASA]. What part did he play?

Little or none, as far as I know. I read his papers; I didn't know him. He was one of those people who calculate how likely stars are to have planets, and so on. The consequences are technical.

And Otto Struve? [astrophysicist; director of the Green Bank Radio Observatory.]

Struve was then Drake's boss. His sympathy made the whole thing possible for Drake. I didn't know him.

And Cocconi?

Whatever I said about myself, the same has to apply to Cocconi. For the first year anyhow; then he dropped out of it. But everything I did, Cocconi did also. We were full partners in writing that paper.

There are some more people: Charles Seeger at Ames.

All the people at Ames—Seeger, Billingham, Wolfe, and a few others who were in the position of scientists in civil service, carrying out their tasks at Ames—they are very important. Eventually, that is what will make it work; when there are people who really are doing it in a professional fashion. So their contribution is the most valuable.

And Bracewell?

Bracewell is another independent enthusiast. He's a radio astronomer who was converted, to some degree, to the possibility of SETI. For a while he took the view it was not so likely. Finally, he came to see it as more likely. He wrote a good book on it. He's a very original person.

There's no question that the coincidence that Oliver and Bracewell and Ames Research Center are all out there within twenty or thirty miles is another important element—more important than the rather minor Cornell connection between Drake, Cocconi, and me, because ours was not a working connection. But it was a working connection among those Peninsula people; they reinforced each other a lot.

When looking at the early days of SETI, let's ask ourselves what we would like to have known about explorers and scientists of the past, but which we probably never will know because that information was not preserved for us. Pick out any people who appeal to you—Columbus, Galileo, Copernicus, etc.—Ask yourself what you would like to know about them, and then tell me what I should record now so that future generations will have similar information about you and the early days of SETI.

That's a very nice question, when you think about Columbus and Copernicus and try to see what was the nub of the whole thing. I think what we're seeing is the difficulty of innovation. What we really ought to know about are all the people like Columbus who had his plan but didn't get it financed, didn't send the three ships to the Indies. I'd like much more to know about them. I know about Columbus. His venture went on. So did Magellan's. So did Galileo's, and so on.

What you really want to ask is: what is our knowledge of the many others, or were there many others? What happened to them? Why did it not work in their hands, and why did it work in Columbus's hands? SETI has not worked in our time. It hasn't properly begun yet, so we're in the position of these people we've forgotten about—the predecessors of Columbus, who *didn't* get to India.

So you're likening yourself not to Columbus but to the non-Columbuses?

That's right.

Not to Galileo but to the non-Galileos?

Exactly; that's much more parallel. If SETI starts going, then there's no problem. But the point is it hasn't quite caught fire. It hasn't accumulated a coterie of strong supporters all over the world. There is probably more popular interest than scientific interest. It's just short of being enough to make it work.

As a non-Galileo, what do you think people in the future should know about you and your involvement with SETI?

If it doesn't start they won't know anything about it. If it does start, they'll know enough. It's that kind of thing.

But your article is established in the literature and people will be referring to it.

No, not if a really successful venture occurs. I think what will happen is that SETI will start in Timbuktu or Moscow or Green Bank or Harvard Forest or whatever, and will go on for N years and then—let us be optimistic and say it succeeds—that success will be recorded very well. The fact that there was a little pioneering going on before might be known by historians but really wouldn't make any difference, just as we don't know whether there was somebody else before Columbus. There probably was!

The interesting thing would be to try to find out about other people who had this idea and did not send it off to *Nature*. One is Frank Drake. While Frank and Otto Struve and Giuseppi and I probably had some vague common origin in the interest at Cornell in the big dish, it wasn't any more than that. I might have seen Frank, but I never spoke to him. He says he didn't have me for a class but his roommate did, and so he knew about me but I didn't know about him. We were all at Cornell and the atmosphere was similar for all, but Frank did not publish.

However, he was at a radio telescope. He had the approval of the director, did make a receiver (of course for other reasons; he had to make a receiver anyhow for the twenty-one centimeter line), he did listen to a number of stars for ten days when they were just getting started. I'll bet you that there were many other people who did the same thing, but didn't get quite that far. They are interesting, too.

As for Cocconi and me, we didn't understand radio very well. But we learned, once we asked the question, that technology had progressed to the point at which we could communicate freely across interstellar space. That was the big surprise for us. If we hadn't found that out, I don't think *Nature* should have accepted our paper. I'm not even sure we would have sent it in because it would have lacked any clear result.

Suppose that the Arecibo dish had been unthinkable—way beyond our persent capabilities, possible only in the far speculative future. I don't think *Nature* could publish a paper on that basis. It wouldn't be so crisp. It would have involved *two* speculations about our technical capacity, as well as about ETI. The surprise and the revelation of our paper was that we humans had already achieved, or potentially achieved, the technological capability.

SETI was the inevitable consequence of the development of radio astronomy. Once it became plausible it was picked up. That happened about 1960, when the Arecibo dish and good narrowband receivers were built. Perhaps five years before, it could be seen coming, but ten years, no.

How curious it is, how paradoxical, that having reached the stage at which we now could search quite easily, that advanced and imaginative scientific minds say "No, it is not worth doing because if anyone was out there they would already have done it."

What worries me most is that, without any empirical test, SETI might be abandoned on the grounds that it must not be possible!

I think that won't happen now. Carl organized a scientists' declaration on that very point, helpful and timely.

What are UFOs?

I believe they are simply mistaken interpretations of many different phenomena, by many different people in many circumstances. Some of them are perhaps even self-seeking deliberate inventions, but most of them are unusual sights which are not easy to interpret. The trouble is the rubric has gathered everything together into one. There are certainly lots of interesting things up there, but I think none of them are disc-shaped craft flown by green fellows from other planets.

What should a scientist do upon receiving convincing evidence of extraterrestrial intelligence?

The main thing you have to do is make it available for confirmation by others, by giving the particulars which enable others to reproduce the results.

What steps might that involve?

I suspect that first would be consultation with his colleagues and friends and so on, to make sure that he is not completely misled.

People have been misled in the past. They thought they had something and they were just wrong. That's what he'd have to do to satisfy himself and a party of people who were competent in the details, but who were not too closely related to hoax successfully. Once he's beyond that I think he should publish it in an appropriate way. I can't give criteria of what the appropriate way is, but there are many general announcements one could make.

Would this first be in scientific journals and then public media: magazines, newspapers?

The nature of the thing is such that if you sent it to any scientific journal it's going to leak out to the rest of the world very soon, so I don't think it matters what he does to begin with. It depends more on his circumstances. You do one thing if you're off in the Sahara or Timbuktu, and quite another if you're in New York City.

What would be the public's response to this information, about the existence of ETI?

That's not an easy question to answer. It's certainly not a question radio astronomers are in a position to answer. We have talked and thought about it a great deal, but I don't think we can say we know. My view has generally been that it would be a matter of considerable, long-lasting impact. There probably would be a great flurry of excitement for a while because of the media but that would die off in a short time. Then there would be many-year-long gradual recognition of the importance and interest of the whole thing, which would permeate the thoughtful community but would not be a headline matter. I think the headline matter would be over rather soon because the headlines just don't have enough sustenance to keep it going for a long time.

What would be the long term effect upon humanity?

The effect is not the question. It's the facts of the case that matter. I believe it will certainly affect us if we decide that we are not unique. Up till now a great many people have the happy view that we *are* unique, the green footstool of creation, and that there is nothing else like us. I don't think that shaking that view matters unless it is genuinely false, but if it *is* false, then that will have to be absorbed by a great many people, in many different ways. It will have an impact over the long run comparable to the notion that the Earth is not the center of the Solar System.

What effect would this realization have?

That's not an easy question to answer. As I've said, when Copernicus claimed that he had arguments to show that the Earth was not the center of the Solar System, and that everything revolved around it, including the Sun, that had a big effect which is not over yet, though we now accept it fully, and we comport ourselves in that way. We say, "Yes, the Earth is one planet among many, and that's the way it is."

People can make whatever judgment they want of that themselves, and after a while they manage to fit that to any kind of structure of thought, but it still is there, and it leads many people to view the world differently from what was the case before that was pointed out. I think exactly the same thing will happen here, but perhaps even more strongly and maybe even more rapidly.

Giuseppe Cocconi

PHYSICIST

Born October 3, 1914

Como, Italy

Guiseppe Cocconi, recently retired, was the director of the proton synchrotron at the Centre Européen Recherche Nucléaire (CERN) in Geneva, Switzerland. In 1959, while at Cornell, he and Philip Morrison were the first scientists to call for a coordinated search for extraterrestrial signals. They noted that hydrogen atoms in interstellar space radiate at a frequency of 1420 megahertz, and suggested that this would be a natural channel for interstellar communication. Their report in *Nature* launched the SETI movement.

Cocconi received his doctor's degree from the University of Milan in 1937 and became professor of physics in 1942. He taught at the University of Catania in Italy until 1947, when he was invited to join Cornell University. He remained there, teaching and doing research, until 1962, when he returned to Europe to work at CERN as senior physicist. His interests include cosmic rays, particle physics and astrophysics.

Interviewed February 1983 by telephone, Geneva, Switzerland

Where did you live during childhood and when you were growing up?

In Como, Italy. It was a small city of fifty thousand, on the other side of the Alps from Geneva. I had my schooling there and then I studied at the University of Milan.

What were your childhood interests?

Until I was fifteen, I was very reluctant to study. But then I started to study astronomy on my own, and I liked it very much. When I was in high school I spent a good deal of my time looking at the sun and the stars with a very small telescope I found in the house, and I wanted to become an astronomer. I got acquainted with an astronomer in Bologna, Luigi Jacchia (now at Harvard). He told me, "If you want to become an astronomer, you must do physics first."

So I studied physics at the University of Milan, but never became an astronomer. Astronomy in Milan at that time was only Newtonian, not physical astronomy: not very challenging. So I stuck to physics, but I remained always very attached to astronomy. I still am. In fact I follow quite closely all astronomical results.

How was your elementary schooling?

At that time I liked much better to play, so I tried my best not to do anything in school. Until I went to high school.

Tell me about the high school.

I had to study eight years of Latin and five years of Greek, and these I did not like very much, but there was a good physics teacher, good mathematics. They encouraged me.

During your college years were there any ideas, books, teachers, classes, other students, or events which stimulated you to think about ETI and the possibility of searching for or communicating with it?

No, not specifically. There was a good physics professor, Giovanni Polvani. He was brilliant. The rest was standard, not particularly enchanting, but one learned the trade.

I graduated in 1937. Soon after, an examination was given in Rome for young graduates to become "university assistant" and as I went there, I had the good chance of meeting some of the physicists of the Rome group, in particular Edoardo Amaldi.

He liked me and invited me to work in Rome for one year. That was in 1938, when Enrico Fermi was there and was interested in cosmic-ray phenomena, just before he left for the States. So I worked with him and with Gilberto Bernardini in building a cloud chamber to study the decay of the meson. That was really the starting point of my career.

A new world opened up for me. Fermi was such a challenging man. But more than challenging he was a fertilizer. He was such that anybody around him got uplifted and managed to produce his best.

The year spent in Rome got me acquainted with cosmic-ray physics and, back at the University of Milan (as an assistant, then as "Aiuto"), I followed that line of research, cosmic rays and particle physics, with special interest for the phenomena of extensive air showers.

In 1942, while under military service, I won a nation-wide competition to become University Professor, and I had a chair in Catania. As the war ended, I spent three years in Catania, teaching and continuing cosmic-ray research.

In 1947 I accepted an invitation by Professor Hans Bethe to spend "a year or so" at Cornell University. There I remained fifteen years, till 1962, eventually as full professor.

When did you first think about the possibility that extraterrestrial intelligence might exist?

I cannot quote a date. It was something I took for granted. It was more or less obvious that evolution could take place elsewhere in the universe.

What about communicating with, or at least searching for, ETI? When did you first think about this possibility?

In 1959 I had many discussions with Phil Morrison and other people at Cornell about gamma rays, about the discovery of radio waves from celestial bodies, and so on. It was all a novelty that so many radiations outside the visible were emitted by celestial bodies. At the same time I was studying gamma rays radiated by electrons in the Cornell synchrotron.

I remember one conversation with my wife, Vanna, who was also a physicist at Cornell. We were thinking that a narrow burst of gamma radiation could be a signal that can travel far and straight in galactic space and be peculiar enough to be recognized. And that was the triggering.

Then, I think just the following day, I went to see Phil Morrison and discussed the possibility of extraterrestrial communication using radiation. The problem was: which is the best wave length?

Morrison said that, probably, the longer wavelengths were better. Considering the noise from the stars, and where the minimum interference was, it turned out that the minimum was near the "water-hole". Then I started to put down numbers to see whether we could, with radio waves, have power enough. And to my surprise I arrived at the conclusion that the Jodrell Bank Telescope was nearly at the point of going as far as the closest star. But that came a few days later.

At that time, September 1959, I had an invitation to spend my sabbatical at CERN in Geneva. CERN was very young at that time. No large accelerator was running, but they wanted some physicists to prepare an experimental program for the synchrotron in construction.

Morrison also had to come to Europe for some period, so we met in Geneva and wrote that first paper here, not far from the house where I live now. Before publication we sent a preprint to P. M. Blackett to see whether he liked the idea, and then I also sent it to B. Lovell of Jodrell Bank.

Blackett was enthusiastic about it. Lovell answered, rather coldly, but saying that he could find no fundamental mistake in our numerical evaluations. So we sent the paper to *Nature*, and that was it.

Has your interest in ETI, expressed in that paper, had any effect upon your career as a scientist?

No, I don't think so. I continued to be interested in particle physics and to follow the astronomy news.

What do your colleagues think about your interest in SETI?

My colleagues here in Geneva?

Yes.

Oh, they barely know that I am involved. Sometimes somebody comes around and says, "You know, I saw your name in some book or some magazine. Are you really the one who did it?"

What about your friends? What do they think about your ETI interests?

They don't think very much and I don't encourage them.

It seems to me that the thing must mature. At a certain point the technique will come to a point where it's very easy and obvious to work on SETI, and then it will be done.

What do you think the public feels about SETI today, about searching for extraterrestrial intelligence?

The public is so mixed up. Probably nobody is astonished at anything now, but the real meaning, the real impact on our civilization, I think, is generally not fully appreciated.

This probing of evolution is really a fantastic thing because we know only of one way evolution worked, on Earth. It is one history, our history. But there are probably millions of other roads equally interesting or even more interesting—or less interesting.

From your view in Europe, who are the outstanding people in the search for extraterrestrial intelligence today and during the past twenty years?

Not very many here, apart from the Russians. In Russia there are Shklovskii and other people who are quite keen about it. But in Western Europe practically nothing, nobody.

What about for the world as a whole? Who do you see as the outstanding people, either today or during the last twenty, twenty-five years?

In what sense? Accomplishment—there's nobody.

I'd just say general interest or activity with respect to SETI.

I think that the best help we had so far has come from *E.T.*, the movie. Because there is a message there. I found it appealing just for the things which are not said but which are obvious. It is the best help, apart from what the people in Silicon Valley are doing, developing the electronics to make powerful multichannel analyzers.

A few final questions. What scientists or other people in the past do you admire?

One can always speak of Galileo and Newton. But of the people I met personally I think Fermi was the most extraordinary person, both as an experimentalist and as a theoretician. He was like Niagara Falls: a beauty, a power of nature. But then, happily, at Cornell I met several stimulating people: Hans Bethe, Dick Feynman, Bob Wilson, Ed Salpeter, Tom Gold, and, of course, Phil Morrison.

What besides physics are your other hobbies and recreations? You enjoy skiing?

Yes, and I like walking in the mountains. Here it's easy because there are so many mountains nearby. Fortunately, I'm still healthy so I can do it.

That's wonderful. What would you say was the most memorable moment or event in your life?

I don't know how to answer. It is too difficult. I cannot adjust to so simple a definition.

Well, finally, is there anything else?

I told you more than I should.

I really appreciate that.

Frank D. Drake

ASTRONOMER

Born May 28, 1930

Chicago, Illinois

Frank Drake is professor of astronomy and astrophysics at the University of California, Santa Cruz. In 1960 he conducted Project Ozma, the first search for radio signals from extraterrestrial civilizations, "listening" for 150 hours to two nearby stars resembling our sun. In 1974 he directed a message at M13 in Hercules, a globular cluster of 300,000 stars 24,000 light years away. He also formulated the "Drake Equation," an algebraic statement for estimating the number of extraterrestrial civilizations.

He received a Bachelor of Engineering Physics degree from Cornell in 1952. After service in the U.S. Navy he earned an M.A. in 1956 and a Ph.D. in 1958 in astronomy from Harvard. For the next five years he was at the National Radio Astronomy Observatory in Green Bank, West Virginia, where his observations revealed the Van Allen radiation belts around Jupiter, and where he conducted the historic Ozma experiment.

After a year as chief of the Lunar and Planetary Sciences Section at Jet Propulsion Laboratory, he joined the faculty of Cornell University in 1964, where he was professor of astronomy and director of the National Astronomy and Ionosphere Center, which includes the Arecibo Observatory. In 1984 he became Dean of Natural Sciences at U.C. Santa Cruz.

He is author of *Intelligent Life in Space* (1962) and many articles for the general public. His numerous technical reports deal with such topics as "A High Resolution Radio Study of the Galactic Center," *Astronomical Journal* (1959); "Interstel-

Photograph by Don Fukuda. By courtesy of Instructional Services, University of California.

lar Radio Communication and the Frequency Selection Problem," *Nature* (1973); and "A Comparative Analysis of Space Colonization Enterprises," *I.A.U. Proceedings* (1985).

He has served on advisory committees for numerous scientific and governmental bodies, ranging from the Very Large Array, National Radio Astronomy Observatory, Kitt Peak and Cerro Tololo observatories, to awards committees of the National Science Foundation and the American Astronomical Society.

He is president of the Astronomical Society of the Pacific, and the SETI Institute. In the International Astronomical Union he was president of the Commission on Bio-astronomy. He has chaired the Astronomy Section of the American Association for the Advancement of Science, the Division for Planetary Sciences of the American Astronomical Society, and the SETI Advisory Committee of NASA's Office of Aeronautics and Space Technology.

He is a member of many other organizations, including the National Academy of Sciences, the Extrasolar Planetary Foundation and the Explorers Club, and has served on the advisory board of the *Astrophysical Journal* and other publications.

His honors include the Harvard University Loeb Lectureship, the Glover Medal of Dickinson College, and the naming of the Frank D. Drake Planetarium in Ohio.

Interviewed June 1981 at Ithaca, New York

Where did you live during childhood and when you were growing up?

Chicago.

What were your childhood interests?

I did all the usual, standard things; nothing of note or unusual about what I did. I used to make little electric motors, little tiny radios, stuff like that which a lot of kids do. I did a lot of amateurish experimentation with a toy chemistry set.

What incidents involving science do you remember from those early years?

There was a very fine museum which is still there, the Museum of Science and Industry. And I did have a friend all through childhood who was also very scientifically interested. We were close friends, and our most common way of playing was to get on our bikes and go to the museum. We knew every square inch of that museum. We used to go there a couple of days a week, years on end, to that museum down the street.

Do you remember comic books, newspapers, anything of this sort?

I remember the comic books but they didn't have any influence.

What about during adolescence?

In my teenage years, when I became interested in astronomy, I turned to the making of optical telescopes. I made three telescopes, one a very simple one with a lens from an old slide projector, and then two larger ones, reflecting telescopes, where I ground the mirror and built the telescope mount and all that sort of thing.

Did your elementary schooling give you helpful preparation for a scientific career?

It was actually quite a good elementary education. More good was done in high school, where I had some excellent courses and teachers in physics and chemistry and mathematics.

Were you oriented towards astronomy at the time you entered college?

No. Actually I lost interest, years before. All through high school there was no course or instruction in it, nothing to excite my interest in the universe which had been kindled at the age of eight.

When I was a sophomore here at Cornell I was allowed to take an elective so I took the elementary astronomy course. I found it absolutely fascinating. I loved every minute of it. At one stage they took us to the little old telescope over here, which is still used by the students, to look at things. I remember looking at Jupiter and being just thrilled to see there really was an object with four satellites going around it. You could see the cloud belts on it and all. It made the whole business seem very real. It hit me and I was hooked from that time on.

When did you first think about the possibility that extraterrestrial intelligence might exist?

Since I was eight years old I've wondered about the origins of people and whether there might be others elsewhere in the universe.

Who or what suggested this possibility to you?

Nothing and nobody suggested it to me. It just occurred to me as I learned more of our world and the way of life we have. As I grew up it just occurred to me one day that the whole thing was somewhat arbitrary and didn't have to be the way it was. And I wondered if there were other places like the Earth—where similar things, but not identical things, with regard to the development of intelligent life—had occurred.

You began thinking about the possibility of ETI when you were a child. What was happening at that time that focused your attention on ETI?

A strong influence on me, and I think on a lot of SETI people, was the extensive exposure to fundamentalist religion. You find when you talk to people who have been active in SETI that there seems to be that thread. They were either exposed or bombarded with fundamentalist religion; in my case, and at least in several of the others.

In my case it was when I was eight years old, I learned enough of science and astronomy to recognize that there was a conflict. Something was not jibing. Sunday school just did not fit in with what the universe seemed to be. That came as quite a revelation. There might be quite a different explanation of our world, our people, and what life is all about besides what I was being told in Sunday school. It was an eye-opener that there were alternatives. In fact, there was a scientific basis for other alternatives. It wasn't just stuff written on old scrolls and things, which I was always suspicious of anyway. So to some extent it is a reaction to firm religious upbringing.

What faith were your parents?

Baptists. Very strong Baptist. Sunday school every Sunday. It should have been wonderful. I remember this because it was really such a puzzlement to me as a little kid. The instructors were all professors from the Oriental Institute at the University of Chicago, which is one of the outstanding archaeological institutes in the world. The guys who were teaching our Sunday school were the world's foremost archaeologists, Middle East experts. They were coming in and telling us little eight-year-old kids all about the Hittites and the history of the Middle East.

It was very difficult to understand, but my reaction was "Why should any of this have any bearing on how we live in the twentieth century in Chicago?" I couldn't see the connection. It all seemed crazy, these great men carrying on and on about this.

They used to take us over to see the Egyptian mummies and stuff. The Oriental Institute was just a block away from the church we went to. It should have been very effective, being taught by the professors with endowed chairs in the world's greatest archaeological institute, but it had the opposite effect. There was so little correlation with the real world as I saw it. I discovered there was a scientific explanation of the real world.

Were these eminent archaeologists invited in hopefully to reinforce the biblical fundamentalist interpretations?

Yes. The idea was they came in to teach that there was archaeological evidence to back up the Bible. That's what they were doing. So they would describe the archaeological digs which provided information which was correlated with what the Bible said about the histories of ancient peoples.

They were involved because, while very few people know it, the University of Chicago was dominated by the Baptist church. The faculty was full of Baptists. So there was that Baptist influence, and our church was connected with the university, informally. It was right on campus. Since that time the university has backed away from any kind of religious influence. Nowadays you would never be able to discover that it had a Baptist founding body. But in those days the Baptist influence was still very active in that university.

Was the level of ideas that your parents and the Sunday school teachers were trying to get across that the world was begun in 4004 B.C.?

Yes, it was right down the line with the biblical version of how things are.

Did you discuss the possible existence of ETI with other people?

No, I didn't. There wasn't anyone I could discuss it with.

Let's clarify the development of your ideas about SETI, because you are a key person in its origin and development.

I became interested at a very early age in the nature of other planets and whether there was life on them. I didn't have much opportunity to either learn any more about it or to do anything about it until I was in college. In college I took that elementary course in astronomy, which kindled a great interest in astronomy in general, though there was nothing in it about ETI. It was very stimulating and certainly got me started in astronomy, but didn't mention ETI. In fact, I was interested in optical astronomy rather than radio astronomy.

During your college years were there any ideas, books, teachers, classes or other students which stimulated you to think about ETI and the possibility of searching for it?

No. Then, when I was in the Navy, I got all this training in electronics, so that when I went to graduate school I was edged into radio astronomy, even though that was not what I had in mind. But once I was in, it was obviously very attractive. Still, my primary interest, really the only interest, was using radio astronomy for conventional astronomical research.

What about communicating with, or at least searching for ETI? When did you first think about this possibility?

I only thought of that when I was a graduate student in radio astronomy and learned the theory and practicalities of radio telescopes. And in particular, learned how to quantify their sensitivities. And from that, to be able to make numerical estimates of what signals they could detect from what distances. Only then did I recognize that this was a promising way to search for other civilizations.

Who or what suggested this to you?

Nobody. I've always been interested in life in space, and had thought about it over the years. I've read some science fiction, although I wasn't a big science-fiction buff. And so whenever I contemplated an instrument, an optical or radio telescope, I would as an aside ask myself, "Could this be used to search for life?" The answer was always "No," until we came to the modern radio telescopes.

I don't remember exactly when the possibility of detecting other civilizations came to me, but I do recall something that happened when I was doing the observations for my Ph.D. thesis. I was measuring the radio spectrum of the Pleiades, which I had done many times. I knew what the spectrum was, and I was doing it over and over in slightly different positions in the Pleiades, and it always looked about the same.

But on one occasion there was a strong narrowband signal added to the normal hydrogen spectrum from the Pleiades, and I had never seen that before. I had it on the record. It was late at night. I started thinking about it and I measured its velocity. It turned out that it was on a frequency which was what hydrogen would be radiating at if it were in the Pleiades and had the velocity of the Pleiades star cluster, which is quite a large velocity. So it was an anomalous frequency but it wasn't hydrogen; it was clearly an intelligent signal; and that just rang a bell. Could it really be an intelligent signal from the Pleiades? It was in the right frequency. It was very exciting.

I worked very hard on it and finally decided it wasn't coming from the Pleiades, but from Earth. But from then on I always had it in the back of my mind. And when I was at Green Bank I did the calculations to see what we could detect, and it turned out that we could detect something reasonable from Tau Ceti and Epsilon Eridani. And I went on from there.

Since you completed your schooling and began working as a professional scientist, who or what has influenced your thinking about ETI and SETI?

There's been a lot of people. I would say people of special significance out of many have been Barney Oliver, Ronald Bracewell, Sebastian von Hoerner, Carl Sagan. I think that's about as lengthy a list that I could give without giving many, many more names.

By 1960 or so, these people and others knew about your Project Ozma work. How did they find out about it? For example, you mentioned one time that Bernard Oliver just showed up at the Green Bank Observatory.

I don't know how he found out about it, to this day. But there was a very small network of interested people, and it really consisted of Morrison and Cocconi, though Cocconi always was peripheral actually. He talked about it but he didn't really do anything serious about it. Morrison was interested in doing something serious. Oliver was and Sagan was; although Sagan never had the chance, he was part of the network.

I'm talking of Project Ozma time, '59–'60. Before that there was no network because the subject was not discussed; there was no way to know who was interested; just no way to make contact or to learn of other peoples' interests. But with the existence of Ozma and the publication of Morrison and Cocconi's paper, correspondence started to flow. People knew who they could write to to find out who was interested.

Essentially the whole network is the ones who attended that 1961 Green Bank meeting. It wasn't limited to eleven because that was the maximum number we had room for; it was merely that that was everybody we knew of who was seriously interested in it.

Of that eleven, several, like Lilly, dropped out of the picture?

They went into other things. Lilly actually is sort of back in it today. He's back studying dolphins again after a hiatus. He's out in California doing the old experiments again on dolphins' intelligence. Some got more active, as Oliver has done.

This is a good time to discuss Project Ozma. What are some of your memories about it?

When I think of Project Ozma, I recall how cold it is at Green Bank at four in the morning. I have two vivid memories of those days. First, of the battle against the cold each morning as I climbed to the focus of the dish to tune the parametric amplifier. Then, of the moment on the first day of the search when a strong unique pulsed signal came booming into the telescope just as soon as we turned it towards the star Epsilon Eridani.

Green Bank was an exciting place in 1959. We had been given the charter, and practically unlimited funds, to build the best radio observatory in the world. We had started to build a very costly 140-foot telescope, which would take years to complete.

We knew that we should build a smaller telescope first to get the place into operation, and to gain some experience and momentum.

So we built an 85-foot telescope, and by early 1959 it was finished, a striking anachronism in those primitive mountains of West Virginia.

Since childhood I had wondered about the origins of people and whether there were others elsewhere in the universe, so it was natural that one day at Green Bank I calculated just how far our new 85-foot telescope could detect radio signals from another world, if they were equal to the strongest signals then generated on the earth. It turned out to be about ten light years. Within that distance there are a few stars very much like the sun.

Every few days the small group of scientists at the Observatory had lunch together at a roadside diner five miles away. We christened it "Pierre's" or "Antoine's" although "The Greasy Spoon" was more appropriate. One snowy day late in winter we all drove down to this forlorn place, and during lunch I mentioned my conclusion that the new telescope might be able to detect intelligent radio signals from some nearby stars. I suggested that we put together some simple equipment—it would require a narrowband radio receiver, something we didn't usually use in radio astronomy—and search some nearby stars for signals.

The director of the National Radio Astronomy Obsevatory at that time was Lloyd Berkner. He liked the idea so, as the last greasy french fry was washed down by the last drop of soda, Project Ozma was born.

I decided we should build our equipment to operate at the 21-cm line frequency. The narrow-band radio receivers to be used in Ozma would be just right to search for the Zeeman effect in the 21-cm line of neutral hydrogen. So if we set up our equipment for that wavelength, we could use it for that important experiment. It would also head off any criticism that we were wasting money on the equipment. In the end we spent only about two thousand dollars for the unusual parts of the receiver, and no one ever complained.

Around that time we received a visitor from England: Ross Meadows. He was an electronics expert so he was given the task of doing all the dirty work of putting the Ozma receiver together. By today's standards it was a simple receiver. It had only one signal channel, and the simplest of outputs, a chart recorder. We also planned to connect an ordinary audio tape recorder to the system just in case something did come in from outer space.

The receiver had other important special aspects. It switched between two feed horns to allow us to distinguish a signal from space from a terrestrial signal coming in the side-lobes of the antennas. The same approach has been used in some form in just about all the searches since.

Also, there was a reference channel to which the signal channel was compared; this was a standard technique in those days and now, and was used to eliminate receiver gain fluctuations and nonlinearities. Because our bandpass was to be 100 hertz, the oscillators used in the receiver had to be more stable than usual, although nothing very challenging. After a while Kochu Menon, a friend and colleague from Harvard, came to Green Bank and worked on the receiver also.

We had been working on this system at a relaxed pace for about six months when an important event occurred. Cocconi and Morrison published their deservedly famous paper in *Nature* in September. They made the same calculations I had made, pointing out that mankind could detect other civilizations with existing radio telescopes. They suggested the 21-cm line as the most promising band to search for signals, and they offered a good reason: the unique status of this fundamental spectral line of the most abundant and fundamental atoms of our universe, and not the practical reason which had influenced me. We felt good because now there were further arguments for what we were doing.

At that time I didn't know Morrison personally, though as a student at Cornell I had been awed by his superb lectures, and he was a great help to one of my best friends. But I had never met him.

By this time Otto Struve, the world-renowned astrophysicist, had become the director at Green Bank. Despite his conservative background, he was one of the few senior astronomers of that era who believed that intelligent life was abundant in the universe, and he felt that everything possible should be done to support any feasible searches for signals of extraterrestrial intelligent life. So, he was for Ozma from the start, and was urging us to hurry with the project.

From his long experience in the real world of astronomy, Struve was also aware of the importance of getting credit for ideas, discoveries, and good research whenever possible. He knew it paid off when it came to getting funding. So, to my surprise he was very agitated and frustrated when the Cocconi-Morrison paper appeared. He was very worried that Green Bank would lose the credit for what he thought was an important idea.

Actually, from the beginning of Ozma we had expected that any public announcement of it would bring a horde of reporters down on our heads, and so we had kept the whole thing as quiet as possible. Now Struve was very upset that we had done that, and he chose to publicize. He had been scheduled to give some prestigious lectures at MIT about a month later, so he used the occasion to announce the existence of the project, ballyhooing the ETI search going on at Green Bank. That did bring the press down on us.

The cat was out of the bag. Looking back now, only good came from letting it out. The first thing that happened was that we were offered the use of one of the first operative parametric amplifiers in the world. Electronics had recently been turned topsy-turvy by two inventions: the solid-state maser, and the parametric amplifier. Both gave receiver sensitivities as much as ten times better than what had been in use. But they were laboratory devices and could not be used in practice in the field, on a moving radio telescope.

So we were thrilled when Dana Atchely offered us the use of a working parametric amplifier, probably the best in the world. He was simultaneously an avid radio ham, an intelligent life buff, and the president of Microwave Associates, one of the most sophisticated American electronic companies. Not only would he provide the amplifier, but he would also send his chief engineer with it to install it.

On the appointed day, I got a call in my office that the chief engineer of Microwave Associates had arrived with the amplifier. Going downstairs, I got a real jolt as I saw before me: 1) A British sports car, top down, made by Morgan. Cars used to be made of wood and this was the last of them, complete with leather straps to hold the hood down. 2) In the driver's seat, a fellow with a long flowing red beard, wearing a red tam-o-shanter. And 3) in the passenger's seat, the parametric amplifier which had bounced all the way from Boston.

Sam Harris was the driver, known to every radio ham as a radio amateur magazine editor, and known to many and soon to me as an electronics genius. He had designed the parametric amplifier, and was the only one in the world who could make it work. And it really worked.

He proceeded to install it, make it do its magic. Then he taught me how to tune it, the task which became my four-o'clock-in-the-morning starter for the day. When all was well, he climbed back into his Morgan and drove off. I didn't see him again until one day in 1966, I met that red beard again. This time he was on the staff of my observatory at Arecibo—I had nothing to do with this improbable event. He has been there ever since, doing his magic.

We had much urging from Struve and Dave Heeschen [assistant director of N.R.A.O at that time, and later the director] to get on with Ozma, because the press and the scientific community were now harassing us. We finally had all the equipment built in the early spring of 1960. In April we began the actual observations.

On April 8, the first day of Project Ozma, I set the alarm clock for three, got up groggily, and went out into the fog and cold which was to be my regular morning greeting for about two months. At the 85-foot telescope, the operator would turn the telescope so that I could climb into the metal can, not much larger than a garbage can, which was at the telescope's focus.

I would sit there for about forty-five minutes twiddling the micrometer adjustments on the parametric amplifier, talking to the telescope operator, as we set Sam Harris's gizmo so that it was doing the right thing. In the beginning we had to do this several times a day because the changing temperature upset the tuning, but as time went on we found ways to evade that problem.

After the amplifier was all tuned, I climbed down from the focal point, went into the control building and set up the Ozma receiver. It was built to tune slowly in frequency so that it shifted its frequency about 100 hertz every minute.

Then we pointed the telescope at our prime target, the nearby solar-type star Tau Ceti. Once the telescope was tracking the star's position, and the receiver was set on the starting frequency, we turned on the tuning motor, the chart recorder, and the tape recorder. Project Ozma had begun!

Whenever you search for extraterrestrial intelligent radio signals, you always feel at the beginning that the signal may pop up right away. The telescope operators and I spent a breathless morning peering at the wiggling pen on the chart recorder, thinking that every time the pen started to deflect up that this was it—only to see the pen go down again, obeying the universal law of gaussian noise statistics. And so it went until noon, when Tau Ceti set in the west.

Then we turned the telescope to point at our second subject, the solar-type star Epsilon Eridani. It was then believed to be a single star. Recently there has been evidence that it has some companions; I wish we had known that then because it would have been more exciting. Again we pointed the telescope at the star, and set up the recorders. We also added a loudspeaker so that we could hear the receiver output. Again we started the chart and tape recorders, and settled back for more of what had already become routine.

A few minutes went by. Then it happened. Wham! Suddenly the chart recorder started banging off scale. We heard bursts of noise coming out of the loudspeaker eight times a second, and the chart recorder was banging against its pin eight times a second. We had never seen anything like this before in all the previous observing at Green Bank. We all looked at each other wide-eyed. Could it be this easy? Some people had even predicted that the most rational extraterrestrial signal would be a slow series of pulses, as that would be evidence of intelligent origin. At that time nobody had any idea of the existence of pulsars.

Suddenly I realized that there had been a flaw in our planning. We had thought the detection of a signal so unlikely that we had never planned what to do if a clear signal actually was received. Almost simultaneously .everyone in the room asked, "What do we do now? Change the frequency?"

The most likely source of a spurious signal was the Earth, and we could check that out by moving the telescope off the star and seeing if the signal went away. So we proceeded to do that, and as we moved off the star, sure enough the signal went away. Then we pointed back at the star. The signal did not come back.

Was it really from the star, or had it been from the earth and it turned off about the time we moved off the star? There was no way to know. There we were, with all that adrenaline flowing and no way to apply all that excitement and energy in a useful way.

Day after day, as we turned to Epsilon Eridani, we tuned to the frequency on which the signal had been heard. We listened for a half hour or so, and then we would go back to our frequency scanning. We also connected a second receiver to a simple horn antenna which looked out of the control room and could pick up interference. A week went by and the signal didn't return.

To our dismay, one of our employees called a friend in Ohio and told him about the signal. The word was passed to a newspaper re-porter friend, and suddenly we were deluged with inquiries about the mysterious signal: "Had we really detected another civilization?"

"No."

"But you *have* received a strong signal with your equipment?"

"We can't comment on that."

And so, aha, we were hiding something. To this day many people believe falsely that we received signals from another world, and that some fiendish government agency has required us to keep this a deep dark secret.

We finally learned the truth about ten days after the *big* day. Sud-denly the signals were there again, blasts of radio noise eight times a second, coming in the 85-foot telescope. But just as strongly, they were coming in the little horn we had poked out the window. The signals had to be man-made radio interference. As we watched them, we saw them grow and fade as though they were being transmitted from a highflying passing airplane.

Were you really certain it was a plane?

Yes, it was something military: electronic countermeasures or some-thing. We've never been sure just what it was. But it was from a circuit being worked out. It had all the earmarks of it.

So we stopped listening as intently to that special frequency while the telescope was pointed at Epsilon Eridani.

As the weeks passed, hundreds of yards of chart paper and tape piled up, all with nothing but noise on them. We were now experts at scanning the records for signals. It even got dull, and I realized that, as exciting and important as is a search for extraterrestrial signals, such searches should only be done in conjunction with regular astronomical research, so that there will be real results all the time to remind the searchers that there are, after all, strange and wonderful things in the sky. So they will keep looking.

We had some special visitors at Green Bank during Project Ozma. One who came for several days was Theodore Hesburgh, then the very young president of Notre Dame University, and an up-and-coming theologian. He felt that the search for extraterrestrial life was an inspiring and a very good thing to do.

Another was John Lear, then the science editor of the *Saturday Review of Literature*. He was a titan among science writers. He'd been the first to expose abuses in the drug industry, for example. He wanted to see history made, and knew that the detection of another civilization would be *history* if it really happened.

A third visitor was Barney Oliver, the vice-president for research of the Hewlett-Packard Corporation, a company which made a lot of oscilloscopes and meters and other electronic gadgets which we and every other observatory used. He dropped out of the sky one day in a chartered plane, full of enthusiasm, to watch the goings-on in the West Virginia wilderness. Actually it was not at all surprising that he was there, because he, too, had thought for many years about the means of detecting other civilizations. He was a successful inventor, electronics expert and physicist, and already knew all about it. He was glad that someone had the opportunity at last to do something.

After one month of searching, we took a break. Then another month, and the whole range of plausible hydrogen line frequencies had been scanned for both stars. We knew that there was a chance that we had looked at the right star on the right frequency, but at a time when "their" transmitter was turned off, so perhaps a second look would hit pay dirt. But no more telescope time could be committed to the project; there was a lot of astronomy to do. And so Project Ozma was over.

Today, with the Arecibo telescope and our 1008 channel receiver, we duplicate everything that was done in Project Ozma in less than a second. We actually do it better. And, everything is warm and cozy. But that does not mean in any way that the effort was wasted. Only by doing the best we can with the very best that an era offers, do we find the way to do better in the future.

Project Ozma contributed to the development of the much better systems of today, and the giant telescopes and computers of today will be replaced with even grander instruments in the future. A ladder doesn't work if some of its rungs are missing. They all have to be there, and you have to step on them all, one at a time.

Since Ozma and the Cocconi-Morrison article occurred at about the same time and both were very significant in the origins of SETI, I want to be sure that I really understand what was happening then. Were you concerned about a possible Darwin vs. Wallace situation?

No. We had very little concern about public reaction or reactions from the scientific community. The reason was that we were running that project on a very low level, with very little expense involved. We did that purposely, to avoid attracting criticism or too much attention. We recognized that if the press became aware of it, they would harass us. We had no facility to deal with that, so we tried to keep it a confidential project. Not military confidential, but we just didn't advertise it.

Our main concerns were really with technology. It wasn't with public contact at all. We were trying some technology that had not been done before. In several respects we were using a new form of sensitive receiver that had never been used on a radio telescope. It was very much a laboratory device, and we were using frequency resolutions which were far finer than had been used before.

We weren't worried about interactions with the rest of the world. In fact, there wouldn't have been any interaction with the rest of the world had it not been that Morrison and Cocconi published their paper, which occurred completely separately. We didn't know they were writing, and they didn't know we were building the equipment. We had the project underway for almost a year when their paper appeared, so the two things were happening quite independently, without any contact. Morrison and Cocconi were thinking and we were building. We happened to be building for the same frequency, 21 centimeters, they had suggested.

When their paper came out my first reaction was, "Fine. Here's some other scientists who think what we're doing is the right thing. We can only take that as encouragement." So to me it didn't matter that Morrison and Cocconi had published, and as far as I was concerned we would proceed, keeping the thing a secret because we didn't want the press harassing us.

But Struve, as I said, didn't like that at all, because he knew that getting credit for things mattered when it came to funding. He foresaw what did, to some extent, happen—that a lot of credit would go to Morrison and Cocconi when in fact we had started our project completely independently. I, as a very young scientist, was naive about that. My concerns at that time were with technology, not publicity.

Were there any other things about the early time, the Ozma period, that we should note here?

The only significant things were: technology was the problem; the business with Struve reacting to the Morrison and Cocconi paper by seeking publicity; and the excitement that went with the detection of a signal which turned out to be terrestrial.

The other director of the National Radio Astronomy Observatory, Lloyd Berkner, approved of Ozma, too?

Yes, he was all for it. His whole history had been to push exotic experiments. He was sort of an entrepreneur and gambler of science. He liked to do that kind of thing.

Is this characteristic of people in a new field that's opening up?

Yes. What happens is the community more or less invites such people to take over because they know they'll be pushers, they'll be brave, courageous and try daring things. That had been Berkner's history. Before that time he developed some important ionospheric measuring techniques, and he had built some of the big nuclear energy centers.

So both he and Struve were . . .

Both were positive and both very supportive. In fact, if we had proposed a more grandiose project, they probably would have supported it. I think it was me that was the conservative in all of this. They were older. Both Struve and Berkner were near retirement age and they were willing to risk. They were venturesome; they weren't worried about their career or anybody saying bad things about them. That was all right at their stage.

This is another issue that comes up with scientists: concern about careers. Sometimes it doesn't seem wise to get into certain topics.

It's not wise to get into really controversial things, but SETI in particular is not good for young people because the chances are very high that you will get no results at all. That's not good for one's career. Around here every year we have a stream of students who want to work on SETI. They're thrilled by it, they think it's the wave of the future and the most significant thing they can do with their lives. They're very serious about it, and they are right, too.

But my advice always to them is, "This isn't good for you because if you work on SETI, you're going to work for four or five years and have nothing to show for it, and then how are you going to get a job?" It's a paradox. What they want to do is basically what the talented human should be doing, but at the same time in our culture you have to tell them that it's professional suicide to do it at this stage in your career. You've got to prove yourself first. You have to have something you can do besides SETI which will justify your future employment, tenure track appointments, and that stuff.

Is there any way out of the paradox?

Not until SETI finds another civilization; then you could do lots of theses on it.

What do your colleagues think about your interest in SETI?

As long as no great demands were made on money or telescope resources, the people were willing to let you do anything you wanted, even if maybe it was crazy.

Did many think it was crazy?

I don't think anyone thought it was crazy. I think many did and do think it's a long-shot experiment with small probability of success. But I don't think there's anyone who thinks that it's just completely outlandish.

So you fortunately didn't have the problems that earlier scientists did?

They didn't burn me at the stake.

Was your colleagues' response to your Ozma project uniformly positive?

It was uniformly positive but not enthusiastic. Again I think it was the fact we weren't investing a great deal of resources. Two thousand dollars was invested. People didn't think it was worth a very careful analysis, but since it wasn't crazy they said: "These guys want to spend two thousand dollars; let them do it."

Would you say it was more tolerance than active support?

It was a little more than tolerance: "It's a long-shot experiment, it's a nice thing to do. Since it doesn't gobble up a lot of funds, we think they should do it." That was their attitude. Now if we were going to tie up that telescope for a year, I think there would have been a lot of opposition.

Do they ever discuss SETI with you or ask you about it?

Well, until recently the answer to the question was "no." But for the last three or four years I've been bombarded with advice, questions, comments, inquiries—the typical question being, "What's going on now in SETI?" It comes from people who were not at all interested in the subject five years ago; now they've built up a genuine interest. There are a lot of crypto SETI people out there who didn't reveal that until recently.

What about your family? What do they think about your ETI interests?

They think it's a very good thing. Of course they've heard so much about it they're believers.

What about your superiors, bosses, and so forth?

They have been supportive. As long as it's handled rationally, and again, no outlandish demands are made on resources, they think it's an appropriate subject for me, a faculty member, to be pursuing, and a subject to be carried out in a university such as Cornell. This has always been true.

What about your friends?

They and other lay public people are generally more optimistic and supportive than are the scientists. That is because scientists are much more aware of the difficulties and all the obstacles to success. The lay public, not being so familiar with the scientific and technical facts, think the whole enterprise is easier than it really is. They tend to be more enthusiastic. That includes not only friends but people in the Congress, people in the press.

What was your father's occupation?

He was a chemist with the city of Chicago. My mother was a house-wife. They both went to the University of Illinois, public schools. Both were Baptists. They adhered strongly to the church beliefs.

Did your mother have occupational skills or work experience other than that of the typical wife and mother of that time?

Just the typical life of an American woman of the middle twentieth century.

How would you describe them?

Reserved and quiet, both of them.

How did they influence your eventual choice of science as a career?

They were moderately discouraging. This was before the space age, about 1951, when I was announcing that I intended to go to graduate school and study astronomy. At that time astronomy was still a very small subject, practiced in few places, with little funding. It was well known that astronomers were poorly paid. So they thought this was all kind of crazy and they tried to discourage me, but not strongly.

But I persevered, and it happened, just about the time that I was finishing my schooling, that astronomy burst forth in a great boom era which was the result of important discoveries—the development of radio astronomy and the space age arriving as it did in 1957. I was in the midst of doing my thesis. So my parents changed their minds.

What was the political environment in your home?

They were very apolitical. Politics was not discussed. They always voted Republican, but there was never any conversation about politics.

So this probably was not an influence?

It was not playing a role in my SETI interest.

Do you have brothers or sisters?

I have a brother and sister. They are both in science. My sister's a bacteriologist. My brother has been in a lot of things. Right now he's more into economics of nuclear energy.

Where were you in the birth order?

I was the oldest.

Did your siblings influence your thinking about science and ETI?

Not really. My sister is three years younger than I am, and she did study biochemistry, but I never recall ever discussing science with her. My brother is eleven years younger than I am so that by the time he was a thinking person, I was already grown up and gone.

You mentioned as a child one friend that you used to go to the museum with.

I had a lot of friends. In fact, there was a large group of us that hung around. I would be considered a gregarious extroverted kid. I wasn't a loner.

What other professional interests do you have, apart from SETI?

Telescopes, planetary science and radio pulsars. Telescope design is one of my things.

What recreations do you enjoy?

I tinker with cars and fix them, although I'm doing less of that now; I hate to get my hands so dirty. Another thing I do is make jewelry, probably my first avocation at the present time.

Was Ozma a full-time occupation of yours at that time?

No, it was just one of many activities I was doing. It had equal priority with other things. At that time we were putting into operation the first large telescope at the observatory. I had other jobs, such as making the telescope work, making it so visiting scientists could use it, and I was conducting conventional research with it. We worked mainly with radio studies of the planets. So the time and effort I was putting into this project was probably 25 percent.

In fact it went very slowly, just because we didn't push it. We didn't feel any urgency, so it took us about a year to put the equipment together, although the equipment wasn't that remarkable in its magnitude. We had two other people working on it part-time.

About what fraction of your working time do you spend on SETI?

These days I still spend about a quarter of my time. Activities in SETI have been building a great deal in the recent years. Right now there's a great deal of planning and some real work going on towards putting together a very large search program. This is the one that's been devised by NASA Ames Research Center and Jet Propulsion Laboratory, and they have science working groups with which I participate. I edit the reports, have meetings to go to, a lot of phone calls, papers to write.

And also I do what I can to help the *Cosmic Search* magazine because it is basically a good magazine and it has been very effective in reaching a small but intelligent segment of the public. It needs all the help it can get because it's doing the right thing and it's doing it well. I usually write monthy articles for them for nothing.

Do you foresee that your time spent on SETI will change much?

I think it will actually increase. There's been a technical obstacle in the search for extraterrestrial intelligence which has been the inability to look at many, many radio frequency channels at once. This has held people back from searching because they knew the searches would not be powerful as long as that obstacle existed.

What has happened in the last year or so is the development of a design and the provision of funds to build a system that will get over that obstacle, which will look at many channels at once. The first such device is now just being finished at Stanford, and it's the prototype of bigger ones to come in the next few years. And as soon as the big ones are built, two or three years from now, there'll be a justification for a lot of serious searching. I think a lot of it will go on and I hope to be involved.

At the Green Bank meeting you wrote your now-famous equation on the blackboard. How did that come about?

The Drake equation was a way of organizing the meeting. I thought we should organize the meeting and categorize topics and establish themes for various sessions at the meeting, and that caused me simply to think about what we needed to discuss and how these things were related. And it was easy to see that they were interrelated in the way that is described by the equation.

Was this before the meeting or after people arrived at the meeting?

It was a few days before. There was really not a great deal of profound thought or anything involved in putting it together.

So it was your outline of the relevant factors?

Yes.

In addition to selecting the basic topics, when did you get the idea of actually expressing them in an equation?

That was also before the meeting.

So not only the basic concepts, but also the idea of linking them together in the form of an equation—both of these occurred to you before *the meeting?*

That's right. The whole thing. I started the meeting by writing the equation and saying this is what we need to know and this is how we're going to talk about it. That was planned before the meeting began.

What are your thoughts about the equation today?

Well, it's still a good equation. People have thought about it a lot over the years and have suggested different ways of expressing it, but the basic thinking behind it hasn't changed at all.

How about the probabilities? In the years since the equation was formulated, have your estimates of the probabilities of ETI changed much?

They've held the same. I don't know whether that's just prejudice but over the years I've hardly changed any of the estimates. I've seen arguments that move the probabilities this way and that but I have never found any of them strongly convincing.

What stages has your thinking about ETI and SETI in general gone through?

It's remained pretty much the same. If there's been any change, it's in the sense that when I first thought about this, I had the idea that other civilizations would follow the pattern of terrestrial civilization rather closely. Now I am of the opinion that there are probably a wide variety of paths of development, and end states, and natures of civilizations. And what we have on Earth is emulated in other places, but is probably a relatively small subset of all the civilizations. In other words, I've become more open-minded and liberal as to what form civilizations may exist in.

In another way my thoughts about SETI have gone through stages, but they depend mainly on the activity of other people. I did what I could do, wrote what I could write, but then somebody else would start a new thrust in SETI and this would get me active again. After Ozma it was really very quiet for nine years, at which time Oliver started the Cyclops study at Ames, and that lighted the whole subject up again and got it really boiling for a couple of years. Then that sort of petered out.

Then Ames started a whole new project, the Morrison study group, which was a much larger thing still. That got the subject going again, got me active and doing some new things with theories and papers and such. Then, about a year ago, there was what might be called the third Ames thrust. It was one aimed actually at building the equipment and doing the searching. It's still going on right now.

Does there seem to be a larger, more permanent, dependable nucleus?

Yes, the last five years, mainly stimulated by this new program, because this new program isn't just talking, for a change. It's aimed at actually doing something, some serious searching. The fact that it's really serious has goaded some people into participating who otherwise wouldn't have.

Do you have any feelings about that, one way or another?

Oh, it's good. More is better.

Are they coming in because the field is serious, or because there's money available?

In every case, these people could get support elsewhere. It's not because they don't have other options; it's because they are interested in SETI. Some of them could make much greater names and a lot more money doing other things, such as Oliver. Peterson at Stanford, who designed the electronics, is in big demand as a consultant to electronic companies and such. He could do much better financially if he skipped SETI.

You mentioned that two thousand dollars was invested in Ozma. If more money had been available twenty years ago, would that have enabled Ozma to go on?

No, but it would've made what was done better. We wouldn't have run it longer because it would've started consuming too much telescope time. As it was, we tied up the telescope for about a month and that seemed to be as much as we could do without starting to elicit criticism. People wanted to use the telescope for their research. The limitation on Ozma was the perceived limit on reasonable telescope time. That perception was never tested. We never queried anybody or tested the waters, but that was what we guessed.

So that rather than finance was the obstacle?

It wasn't the finances. With our finances we were careful to build equipment that was going to be useful for conventional astronomy anyway, so that nobody could say we wasted a penny, which was true. And that was the reason we chose 21 cm, because that was a frequency on which you could do radio astronomy with narrowband equipment, which was what we were building. It wasn't for any profound reason except that one. Later Morrison and Cocconi came around with the profound reason. That was nice.

Do you recall any real opposition to SETI?

No. At the time we didn't receive a single letter protesting or criticizing it. None. Nobody called on the phone, nobody wrote letters. There was no criticism. People often ask me that. I imagine that there were people who criticized it over the dinner table, but never to the extent where they felt they had to sit down and write a letter or phone. There were just none.

Did you get positive letters or other positive responses?

There were positive letters, but they were mainly from the usuals: the Hesburghs and Olivers. A lot of letters from the public.

Do I understand you correctly, that there was not the kind of opposition for you that Galileo met?

People keep hoping that there's some kind of history like that. It just didn't happen. There just wasn't any, and again I think because the project was so modest it would be trying to kill a flea with a cannon. If somebody had come after it they would've felt they were going after small potatoes.

This is different from the responses that were brought forth later by the Pioneer and Voyager missions.

Yes. There were complaints about racial, sexual issues. There were people who felt the money shouldn't have been spent, even though it was very small.

There was an episode which is relevant here. In 1974 we sent a message from Arecibo to the stars. This was written up in the newspapers and was read about by a very senior British radio astronomer, who is now the Astronomer Royal, and he got all upset about our activities. He felt we were notifying the universe of our existence and they were going to come and eat us. He got really uptight. He was writing all kinds of letters and wanted the I.A.U. to pass a resolution condemning all such activities.

Martin Ryle?

Ryle, yes.

Is this emergence of criticism related to the public's awareness of the attempts?

I think these later efforts appear to be larger efforts. They are with regard to resources. No one has criticized the proposed Ames-JPL program but there again the demands on telescopes are very slight. To me that's the trigger of criticisms. If you start cutting out other peoples' opportunities then the criticisms pop up. But until that happens it's live and let live.

We're talking now within the scientific community?

Yes. There was a very rare example of public criticism of SETI. It was in the *N.Y. Times* about six months ago. It was an article on the Op. Ed. page. It's the page opposite the editorial page, and.every day they have some guest comment there, usually by some expert. It'll be on 'was it right to bomb the Iraqi reactor?' or some big thing— inflation and so forth.

One day it was an anti-SETI article. Very strongly antiSETI, by someone nobody's ever heard of, and it was so poorly written that I as a SETI person still don't know what the guy was protesting. I don't know how it got in the *N.Y. Times* because it was so illiterate, but they chose to print it, which said something. And it wasn't triggered by anything. There hadn't been any events or news in the paper. It came out of the blue.

Did the other members of those early days in SETI have a perception of themselves as being somewhat different from mainstream scientists?

They had perceptions, but it wasn't that. Every one of them had excelled. They weren't mainstream scientists. They were all in some way elite in their fields. Morrison was an elite theorist of physics.

Calvin was a semi-god in biochemistry. Atchley had been one of the great successes in the electronics industry. Struve was the greatest astronomer of the twentieth century at that time. Sagan was still very young but was already making a mark as a planetary astronomer.

So every one of them was, in that respect, above the average scientist in their discipline. There was a bunch of different disciplines involved. Oliver was a great inventor of all kinds of things. There was a common thread there, which was superiority—superior performance in a host of different ways.

What brought them together on this particular issue?

Just whatever they did. There was a physicist, mechanic, and so forth. For whatever reason, they were also interested in extraterrestrial life. I never asked. I told you how I got involved, but I don't know their stories. It's an interesting question, now that you mention it, that should've been raised at some after-dinner get-together. "Just what were you interested in? How'd you get started in this?"

One common thread seems to be location. You're at Cornell and Morrison and Cocconi were, too.

That was coincidence. It turns out that a lot of the people at some stage were at Cornell. As far as I can tell, that's all or almost all coincidence. As I mentioned, when I was a student here Morrison was here too, but our paths never crossed; never. I only came on the faculty long after Project Ozma. But Sagan is at Cornell partially because I'm here. Sagan started looking for a job around '67 and I beat the drum to get him hired here. And that's the only fortuitious circumstance that had so many of the people at Cornell.

So there was not a little SETI cell?

There was no cell, no secret society or study group or anything. None. People didn't know each other then.

After something gets started, then the cell can keep going.

Yes. It's self-feeding, which shows itself by Sagan being here. If I hadn't been here Sagan would almost certainly not be here. He would be somewhere else.

And Cocconi and Morrison?

Their collaboration? Cocconi and Morrison were both here when they wrote the paper. That was a result of lunchtime conversations. They were both primarily in physics but had an interest in extraterrestrial life, for I don't know what reason, which they would talk about over lunch, and that led to their paper.

What would you do differently if you were starting over again to study SETI?

If we were doing it again we'd certainly use modern equipment, because the equipment we have nowadays is unbelievably better than what we had then. What we had then was terribly crude. The only other thing I would do differently is to exclude news people. What happened was that just about the time of the experiment I was called by the science editor of the *Saturday Review*. John Lear. He was terribly interested and begged to be allowed to observe the experiment.

I was very reluctant to do it but he was a very eminent science writer even then. He was so persistent and he promised not to interfere that I finally agreed. He came. And by the way, we've been good friends ever since and interacted many times, so basically it was all right.

But what developed was that he was a serious science writer, and despite his promises and his appreciation of the difficulties, he just couldn't bring himself to just remain in the background. He couldn't avoid asking questions, probing around to find out what's going on and why are they turning those knobs now—things like that.

And that was quite a distraction. It was a mistake, despite the fact that it's no reflection on him. He was doing his job and restraining himself as much as was reasonable, but even so, it was a distraction. It made you feel as though you were in a goldfish bowl, and that's not a good way to do science.

What do you think is the best way to deal with a subject like SETI which is trying to get started and in which the public has some interest? Do you think something should be done to generate more public interest, or do you think the public understands about as much of it as you'd like them to understand now?

The answer to that question is complex because there are many things that you need to do. There are three segments of the world that you have to gain the support of. One is the public. The public has to be for it or you're dead. But that's pretty easy to do, particularly with something like SETI, because the public is very enthusiastic about it.

With the public the problem is not to generate enthusiasm but to generate *informed* enthusiasm. The problem is that the public, being somewhat naive scientifically and largely misled by TV and movies, cannot distinguish between what's scientifically and technologically credible, and what's not. They have a very difficult time judging between the credibiity of using spacecraft to go to the stars, between spacecraft coming to us, UFOs, and between radio searches.

To most of the public, all these things are equally sensible, whereas to scientists only the radio searches and similar means of contact make sense. So it's easy to get the enthusiasm of the public, but there is a job to be done, which is to channel that enthusiasm and focus it on what is technogically sensible.

The second segment consists of the bureaucrats who provide the funding. That is, the officials of NSF and NASA. Usually there's no problem convincing them of the reasonableness of the technology and the search strategy. The problem with them is providing them with enough ammunition to head off any criticism. They're very sensitive to any kind of criticism that would make them look as though they're wasting taxpayer's money. What they really need is armor: information they can use to head off critics such as Senator Proxmire.

The third segment is the Congress. They need to be educated, just to head them off from making specious or malicious criticism of the projects. You don't want them coming down on bureaucrats, because that's exactly what bureaucrats are very sensitive to. And so, for adequate funding, you must educate and create enthusiasm in both the funding agencies and the Congress. Neither one of those is going to cooperate if the public is not with you, so that's why you have to include the public also. It's a complex game.

Why is ETI important to you?

It is most important to me because it offers the answer to the philosophical question, "What is the significance of intelligent life in the universe?" To learn what varieties of that intelligent life exist would, to me, answer that question, or go a long way towards answering it. That's the prime motivation.

There is a second one, which is that I'm just curious. I like to explore and find out what things exist. And as far as I know, the most fascinating, interesting thing you could find in the universe is not another kind of star or galaxy or something, but another kind of life.

And the third reason is that I really think that our civilization would greatly benefit and be improved by what it would learn from studying other civilizations. Those are the reasons.

What has been the most difficult aspect of your ETI work?

Getting the time to do the searching.

How do you think most scientists today view SETI?

Legitimate, important, but not worth the commitment of major resources at this time, because of our level of ignorance.

What scientists or other people in the past do you admire?

It's a long, long list. It depends how far back you want to go. I'll list them sort of in chronological order: Newton, Maxwell, Einstein, Eddington, Chandrasekhar at the University of Chicago. He's just about retired now. Certain radio astronomers, J. Pawsey, Peter Debye, Hans Bethe—that's probably enough.

Who are the outstanding people in SETI today, and during the past twenty years or so?

One of them unfortunately is dead. That's Struve. He'd be very interesting to talk to. He goes back farther than anyone.

From the very early days there were the key people, Morrison and me. Morrison's interesting to talk to. He's always been an enthusiast and a thinker, and a very effective supporter, but he's not one to actually build things and search. He's a theorist at heart.

Another one is Oliver. He was very interested in the early days although he was not active then. He picked up the ball later and ran with it. He was a part of it, and when he had the chance to contribute, he did. Sagan was not a participant till very much later. He was an eager spectator in the early days, and it is good to have the views of an informed spectator. Those were the key people, and Cocconi also.

Billingham has been behind all of the work at Ames, all three thrusts. He's been the man behind the scenes. He's the one who put things on the tracks. He's good to talk to if you want to know the modern history, the history since 1970. He's the one who put the ingredients together to make things cook.

Well, that's all on the American side. There's the Soviet side we haven't talked about at all. They're the other active group. They had a small core of enthusiasts, too. One of them is brilliant and that's Kardashev. Another one is Troitskii. He's the vice-director of the radio physics research institute in Gorky, with some three thousand staff members. He's been pushing SETI and that's fine. There used to be another supporter, Shklovskii, a very brilliant person. For some reason which has never made sense to me, he's now believing that we're alone in the cosmos. So he sort of backed off and left the SETI fold.

The Russians have conducted programs almost continuously, whereas the American programs have been sporadic; we run for a few months, then there may be years without a program. The Soviet Union's been running continuously for ten years or so. They've got about five stations going. So they are trying, but at the same time their equipment is very inferior to ours. As a result their searches are, to our American eyes, a waste of time. Just trivial.

There's one other person that should be mentioned. That's Kraus at Ohio State. He picked up the ball and dedicated a big radio telescope to SETI and has been doing it for years. That's all it does now. It came with modest sensitivity and only a few radio channels, but he's poured the whole observatory into SETI.

What is the response of his superiors, supporters or the people who pay the bills for Ohio State? Do they approve?

I'm really not close to it, but the impression I get is they approve. Somehow, I don't know how he does it, but he gets money to keep the thing going. He spends all his time either running his telescopes or doing *Cosmic Search*. *Cosmic Search* he funded with his own money. And it's been losing money right along but he keeps it up.

The only reason he can do that is that he is the author of best-selling textbooks on antennas. The books are used by universities all over the country in electrical engineering departments. So he got a lot of royalties from his textbooks on antennas and decided to pour his royalties into SETI by supporting *Cosmic Search*. He may even fund the observatory privately. I'm not sure. If so, it's pretty amazing. It's certainly a unique case.

How would you characterize your own role in SETI?

I provided two functions. One was I quantified it. I expressed SETI in the form of equations which could be used to justify searches and to optimize searches. And second, I was the advocate for a long, long time, which kept the subject alive, but more important, gave the subject a credibility which allowed other people to enter the field successfully. Because of what I did, people like Jill Tarter can work in SETI. If I hadn't done that, they would've probably found it impossible to get started. So I quantified it and I opened the doors for other people.

What else should future historians know about you?

I don't think I have anything suitably profound for such a question, but two thoughts that come to mind are: one, don't try to make it a one-person show. Don't do it to glorify yourself, because that will simply irritate people and cause them to put up obstacles. It's better to make it a community project and enlist other people.

And the other thing is, thinking about Galileo, you can't know enough about the technology involved. You've got to be an expert on the technology. If you take some new, unknown and untried instrument and try to use it effectively, you're going to be fooled unless you very thoroughly understand it. That applied to Galileo's telescope.

Would this imply that it would be better not to go ahead with a certain step unless someone is able to follow it through and work out all the details?

Yes. I think you'll find that either the project won't work at all, or it will get results that are enigmatic. You don't know what they mean: whether to trust them, what flaws the instruments might have made. I know that from SETI, but I know it also from a lot of things in conventional radio astronomy, where people have made what they thought were great discoveries but it turned out to be completely fallacious. There was something in their equipment: subtle, deep, that when you know about it, you realized it couldn't have made the big discovery.

Don't trust things to other people. That's a corollary. If you want something done right, you've got to do it yourself. A rule I know.

Is there anything else about you that we should note here?

There's two things about me that don't usually make any of the articles and things. One is that I'm quite good at understanding what other people are doing, and offering helpful suggestions for improvements. Anyone can discuss with me an instrument or scientific experiment, and I have good insights about those things and can suggest to them how they can do it better, whatever it is.

The other thing is that I'm very good in dealing with people and encouraging them. I've come to recognize I'm a little peculiar that way. I'm not an egotist and I'm not highly competitive. It has always been to me a good thing to help someone else to be good, which I do by working to see that they have the wherewithal and the time to do things, and to see that they get credit and public attention. Also I do what I can to improve their experiments.

For many, many years I thought everybody should be that way, but I recognize that very few people are. Most people are very competitive and get all upset if their colleagues or subordinates are successes. The best thing that could happen is if someone on my staff or who I'm working with becomes a great success and is internationally renowned. I think that's terrific, and it makes me feel very good, whereas I've discovered a lot of scientific directors and administrators don't really like that. They feel threatened, I guess.

But the consequence is that whenever I've been in charge of something, I've always had very good relations with the people and agencies I've worked with, and they've been pleased and many of them have gone on to do good things. Like I gave Sebastian von Hoerner his chance, really—one example of many.

And that's one thing that falls through the cracks when everybody's writing articles because it's not very exciting; it doesn't make for good press.

You mentioned UFOs. What are they?

I've given it a lot of study. UFOs are, in my opinion, in most cases natural phenomena misinterpreted. That's the great bulk of them, and it doesn't include most of the spectacular ones. Some are psychological failures of the perception system. There's no doubt about that, because in investigating some cases, particularly when we were investigating bolides and fireballs, with the goal of finding a fragment, we would interview witnesses who had seen this thing. We knew exactly what they had seen, which was a very bright meteorite.

It was very interesting what was reported by some of these witnesses, who seemed in every respect completely normal, sane people. To them it had been a spacecraft, they had seen people in it, they had seen people jumping out of it. The misperceptions were very dramatic and very reminiscent of UFO reports. But clearly they had *not* seen that. It was something that their perceptive mechanism in their brain had concocted.

A substantial fraction of UFOs are frauds and hoaxes. We run into a lot of those. There are many people who just enjoy doing frauds and hoaxes. It's kind of sad, but the whole UFO literature is contaminated with that kind of stuff.

Since no UFO has ever left an artifact, there's no evidence that any of them are spacecraft from another world. By artifact I mean an object that was clearly manufactured elsewhere than on earth, or a fact that we did not know but which we could go out and verify. There are all sorts of reports from people who supposedly have been given new information by creatures in UFOs, but none of these things have been anything of interest. Certainly nothing that we didn't know already.

So, given that set of information, my conclusion is that there is no evidence that any UFO is the product of another intelligent civilization.

Realizing that it is only a guess, when do you think contact with ETI will be made?

I have hunches. They're not well informed. Nobody's well informed on this but my usual answer is before the year 2000.

What might ETI be like?

They won't be too much different from us. What I usually say, when people ask me that question, is that a large fraction will have such an anatomy that if you saw them from a distance of a hundred yards in the twilight you might think they were human.

There are reasons why we are bipedal, why our head is on the top, why we have two eyes and why they have to be in the head and close to the brain, and why the mouth has to be close to the eyes. There are arguments for our basic anatomy being the way it is. We don't need five fingers but we need fingers. We don't need a nose but we do need the mouth and eyes and ears. There would be enough similarities so that if you saw the creature in the twilight you would think it is a person.

Assuming that you or your colleagues do eventually receive evidence of ETI, what should be done when you receive that information?

That's the most popular question being asked these days. It's also a question for which we have no good answer. The real answer is, "It depends." It depends on the nature of the signal that is received. If it's a marginal detection of a signal, where you have only enough sensitivity to show that there is a signal there, without being able to extract information from it (which in fact is a good possibility and the most likely possibility), then you must use great care. You probably will ask another observatory to verify the existence of a signal, just to be sure that it's not an artifact of your equipment or a local hoax.

So you would get verification and then you'd announce it to the world. There's nothing else we should do with it. You would inform your funding agency and have a press conference where your group would be represented, and your institution and your funding agency. That's the right protocol.

It's possible that a signal would be received with enough signal to noise that you could actually extract information from it. And that's another ballgame. The information may be unintelligible, which creates a problem. Again, you have to hold your press conference, reveal the existence of a signal and tell something about it, but you're not able to say what it's saying, and that will cause a great deal of awkwardness.

In almost any scenario you'll have to release the news soon because it will get out one way or another. That's because none of the radio observatories have any security setup or any way to inhibit the news from getting out. It would leak.

There's another scenario, where the signal is strong enough that you can get the information from it and actually determine what the information is telling. For instance, if it's really television pictures, or some of these simple pictures like we play with in our coding schemes. Then, of course, it's *very* exciting, and you'd better take a close look at the information to see if it would appear threatening to anybody, and make a judgment as to just what you say. It's hard to say. You can't foresee all the possibilities.

The next question concerns the public's response when a signal is received and announced.

A big range of opinions on that. I think there will be a great deal of excitement, for at least a few days, and then the excitement will subside. I don't believe there will be panic or anything like that, but it will certainly be a hot topic for a long time to come. Any time an observatory comes up with more information, it's going to make a front-page story once again. So I think it will attract a great deal of public attention.

I don't think it will lead to catastrophes like the War of the Worlds broadcast, and the real impact will be seen only years later, as the information gathered starts to affect decisions made on Earth, regarding technology or whatever is affected by information which is garnered from another civilization.

The long-term effect is going to be enormous, more than anyone can imagine. Once we know there is another civilization out there, there will be a very strong impetus to mount much more powerful searches, much more powerful listening systems, because every government will realize that there is possibly very valuable information to be gained which is useful for economic reasons, technical reasons, military reasons. There is prestige involved, so all high-tech countries will pour resources into this thing.

And if there's one civilization out there, there's more than one to be found, so there will be a massive listening effort, and the consequences of that, probably in a matter of decades, will be detection of a number of civilizations and the acquisition of useful information from them. Information as to what technologies are possible. Does it really make sense to colonize space? How much can you alter a planet to suit your purposes? Things like that.

Once we know what is possible, and maybe even what is desirable, we may find general rules of civilization as to what they do with themselves. This will influence people's thinking. Decisions will be made in the context of this knowledge of what civilizations do, so it will affect the course of economic development, governmental systems, how we cope with population growth, energy and resources—all those things will be affected. And this will have an enormous impact on the nature of the life of human beings.

Bernard M. Oliver

ELECTRICAL ENGINEER

Born May 27, 1916

Santa Cruz, California

Chief of NASA's SETI program, Bernard Oliver manages the joint teams at Ames Research Center and Jet Propulsion Laboratory. He was among the first to decide that radio was better than lasers and other means for interstellar communication. He subsequently directed Project Cyclops, the 1971 engineering design study of a large system for detecting extraterrestrial intelligent life. He participated in numerous important meetings, including those in 1961 at Green Bank; many in the United States during the 1970s; the 1971 conference at Byurakan, Armenia; and the 1981 conference in Tallin, Estonia.

He received a B.A. in electrical engineering from Stanford in 1935, and an M.S. in 1936 and Ph.D. in 1940 from the California Institute of Technology, magna cum laude. From 1940 to 1952 he was employed at the Bell Telephone Laboratories in television research and radar development.

He joined the Hewlett-Packard Company in 1952. He was Vice President for Research and Development for twenty-five years, and a director of the corporation. Under his guidance many new product lines were developed, including computers, desk-top and hand-held calculators, mass spectroscopes, laser interferometers and atomic clocks.

Retiring from Hewlett Packard in 1981, he continues there part-time as a technical advisor, and assumed his NASA position in 1983. He is an Adjunct Professor in the Astronomy Department at the University of California, Berkeley.

He holds more than fifty patents in electronics, developed over a forty-year period, and a similar number of publications. Examples of his reports on SETI-related subjects include "Radio Search for Distant Races," *International Science and Technology* (1962); "Proximity of Galactic Civilizations," *Icarus* (1975); and "Signal Processing in SETI," *Computer* (1985).

He has been a consultant to the U.S. Army Scientific Advisory Panel, and a member of the Congressional Review Committee for the National Bureau of Standards, the President's Commission on the Patent System, and the NAS/NRC Space Science Board. He is also active in regional affairs, serving on the Palo Alto school board, and engineering advisory councils for Stanford and the University of California. In addition, he was a consultant to the Bay Area Rapid Transit system.

Among his awards are the National Medal of Science, the Lamme Medal of the Institute of Electrical and Electronic Engineers, and California Institute of Technology's Distinguished Alumnus Award.

He has served on the board of directors of the Institute of Radio Engineers, and was president of the Institute of Electrical and Electronic Engineers. He is also a member of the National Academy of Engineering and the National Academy of Sciences.

Interviewed June 1981 at Palo Alto, California

Where did you live during childhood and when you were growing up?

Soquel, California, on a farm there, near Santa Cruz. We always called it "the ranch," but it was really a farm.

What were your childhood interests?

My childhood was a little unusual in two respects: I had nobody to play with at home—I had no siblings—and we were a long way from any neighbors with children. So except for those occasions when we visited my cousins in Santa Maria, or some other relative, I was essentially alone, which meant that my play was kind of self-invented. On the other hand, I did go to school rather early.

My mother was drafted into teaching because of the great shortage of teachers following World War I. She started teaching at Aptos, a little town south of Santa Cruz. I was four years old at the time, and she asked the superintendant of schools, "What'll I do with him?" He replied, "Take him along."

So I was bundled up every morning at 5:30, and we rode in a horse-drawn buggy four miles to Aptos and school began. I started in the first grade that year. I remember I did all right.

At four years of age?

Four years, yes. It was originally a one-room school. Then it split into two classrooms, and so they had a primary-grades teacher and an intermediate teacher. My mother was the principal and taught the intermediate grades. I don't know whether it was to get brownie points or what, but the primary-grades teacher had me skip third grade. As she put it, I took the third and fourth grades in one year. It advanced me another year, which put me seriously out of step with my classmates agewise. I was three years out of my peer group. In those ways I think my childhood wasn't an ordinary one.

I remember being interested in astronomy as a very young person. I guess it was when I was about four or five years old, because my parents were somewhat literate as to what the stars were all about. My dad had a theodolite which he would set up, and we'd look at the craters on the moon, look at the satellites of Jupiter, and things like that. So I got an early introduction. I read very extensively in astronomy, popular works, to be sure, because I wasn't sufficiently educated at that time to read technical literature.

Do you remember any particular things you read then?

There were a couple of books from England my parents got me, called *Splendour of the Heavens*. They had very comprehensive star atlases and treatises on various aspects of astronomy. I also read quite a few things written by James Jeans, who was a great popularizer of the period. Jeans was talking in those days about a barren universe. He felt that life on earth was indeed a unique phenomenon.

I remember finding that point of view rather distressing because the universe seemed so vast, and to have the only consciousness that appreciated it exist only on this tiny planet seemed to me a colossal waste. Those were not very good scientific reasons, but in a less scientific and more philosophical vein I think they had some validity. I didn't like the idea that we were alone in the universe, but who was I to refute Sir James Jeans?

Do you recall reading any comic strips when you were a kid?

Sure. I used to read the comics avidly.

Any particular one?

Alley Oop and the others in the papers we took. I didn't buy comic books. I didn't read any space comics, I read science fiction. One of the more profound influences of my young life was Hugo Gernsback. *Science and Invention* was the first magazine I remember, and then *Radio News*. *Science and Invention* was a popularizaton of science that was a little bit different from *Popular Science*, another magazine.

My father brought *Science and Invention* home to me one time, during the years when I had to amuse myself. It was 1924. I must've been about eight years old. It had a picture of two movie stars kissing each other on its gold cover. I thought that was kind of shocking. The cover article was about how many thousand germs were exchanged per kiss, that sort of stuff. When I opened the magazine and read it, it was just full of all sorts of tantalizing ideas. I read it and read it and read it. Every night I went to bed with it, and I couldn't wait for the next edition to appear.

Then, around 1927, when I was in the sixth or seventh grade, Hugo Gernsback brought out *Amazing Stories*. I read it avidly and still have the first copies. Ferren Cathey and I, who were buddies in Aptos School, alternated. He'd buy a copy, I'd buy a copy, because we didn't have that much money, so each of us had half a complete library of the first edition, the first year of *Amazing Stories*.

Did you run across Buck Rogers?

Well, I knew who Buck Rogers was but I never read him because I thought that was very poor science fiction. The science fiction I was reading in *Amazing Stories* was a different kind of thing. It wasn't a horse opera set in space, which is what Buck Rogers was. We disdained it. We thought it was terrible, prostituting the new medium that way.

Many stories warped the facts. We looked upon science fiction as good when it was truly correct science and led to a situation that was totally new in man's experience. And many of the stories were just exactly that. The best science fiction doesn't warp the facts but tells an enchanting or moving story, sticking to what we know but perhaps extrapolating. I think if you're going to deviate from the facts, you ought to tell one big lie and stick with it rather than abandon all science. I'm not very sympathetic to fantasy.

So Flash Gordon was probably even farther from your acceptance?

How dare you mention his name! No, those things, either through choice or happenstance, didn't get into my orbit. Then I had a long hiatus with science fiction, and I didn't reestablish connection with it until about 1950. All during the thirties and forties I dropped out because I was in college and studying like mad and didn't have time for it, and later because of graduate school. Then I was in Bell Laboratories, then WWII came along, and we were working overtime on automatic tracking radar. So it wasn't until post-WWII days that things slowed down enough and I had enough time. I took it up again. Meanwhile, Claude Shannon had been an avid reader all during this period. So I had his library to go back to.

I was fascinated with radio, which was new when I was new. My father built one of the first radio receivers in California. We listened to broadcasts from KDKA at that time. I remember once sitting up in the wee hours of the morning and tuning in a faint squeal. It turned out to be a 50-watt station on Prince Edward Island, Nova Scotia, which I thought was doing pretty well on the apparatus of the day. This was around 1925.

Then I had some interest in music so I got to making electric reproducing systems for phonograph records. I got in the good graces of the projectionist at the Santa Cruz theater. I'd skip gym class at the end of the day in high school and I'd go up and run the projector for him, so he used to give me all the records he had. You know, a little symbiosis.

When I got to be sixteen years, I then was able to get a license and drive a car. We had trucks on the ranch that I could drive around. That sort of opened heaven's gates to me. I could go anywhere. But in the early period I was pretty well confined. When I was in high school (I entered high school at eleven), I had to ride in the morning with my father and then wait around in his office, then ride home at night with him.

I spent my afternoons then playing with a rotary calculator he had, learning how to extract square roots on it and do other things that were fascinating. In fact, there's the calculator on the desk. That may have had something to do with my getting involved in hand-held calculators.

How was your elementary schooling?

I guess mine was, in an educational sense, better than the average. I think my mother was a good teacher. She was certainly a disciplinarian and she was certainly a person who venerated the idea of learning. She was intellectually inclined, and she considered her task in life to drag these peasant boys and girls up in the world. So in that sense it was good.

In another sense it was bad because she was also my mother and leaned over backwards to prevent any scandal of playing favorites, so she became very harsh with me at school. I gained a teacher and lost a mother.

Did it give you helpful preparation for a scientific career?

We had a book that was what's now called an "enrichment" book. It was called "Thinkers and Doers," and it was about American inventors and other early inventors. It was exciting, a kind of a real historically true Tom Swift series, you might say. In other words, instead of reading books about a fictional boy genius, we read a book about real people, and that was a very exciting thing. I had made up my mind to go into science at that time. I had a chemistry set, and I was fascinated by it. So science was it from the very beginning.

What about secondary school?

Spotty. My physics teacher was terrible. His name was Kazmerek and he was a ham-radio enthusiast. After he took the roll he would go out to his radio shack and talk to hams in other countries while I ran the movie projector for the class, showing films made by General Electric on various topics in science. That was our physics curriculum. It was really very, very poor.

I learned more about physics from the principal of the school, who came in one day and told us all about simple harmonic motion. That was more than we ever heard from the physics teacher. So I had no quantitative introduction to physics until I got into college, and then it was tough. On the other hand, high-school chemistry was a pretty well-administered course. I never took biology or botany.

How would you assess high school's helpfulness to your scientific career?

Not very great. Well, I shouldn't say that. I think our math courses were pretty good; and since that's really a foundation for almost everything else, I would say I did learn algebra and I did learn geometry and trig. I could handle those subjects very well, but we had no calculus in high school, and that's the math that really opens your eyes; that's what math is all about. When I got to college I remember that calculus and freshman physics were a revelation to me.

I gravitated toward radio communication, electronics, and electrical engineering and graduated in '35 from Stanford and then went on to graduate training at Cal Tech. I left there in late '39 and was hired directly by Bell Laboratories and completed my degree while working for them. I wrote my thesis and got my degree actually in '40. I worked very happily there for twelve years and then in '52 I was persuaded by Hewlett and Packard, whom I had known at Stanford, to come out and join their fledgling firm. I always had kept track of it in the intervening years. In my summer vacations I'd drop by to see how things were going. They finally persuaded me to join them.

When did you first think about the possibility that extraterrestrial intelligence might exist?

Oh, probably at the age of four. Very early.

Who or what suggested this possibility to you?

Well, you've got to cast your mind back to the pre-NASA days when there was a real question as to whether or not Mars and Venus had life on them. People speculated in the early part of the century, following Percival Lowell, that Mars might be inhabited. Venus was a mystery planet. We didn't know the surface conditions as we now know them. It seemed possible at that time that Venus might be covered with water and might be a very steamy jungle planet, with saurians and things like that. So speculation was rife in those days, and I can't say when I first thought of the idea, or had it presented to me.

Speculation then was about life on other planets of our solar system. There was a common belief that the solar system was unique. Sir James Jeans was saying the rest of the universe was barren because, he felt, the chance of formation of a planetary system around stars was negligibly small.

The reason he was saying these things was that the earlier nebular hypotheses of Kant and LaPlace, in which they visualized the stars collapsing from a rotating cloud of matter and spinning off planets, had been called into question in the middle of the nineteenth century, through the discovery that most of the angular momentum in the solar system, something like 99 percent of it, is in the planets, whereas 99.7 percent of the mass is in the sun. This maldistribution of angular momentum is inconsistent with the statistical mechanics of formation of stars by that mechanism.

So that led the whole scientific community astray for a long period, and they considered all sorts of other possibilities. No one could think of a way to put brakes on the sun. If the sun were spinning too fast initially, which the theory predicted, how would you ever slow it down? That was the question. So then they went astray on these catastrophic theories of planetary formation, and those held sway in the late 1800s and early 1900s. James Jeans was part of that movement. Those theories predicted that planetary systems would exist about one star in a billion.

Furthermore, in that era the origin of life was shrouded in mystery and thought by some to be perhaps supernatural. The chances of it coming about out of ordinary chemistry were thought to be minuscule; perhaps a factor of 10^6. By the time you multiply those two, we were indeed unique in the galaxy. That's what James Jeans was basing his arguments on.

What's happened since then is that magneto-hydrodynamics has been discovered. Rapidly spinning stars throw out matter in the form of plasma. The plasma is attached to the magnetic field of the star and brakes the star's rotation. Nobody sees any difficulty now with the original hypotheses of Kant and LaPlace. So that has increased the factor of f_p in the Drake equation by a billion times.

The work of Stanley Miller and others on early planetary atmospheres, the synthesis of amino acids, the work of Schopf and Barghoorn in detecting early microfossils back as far as three and a half billion years, shows that life started very quickly, probably out of the prebiotic chemicals that were evolved by the early atmosphere and by energy sources acting on them.

So there's a chain with one little gap in it: how did the first self-replicating molecule get made? But aside from that gap, there's a fairly unbroken chain of cosmic evolution from the Big Bang to us. In Sagan's words, the best assumption we can make is that we are mediocre: the earth is a mediocre planet and the sun is a mediocre star and we are mediocre forms of life.

Did you discuss the possibility of ETI with other people?

Well, as a sci-fi fan we youngsters believed that it existed. You had to believe or you wouldn't be a member of the club. So I didn't have to convince anybody. If they read sci-fi, they *knew.*

What about communicating with, or at least searching for, ETI? When did you first think about this possibility?

Post WWII, or about the same time that Frank Drake actually set up his experiments. I was in radar during the war at Bell Labs. In radar you have dishes and microwave transmitters and send out a bang of energy. It is reflected off objects and the reflections are picked up by the same antenna that radiated the bang. The radar range in WWII was on the order of twenty or thirty miles, reflecting off an airplane target.

Just for the hell of it, I made some calculations one day to find out what the range would be if you *didn't* reflect the signal off something, but picked it up with another antenna without those intermediary reflections. I was astounded with the results. I found that you could communicate to Mars or distant planets with a watt or so. This was startling. So that immediately said, "Well, let's make a big antenna and we can perhaps get signals from other stars." We put those numbers in and they were rather encouraging.

So I did that and, when lasers came along, I jumped at the thought that maybe *that* would be the way to do it. I did some calculations using lasers and found to my surprise that the laser was inferior to the radio channel. It's just that you've got more energy per photon, so you take more power to get a few photons. You can focus the light easier than you can radio waves. But finally, I drew up some comparisons, and it was that that caused me to be invited to Green Bank in 1961 for the first SETI conference there; I think it was that. Anyway I gave a paper on it.

During your college years, was there any stimulus to think about ETI and the possibility of searching or communication with it?

No, I don't think so.

After you completed your schooling and began working as a professional scientist, who or what has influenced your thinking about ETI and SETI?

Around 1960 I read about Ozma in *Time* magazine. Shortly afterwards, on some Hewlett-Packard business, I found myself in Washington with a meeting on Monday and a meeting on Wednesday and a free day in between. I called Frank Drake and said, "Would it be possible to visit you and see what the apparatus is you're using on Ozma? I've read about it a great deal."

He said, "I'd be very happy to have you, but you can't get from where you are and back in a day and have any time left."

I said, "Don't underestimate me. I happen to have an airplane at my disposal. One of our salesmen here has a Mooney and is interested in flying out himself."

So we did. We took off in the morning from Washington National and headed over Fredricksburg. On the map there was a radio range at Elkins, West Virginia, and the south arm of that crossed our path about a minute before we should get to Green Bank, which we estimated in twenty minutes.

Sure enough, at nineteen minutes the radio range went steady but down below there were solid clouds. We carried on for about another minute and I saw a hole in the clouds and I motioned to Bob. He did a wing up and we went down through the hole. We were right over Green Bank! I saw a car start up and head out to the little landing strip there, and that was Frank Drake. We landed safely and had a lovely day.

Drake pointed out that we ought to take a look at Sugargrove on the way back, which we did. It was supposed to have been a 600 ft. telescope. At that time it was still under construction; it was later abandoned. But I saw Frank's apparatus and we kept in correspondence after that.

He was gracious enough and kind enough to invite me to the National Academy of Sciences meeting that took place in Green Bank, the famous one where the Order of the Dolphin was formed. That whetted my interest in the subject. I began to think about it, and decided that before we ever made contact, we probably would have to increase the size of our receiving antenna substantially. So I began to think about how one could produce arrays of antennas that would be effective, what the requirements on the arrays might be, and what the problems would be if used to form an image of sky and various things of that nature.

I gave a series of talks both before and after the time I was presi-

dent of the IEEE. People would ask me to give a talk and I would talk
on the potentialities of this happening. Well, me and my big mouth.
That's what got me chosen to direct Project Cyclops out at Ames. John
Billingham and Hans Mark decided this would be a good Summer
Faculty Fellowship Program for Ames and they invited me to head it
up and I accepted. That, of course, led to a greatly increased rate
of thinking on the subject and realization that this might all come
to pass.

Since then we tried to keep the flame of SETI alive at Ames
through several rather lean years. It's still alive, not flourishing exactly, but still alive.

To get back to the thread of what I was trying to say, the realization
that scientific thinking had changed and that what was in my childhood a barren universe has become a potentially highly populated
universe; the realization that physical space travel, despite the science
fiction writers, is not a possibility in a human lifetime; and my earlier
expertise in electronics—all combined to make this a fantastic application of electronics. It was the fact that I was familiar with the technology that might be the key to a revolution in man's intellectual and
social history that impelled me on.

My chosen field is still developing its potentiality in this direction
because of the LSI revolution. In 1971 we had to resort to optical data
processing in the Cyclops system. This data processing can now be
done with integrated circuits and will be done much more powerfully
a few years from now.

Now is the time to really be thinking about the design of the ultimate data processor for SETI. What we really need is something that
sifts this cosmic haystack for the tiny little needle of signal, that may
be present at any frequency in a very wide band at any time, and find
it. We need an electronic brain with that one purpose in mind: to find
an artifact in all the noise and hiss we get from space.

*What effect, if any, has your interest in ETI had upon your career as a
scientist?*

It's diverted time from my other, more mundane activities. I took
time out in '71 to run the Cyclops study. And I've spent time off in
giving talks and preparing for them, and so on. So it has been a little
drain on my energy and my time.

What do your colleagues think about your interest in SETI?

I think there was mild interest. I don't think they exactly did hand-
springs, but they were interested in it and were very tolerant of my
taking time out and giving a talk on the subject, even though it didn't

bear on Hewlett-Packard business. It's a subject that takes most people a little while to get used to. Then I think that most people who have any technical training become interested. At least that's been my experience: skepticism at first, followed by enthusiasm later on.

Some of my colleagues, having been exposed to talks I've given or things I've written, have come back and said, "Hey, you know there really *is* something to this. I didn't think so, but I do now." I've had that happen four or five times, but that's not a large number.

So they ask you about it and discuss it with you?

Yes.

I don't mean the Billinghams.

Yes, I know. The others.

What about your family?

You mean my immediate family? I don't know. They don't scoff openly at it. I think that they're rather intrigued with the idea.

What about your superiors, bosses, etc. What do they think about it?

To cast myself back a couple years, I think that Bill Hewlett and Dave Packard both were interested, but they had a kind of a pragmatic way about them. They'd be more interested if there were any direct proof that there was something out there; inference was not quite enough. In SETI we must really go on the basis of dead reckoning, so to speak. As the Cyclops report says at the very beginning, "Absence of evidence is not evidence of absence." That's our motto.

What about your friends?

A lot are very positive. I don't know of any that have a negative view on it. If they do, they don't tell me.

Before going on to discuss your ideas about SETI, let's get a little more background information. What was your father's occupation?

He was a civil engineer. He worked in the county surveyor's office in Santa Cruz for most of his adult life. He had done some private surveying for a few years and then joined the county staff and stuck with it all through the twenties and thirties and into the forties. He died of a heart attack just prior to his retirement.

So radio was not really part of his job, or part of his occupation?

No, it was everybody's toy at that time. You got some buss bar and some sockets and things, and instructions to tell you how to lay the thing out, and you wound so many turns of such-and-such a wire on such-and-such a diameter tube, connected it up to this thing called the variable capacitor, or "condenser" we called it in those days, and away you went. We had a long time getting the doggone thing to work. I forget what was wrong with it. Something had been miswired. But that straightened out and we had a radio, a Lawrence M. Cockaday five-circuit tuner.

How much education did your father have?

He was a civil engineer student at UC Berkeley, but his father was ailing and he had to come back and take care of the ranch, and he didn't have enough money to finish. So he never got his bachelor's degree.

What about your mother; what formal education did she have?

My mother, on the other hand, who started out contemporaneously, was able to finish. She was a school teacher. She graduated from the University of California in the class of 1905. She went back for her master's degree and got that in 1907 and went into teaching. She taught high school in Sonoma for several years, and then taught elsewhere for a while. Finally she and my father, who had known each other in college, were married in 1915.

She was a kind of fire-breathing young socialist at the time. She was what would be looked upon today as a member of the Fabian Society, except that I don't think that she ever carried a card. My mother was a very dominant sort of person. She was forensic, I guess is the word, except that's come to mean other things these days; she loved to argue.

She felt an evening was wasted unless there was an argument on some intellectual subject. She was an intellectually oriented person, very, very fond of ideas, but not very scientific, although she thought she was. In other words, she didn't require that her truths be proven; she didn't subject her ideas to rigorous tests. She shared a quality of gullibility with her brother, my uncle, whom she disliked.

She got involved in spiritualism and automatic writing and that sort of thing. I accepted those as a child, and then later had to reason my way out of that. She was a great believer in survival after death. She really felt we went on. She did a lot of automatic writing. One of my earliest memories was of seeing her lying in bed and writing automatically with her eyes shut—great long manuscripts allegedly from Jack London who had died and whom she had known in Sonoma.

I became disenchanted with that stuff. I kept asking her for tangible evidence. I had a cousin, Rebecca Bell, who was supposed to have telekinetic powers. I kept asking to meet her but somehow this never came to pass. Nothing ever materialized for me. I finally just turned against it and said, "This is junk," and I clawed my way out of the miasma of swampy vapors that constitutes the spiritualistic world. The sciences felt so good I never wanted to go back.

My father was perhaps a more level-headed sort. He was a very kind man, a very quiet person. He was good natured, thoughtful.

How did they influence your eventual choice of science as a career?

Not at all. They were only helpful.

Were there other people we haven't mentioned who influenced your ideas about science and ETI?

Well, science generally, of course lots of people come into play there. I had various professors, like Terman at Stanford, who were instrumental in leading me along. I had some good ones in my graduate work at Cal Tech. I had Smythe, who was pretty hot on electricity and magnetism, a tough course. And Lindvall, who taught an interesting course in engineering, all the kinds of engineering mixed together.

Yes, a lot of people have influenced me, and I suppose that influence didn't stop at college; it went right on to my associates in Bell Labs. I found myself in a very big pond there with a lot of smart frogs in it—people like Shannon.

What religious beliefs did your parents have?

They belonged to no orthodox church of any sort. I think my father had been christened a Congregationalist by his mother when he was very little, but he never went to church; it didn't interest him. My mother, however, had this strong interest—a philosophical interest, let's say, in life: what was life? And she believed that there was a soul. And the reason was that material things were far too gross to in any way hold this marvelous quality called life.

She shared the ignorance of her day about how complex that matter really is, what a marvelous and intricate thing the brain is. So she was a victim of the mythology that has endured for a couple thousand years. She never went to church; she felt herself a cut above that.

About the time I entered college, I decided that I couldn't hold some of her beliefs. My early acquaintance with science and with astronomy made it appear that hers weren't tenable beliefs. And so we used to argue a great deal about it. We developed into a polarized situation in my later years. She tried to convert me to her ways, and I had given up on doing the same to her. I could see that she was too old to change.

If you had asked me about *my* religion, I am certainly an atheist. It's hard enough to explain the universe; don't ask me to explain God, too.

Frank Drake brought up something similar which led him towards science, reacting against the fundamentalist religion he was exposed to as a child.

I think perhaps my early familiarity with the cosmos had more to do with it. It was very easy in the ancient world, when the earth was the center of all things, to pretend that there were gods in the sky, because nobody had ever been up there. You didn't put them in the next county because you could go to the next county and see them if they were there. So you put them somewhere inaccessible. Well, now we've been to all those inaccessible places and thoroughly explored them, and we haven't found any such things.

You've already mentioned that you didn't have any brothers or sisters?

Yes, I was an only child. That perhaps is not insignificant, although in what way it affected me I don't know. I was forced to rely on my own ingenuity to entertain myself after school. There wasn't time to worry about it before school. When we got home after school it was usually about 4:30 or so, and dinner would be 6:30 or 7. So I had a couple hours in which I had to play by myself because there were no neighbors with children. I grew up in a very isolated way and was left to my own thoughts a lot, and I suppose that does tend to make a person pensive.

What were you like during childhood and adolescence?

I was always kind of a baby. It didn't matter too much in grammar school, but in high school it made a big difference because I wasn't adult sexually. I didn't date. I thought girls were nice, but I would never get near them. That social handicap carried over right through college into my graduate years. It wasn't until much later that I overcame it. Some may think I never have and others may think I have too much.

I had some very close friends in grammar school. It was a very sad day when graduation occurred because I thought I would never see them again, and I was really despondent about it. But it turned out I did see them again.

I had a lot of friends in high school, mainly because we shared an interest in the technology of the time, mainly radio and other things of that nature. When I got to college, I began to broaden out socially. I would take gals that I had known in high school to the beach and other places during vacations.

What other professional interests do you have, apart from SETI?

A rather broad interest in technology and the potentialities of looking at nematodes for their possible utility as a biological pest control. I'm starting a small business doing that. So that's a little off-center, isn't it? Does that make me eccentric?

What recreations do you enjoy?

I used to do quite a bit of photography, and I used to ski when I was younger, in Germany, but I gave that up when I lived in New York because it was too hard to get to. And I never took it up again. I used to play badminton at Cal Tech just because they had a court and it was the going thing there. I never got into tennis.

I did play football for a couple years until I dislocated a knee. I did some in high school, then I did a year and a half in college, then I banged up my knee and I had to be on crutches for a month. When I came back, the coach chewed me out because I hadn't been attending all the sessions; I'd been using the time to catch up in my work. So I decided that I didn't really see eye to eye with that. I wasn't really there to play football. I've never been heavily sports-oriented, though I like to swim and I like to do action sports of one kind or another, or I did before I lost this other knee.

I like to cook. I guess one of my problems is that my foods are too favorite—too many of them.

What has been your greatest satisfaction in life?

That's a question that one will answer very differently, depending on the mood you're in. There have been many satisfactions and many rewarding things. It was a great satisfaction to me, when I first went to college, to understand calculus and physics. It was a thrill, it really was, because I worked like hell. My physics preparation, as I told you, had been poor in high school, so I had to work like a son of a gun to understand it, but I did, and it gave me the greatest sense of satisfaction to be able to visualize vector diagrams in my head and know how the forces were acting in a certain situation, and know what the motion they would produce would be, and the dynamics of things. That was just great. The biggest acceleration in my life was that year of learning.

Falling in love was a great experience, and certainly my family has been at times rewarding.

In general, I have not contented myself with learning in isolated disciplines, but I've tried to integrate everything together. In other words, I refuse to hold what I consider to be mutually conflicting views. If I believe one thing, and another belief is impossible because of that, I have to choose between them. I have tried to find the way the world works.

And I think that this has been rewarding only over the last twenty or thirty years because so much progress has been made; not by me, but by everybody. It's been just great to follow all these sciences and see how everything is tying together. Incidentally, that's not unrelated to SETI. All of these things have converged to change people's ideas about the population density of the universe.

What has been your greatest disappointment in life?

I must confess I'm weary. I'm weary of nothing happening in SETI. As you point out, it's been twenty years since Morrison and Cocconi, and I frankly would've expected more to have happened.

The thing that keeps annoying me is that there is so little discernment, or discrimination in the positive sense, in our government that when budget cuts are made they take the coward's way out by cutting everything X percent. Nobody gets favored and everybody is equally unhappy. I think there are many things our government is engaged in that should be cut 100 percent and other things that it is engaged in that should be expanded a few percent.

I don't have to tell you that the whole NASA budget is 5 percent of the federal budget and the planetary exploration is about a tenth of that, and SETI is about a hundredth of that: about one cent per year for the average taxpayer. It's so minuscule that nobody would even feel it if it were augmented by a factor of 10. Yet that argument doesn't seem to have any force because everybody's got his pet project about which he feels that way. I could contact extraterrestrial civilizations on what we're wasting on food stamps, do a good job of it, just to draw a kind of a staggering comparison.

What was the most memorable moment or event in your life?

I don't know; there've been a lot of them. I guess first love is probably the most memorable thing; you remember those emotions longer. Or even second love. I had a wonderful time when I was an exchange student in Europe, in Germany during the 1930s. I met an Australian girl on the trip and we saw quite a bit of each other, and I traveled around Europe with her afterwards. It was certainly memorable when I had to say goodbye to her. I can see her standing now at the dock in Southampton.

What about SETI? What would you say was the most exciting moment or occasion in your connection with SETI?

Well, it was very interesting and exciting for me to hear the discussions of the 1961 Green Bank Conference. Because I had never met Phil Morrison, for example, before. He was a most eloquent expositor. I never met Carl Sagan before; he was there. There were lots of interesting people. And so the whole conference was intellectually exciting. Melvin Calvin, incidentally, learned while he was there that he had won the Nobel Prize.

I suppose the next landmark was Cyclops. I don't know whether I was excited about it; I was pleased to be offered the opportunity to direct Project Cyclops, which said to me that the government was at least willing to give $120,000 to consider this thing a little more seriously.

But in the seventies and into this decade there haven't been any really outstanding periods of great excitement. One of the temporary ones was when, at one point, we were promised funding escalating to 5 million dollars a year. But it turned out that it just didn't happen. One problem with NASA is that the cast of characters keeps changing, and every time it does you have to re-educate them all over again. Fortunately, that isn't the case this time. Hans Mark already knows about SETI; he was at Ames when we did Cyclops.

SETI has moved far less rapidly than I had hoped it might, and it's been a tedious twenty years. I hope the next twenty years will see a great deal more activity.

About what fraction of your working time do you spend on SETI?

This is subject to change in the immediate future. Right now I'm spending very little of it, 5 percent or something. It's going to go up to 20 percent.

What stages has your thinking about ETI and SETI gone through?

Well, there has been change in the fine, but not really in the large. My belief in the optimum methods of contact has stayed pretty fixed, but I have certainly broadened out my thinking as to the putative signals you might detect. I'm certainly not limiting them any more to CW signals as we originally did, because there are advantages to pulsatile signals.

In the same way that a lighthouse attracts attention better with a flashing light than with a steady glow, so does a beacon attract attention more easily if it's pulsed properly. So we've had to redesign our apparatus to handle pulses, too, and that's a major change, but I don't see that we've had any huge revolutions. It's been an evolving thing.

What alternatives, what other possibilities about ETI have you considered?

I'm not sure I can get the chronology quite right on this, but it seems to me that the laser was invented about 1960. I was involved with Wescon in those years, the Western Electronic Show and Convention, and I made some calculations about what this new toy, the laser, might be useful for. We called them optical masers then. I talked about such things as eye surgery, and actual transmission of power over distances, and the development of huge electric fields in the focal point of the laser.

I also considered its potentialities as an interstellar communication device. The idea was that you can sharply focus a beam of light, so maybe that's the way to do it. After having seen Ozma I did these calculations and found that lasers were nowhere nearly as good as radio for interstellar communication.

So I came in through the laser route, rejected it, and landed in the microwave region along with everybody else. Other people later have been suggesting lasers over and over again, for the wrong reasons. I made that same initial circuitous approach. I made some erroneous early calculations which indicated that if you were signaling from near a given star you ought to be able to make that star change its apparent brightness. I neglected a couple of factors of 10 to the 6th or something, but that would be an ideal signaling method. Just let the star flicker at you.

Once for 'yes' and twice for 'no'.

Lo and behold, not too long thereafter, AC stars were discovered, called pulsars.

There is a problem with SETI, which, if solved, would change the whole picture, and that is the question of determining from some natural event or synchronization factor where and when to look. People have given that a lot of thought. There's a fellow who keeps calling me all the time with his proposal. Whenever a supernova or some other cosmic event occurs, we radiate. Other intelligent beings elsewhere in the universe will also be looking in that direction, so we radiate signals in the opposite direction.

The only thing wrong with that is, if you use the kind of high gain antennas we're talking about, which you have to do to get across the miles, you have to point them in a million directions to cover the sky. So you'd only be sampling one-millionth of the sky if you point it in the directions he suggests. Therefore, if there are N races in the galaxy, you'd have something like 2 million over N events that would have to transpire before you got 63 percent of the directions covered. And those events aren't that often. I'm not willing to wait 2 million centuries for this to happen.

If we had some way of knowing preferred directions as well as times, it would change the whole picture. That's the weakest link in our whole reasoning.

Otherwise our reasoning is, I think, pretty solid. It's really a philosophy of minimizing energy, because that is a commodity that is going to be pretty similar throughout the universe, and it's going to cost you something. We've tried to arrange it so that we signal in the way and in the portions of the spectrum that will take the least energy. The radio search is about 10 billion times as efficient as a spaceship, which takes an enormous amount of energy.

What would you do differently if you were starting over again to study ETI?

Well, that's one of those hypothetical questions that you can always ask: What memories do I retain? What do I know when I get back there? Am I ignorant as I was then, or do I know what the world knows now? You have to tell me what the ground rules are.

For example, Frank Drake said, in response to this question, that he would be stricter in dealing with the press because they got in his way and made things more difficult, so if he could do it over again, next time around he would handle the press differently.

Well, it did come up in his case, I guess. He had the misfortune of starting this as a real pioneer, getting the publicity very early on, when the scientific community itself was not prepared for it. So I think the publicity which the newsmen gave him reacted unfavorably on some of his colleagues. They felt, "What's he trying to do here? He's supposed to be an astronomer. Astronomers don't do that kind of thing."

In my case that never happened. So the only thing I can say is that if I wanted to pursue a career in SETI, I would equip myself with some other subjects than I have taken. I would take a lot of the same things, but I would go in more now for computer theory because I know from the outside what can be done, but I don't know from the inside how to build a computer quite the most effective way. If I did, I think I could design some of the processing equipment a lot better than I have. So that's what I feel most.

There are two things that distinguish the SETI system: One is the most sensitive receiver, which means large collecting area and low noise. And to get the noise down you have to be in the right frequency band. And, two, a data-processing system to tie to that receiver, that sifts through this haystack of noise for putative artifact signals of one sort or another. It's really a brain that scans all this stuff in a very rapid fashion and discards all but a minuscule fraction of the signals, those that we'll make further tests on.

You have to do this in an automated way because the human cannot stand month after month of negative results.

So I would say that where I am weak and would've changed my

training would've been to get a lot more teaching into my noggin on how to design hardware and software structures.

Why is ETI important to you?

I guess because I have a philosophical yearning to answer the question which was brought up—and this might be significant—by my many arguments with my mother. My mother was involved in quasi-religious or metaphysical things.

The question is really, "Is life a negligible and extremely rare phenomenon in the universe—intelligent life, that is—or is it so prevalent that the universe can be considered to be somewhat efficient in producing it?" In other words, I wonder if life is just a fluke on Earth, or whether in fact it is prevalent enough so that stars are likely abodes or places where other life is?

In the latter case, life could, in the course of time, become an important force in the late evolution of the universe. I can imagine, though I can't tell you how, that this life, in a network of communication, could form a sort of super-consciousness throughout the galaxy that, in ways we can't foresee now, might modify the history of it.

In any event, I cannot really warm up to the idea of life existing in lots of places in the galaxy and remaining forever in isolation. And in particular, I don't want my own team to do that. I don't want the human race to be an example of a galactic recluse that never tries to find anyone else.

It's these things that intrigue me. Another thing of course is the sci-fi background, which has long produced an interest in other worlds, coupled with a sober realization that we're not going to *go* there. It says, "Okay, if we're not going to go there, then this is the only way, so let's get to work on it." That's another thing that makes me interested in it.

If successful, SETI would be hailed as the landmark, a great watershed in human history.

What has been the most difficult aspect of your ETI work?

Getting funds. Everything else is simple in comparison.

How do you think most scientists today view SETI?

There are all shades of feeling there. A great many of them think it's fascinating. Others feel that there are more important mundane problems to work on; we shouldn't be wasting money on this. Others feel that, "Well, yes, it's ultimately important, but maybe the time isn't right yet because we don't know how to do the job best." A lot of our thinking in the last decade has gone toward trying to provide answers to such positions. In other words, why should we want to do it now?

Well, we have some pretty good answers. We'll never be able to make more sensitive receivers than we can now; we're up against natural limitations there. It's true that the prices of data-processing equipment are coming down; they've come down enormously already. Things that were not within our grasp in Cyclops days are now easy to do in solid state. You don't have to resort to photographic processes as was suggested in Cyclops.

And the spectrum is getting filled up. In other words, we're dirtying our window on space with other signals that will prevent us from seeing through, unless we go to a great entry cost later on, of putting something large in space on the far side of the moon, to shield ourself from man-made noise. To prevent a high entry price later on, I think we should get going now.

What about public attitudes toward SETI?

The concept is viewed much more enthusiastically by lay audiences than by scientific audiences. Scientists tend to think of all the objections, but lay audiences assume that science has the answer to everything, and this is a damn good thing to be working on. I have never encountered significant criticism in a lay audience to a talk I gave. Questions are asked but generally they are quite supportive and the applause has been warm. This includes audiences of 1500 people. SETI has enormous popular support.

I think that support is underestimated by everybody but the press and television. Congressmen underestimate it, NASA officials underestimate it. I don't think NASA officials really realize why we're in space. We're in space because of the early science-fiction writers and the people who speculated upon life on other planets in the solar system. If there hadn't been the idea in the popular mind or in President Kennedy's mind that there might be life elsewhere in the solar system, I don't think NASA would have done Apollo.

Why do you suppose that legislators and others underestimate the public's support for SETI?

I think it is a fear of ridicule. I think that politicians are afraid to take a position which is (a) unconventional, (b) cannot be proved to be right. It opens them to criticism by people like Proxmire. "You get the Golden Fleece award for that sort of thing, friend." So they're really gun-shy and afraid of it. NASA is also afraid of appearing to veer toward the UFO buffs, who believe quite other things: they believe "they" are already here. NASA is very scared of the allegation that they've succumbed to the UFO believers.

So there are those fears that affect their actions, but also they don't have as good news sense as a journalist does. You don't have to explain to a journalist that if there had been one worm on Mars, it would have been the greatest news story of the century. You don't have to explain that to him—he knows it because he knows how *he* feels. He's Mr. Public. His job is getting the public stories that *he* likes. A good editor is not a person who *thinks* about things. A good editor *feels* things. He's just like his customers. He knows how they feel because he's one of them.

So we don't have enough editors in Congress?

That's right. All we have are lawyers who never looked at the stars.

Realizing that it is only a guess, when do you think contact with ETI will be made?

Well, that's so dependent on funding. I don't expect to make contact with any of the things we're doing now. I think that what we're going through is a dress rehearsal, learning how to parse these signals apart and look for the things that might be there. And we can test our equipment in various ways. We're learning how to automate it. We will have a field test.

We're going through all this stuff, but we will not, probably, find anything unless we're extremely lucky, because our sensitivities are not high enough. Sure, there's always a chance that maybe there is something on Alpha Centauri, four light years away; if there were, we could hear it. But that's a long chance because that's one star, and we probably will have to sample thousands before we're successful.

So I really feel that it's a question of increasing the sensitivity through increasing collecting area, after we've made our first round of tests. You don't want to do that first, because maybe you don't need the area; that's the expensive part. So I think we'll probably go to a small array as a next step. Then maybe either a larger array or something in space as the second step, depending on the economics and the politics of it all.

Now if that support were there, contact might happen sometime in the next twenty to thirty years. If the support is slow in coming, it'll be dragged out.

What might ETI be like?

That's a delicate question. Nobody has that answer. There are two schools. There are the Divergenists, who believe that evolution is very sensitive to physical conditions, and what happens is the result of a concatenation of accidents that leads to a certain species, or an order of animals or plants. They say, therefore, if the conditions were just a little different on earth, the whole flora and fauna would be very different. Different accidents would have happened. Different species would have dominated at different times. The whole evolutionary chain would be very different.

There are the Convergenists, who say, "Yes, but there is an optimum design for a creature if it is to do certain things. So what happens is all these divergences occur, but then the branches that are too far out get pruned off and eventually only those that point towards the desired end survive."

It all depends upon whether our thinking is dominated by the richness of mutation, or by the natural selective process.

I think I'm a Convergenist, but that doesn't mean that I think all intelligent societies are erect bipeds that suckle their young. I certainly think that there could be other, very different types of animals. But if there is going to be cultural evolution, there must be a mechanism that maintains contact from one generation to the next. It's very hard for me to see cultural evolution taking place with turtles, who abandon their eggs, for example. Teaching the next generation what the previous generation knew seems important. It could be replaced by great individual longevity, but I don't know how much longevity is possible. What makes us wear out? We don't even know that yet!

It's been known for a long time that the evolution of life on Earth has been seriously disturbed at many times. The tree of life looks as if it had been chopped off by a vicious gardener, or several gardeners of different heights. There are branches which come out and are suddenly terminated while others go on and then branch further.

Most recently the asteroidal theory of extinctions has come up and appears to be a very satisfactory explanation. So now the question arises: Is this typical of terrestrial-type planets? In other words, do these extinctions typically occur or is the solar system unusually dirty? You might have a sister solar system in which the dinosaurs got a footing and never were extinguished. Then, you ask, would interstellar communicators be chatting with dinosaurs or would they have evolved at all? These are serious questions.

Dale Russell, who has done a lot of studying of dinosaurs, feels they would have evolved. He points out that at the end of the Cretaceous there had developed a cute little dinosaur that was much smaller and more agile than the previous specimens, had a larger brain-to-body ratio, and had opposed thumbs that could manipulate things. He could pick stuff up and handle it. So there's evidence that even if the mammals hadn't come along, something might have changed. The dinosaurs could've evolved, but we only have the one example to look at.

I rather suspect that asteroids and meteoroids are pretty common. Cometary matter is certainly common. So this may be nature's way of avoiding dead ends. Let's assume that Russell was wrong and that the dinosaurs would never have done anything. Then the world would have been a pretty stupid place. The dinosaurs were around for 130 million years, which is twice the time since their extinction. They were in fact a very successful family, more successful than we are, perhaps. If something hadn't wiped them out, they might have stuck around for another few billion years. Nothing would ever have happened on earth. So maybe these extinctions are ways of giving other twigs of the tree of life a chance.

As one particular twig on the tree of life, I'm interested in not becoming extinct. Have you heard about Project Space Watch? That's another project I've been engaged in starting with NASA.

John Naugle has organized a couple of summer meetings at Wood's Hole. The last one was just following Alvarez's paper so I suggested to Naugle that he invite Alvarez and Gene Shoemaker and others to the conference, which he did. The result of our discussion was Project Space Watch, which NASA has taken rather seriously.

The idea is that you set up one or more observatories that are automated, and you search the sky in a methodical manner, looking for something moving. You do this by taking a picture. You don't use photographic film, you use CCD arrays now and you store the pictures in a data base. At an appropriate time later you take a second picture. You subtract the two. If anything has moved, you get a positive and negative pulse in the difference picture. All stationary images cancel. This way you automatically detect motion.

It gives you the velocity vector right away, so you know where to look for the object the next time around. And you track it long enough to determine whether you already know it. You can tell that very quickly. Or if it's a new object, then you track it long enough to determine its orbital elements. Then you go back to the search. You store all the orbital elements in the data base. Periodically you compute the ephemerides of all the objects for fifty years in the future, and you see if there are any imminent close encounters of the first kind.

If there are, if you find something that's on an intercept path with the Earth, then you refine your data on it. And at an appropriate time you launch a mission to rendezvous with the object, ride it down to perihelion and give it a nudge. The nudge isn't all that much. A few centimeters a second will convert a direct hit into a near miss.

We concluded our report by saying: Yes, the dinosaurs were a successful order in all respects but one. They didn't develop the technology to avoid their extinction. We have.

It's an interesting project because the cost is very small, comparable to SETI. Unless there is an imminent collision; then the cost goes way up because you have to mount a space mission. But I don't think you'd have any trouble getting the funds.

You'd have to get the funding enough time in advance to make the preparations.

Yes. I think you appeal to the highest levels, in that case.

What are UFOs?

There are all sorts of phenomena that appear in the sky. I've seen things I couldn't identify but I don't think they're extraterrestrial. There are many things which are so small or under such unfavorable viewing conditions that you can't tell what they are, like weather balloons and other sources of light, and so on. Frequently there are objects that people attach this mysterious origin to. "Unidentified flying object" is an unbiased term, but to say that those are "visitors from space" shows a definite bias. In other words, I don't see them to represent extraterrestrial intelligent activity.

In the forty years since the term 'flying saucer' first appeared there is the meagerest kind of photographic evidence: streaks on developed film and things that looked like flying garbage-can lids; nothing at all interesting in the way of photographic documentation. In July 1972 the Earth did have a visitor from space. It came in over southern Utah and disappeared over Canada. It was in the Earth's atmosphere for about twenty minutes to a half hour. During that time there were ten thousand pictures taken of it, miles of movie film. It was a fireball that came in and flew past Grand Tetons and Yellowstone National Park and was photographed by everybody.

There was an unannounced intruder that gets documented to a fare-thee-well. So I think that if there were something to UFOs we would have better evidence than we do.

What scientists or other people in the past do you admire?

Aristarchus of Samos. I admire him for his very early conception of the heliocentric solar system. I admire Eratosthenes, the first man to measure the diameter of the Earth. Also Hipparchus, who was quite a Greek. He estimated the radius of the Earth and the radius of the Moon, from the curvature of the Earth's shadow on the Moon during a lunar eclipse.

All these people knew that these things were spheres; they weren't flat-earthers, even in those days. There were flat-earthers but not the educated people.

Galileo, of course, and Kepler. Kepler was a kind of a funny mystic, but I admire him for his tenacity and for his refusal to cave in when he found the problem he had been assigned to wasn't soluble in the way he had been told to do it. He was told to see what the best resolution or decomposition of the orbit of Mars was, using circular curves only. And he found the closest he could get was within six minutes of the correct position at times. But he knew that Tycho Brahe had observations within two minutes, and on that factor of three hinged the future of astronomy, for he then went to work and found out the truth. He posed a very tough problem for himself.

Gauss, Newton, all great people. I could give you a litany here of great scientists.

Who are the outstanding people in SETI today, and what role has each played in SETI?

Morrison and Cocconi had the vision of communication in the way we think is most efficient. Drake tried it with the crude equipment at hand, so he deserves a round of applause for that. And he's been very factual in his ideas about it ever since. Billingham has been tenacious in seeing that SETI didn't get dumped. He has a strong interest because it involves life in the universe, and that's what fascinates him; if it hadn't been for him, we wouldn't have a program here. Sagan, of course, did a lot along with Shklovskii, in their book on the subject, and its popularization did serve to build up a positive attitude about SETI. He's been influential in dealing with some of our congressmen, too.

How would you characterize your own role in SETI?

My own role has been the importation of certain engineering concepts to the field. The possibility of using large arrays, the importation of data processing technology to do the job better. I've provided certain arguments for expecting life to be out there and not here. I have had a role organizing some of the effort.

What else should future historians know about you that would not otherwise be preserved?

Well, they should know that I find it inconceivable, if indeed advanced intelligent life is a common phenomenon in the universe, that each of those examples of life should go through its entire evolutionary history in isolation. Looking at the galaxy as a whole, we see hundreds of billions of stars, of which maybe 10 billion have life sites. I see the galaxy peppered with these islands on which life exists. I find it inconceivable that each of those should remain forever ignorant of all the others.

I attempted, at one time, to put some numbers into that feeling and wrote a paper called "The Proximity of Galactic Civilizations." It was published in *Icarus*. It's not a good paper, it needs to have further thinking done on it, but I hoped to open up a new line of reasoning that went as follows: The average separation between advanced cultures may be quite great. It may be a hundred light years or more. But the distribution of these cultures is certainly random and, therefore, there are a few that are fortuitously close. In those cases, they would learn of each others' existence almost without trying.

Suppose there were a star only a light year from us that had a civilization emerging more or less synchronously with us. We could detect their radiation with relatively crude equipment. Our own present radio telescopes with very crude receivers could do it. Or suppose our sun were one member of a binary system and there were planets about both stars. If the separation were 50 AU we would easily learn of the existence of the other civilization, or in that case, we could travel there and see the remnants of the civilization, or see it emerging.

Cases where two civilizations have emerged synchronously and fortuitously close provide the nucleus for establishing a network of interstellar communication. Those two civilizations would begin communication on a long-term basis. The fact that each sees the other and thus knows of the other's existence would then encourage them to look further into space and establish further connections. I can visualize a network of intercommunicating civilizations growing up in this fashion around these nuclei.

There are three reasons for listening first rather than transmitting first. One reason I will dismiss: that is, you should keep quiet in the jungle. The second reason is that if anything is going on, you learn about it right away by listening, whereas you have to wait out the round trip light-time if you transmit.

The third reason is that if something is going on, it might be a concerted program established by a network of communicating civilizations that has picked us out as emerging. They might have been listening to us for a long time, for example, and have seen that emanations have been occurring, non-natural emanations, on the third planet around the Sun. So we might actually have some modest power being beamed at us right now.

When I was invited to become a member of the National Academy of Sciences, I thought this was quite an honor. Then I found very rapidly that all you did once you became a member was work on membership committees to get new members, and I suspect that might be true of the "galactic club," too.

The existence of a galactic club is a charming idea that's been developed pretty well by Bracewell and others, but there's something very serious in back of it all.

One thing I would say about interstellar communications is that, if and when it happens, it won't be like we might think it will be. It will be very, very different indeed. For one thing, if the distances are on the order of fifty light years or more, communication will not be between individuals but between societies and will be adapted to that purpose. It will therefore be different from back-fence chitchat.

Second, we will probably find ourselves in communication with those societies whose longevity is great, simply because they are around. They are the populous ones. They've stayed a long time. That means that some of the benefits may be not simply the advanced science, but the advanced social behavior and political forms that are stable and encourage longevity. We may find out how they survived World War III or kept it from happening.

But even beyond that, if there has been a confederation of societies of the type I described, it will have existed on the order of a billion years or more, because the evolution of galaxies suggests there were intelligent societies at least 4 or 5 billion years ago. Therefore it will possess archival information of great value, have a very long-term, mature perspective of the universe and, in fact, may be engaged in pursuits that we have no conception of.

One of the big questions we could ask is: Is life merely a peripheral by-product of the evolution of the universe, an incidental, interesting aberration of the evolution of all the matter in the universe, or will life, as it continues to evolve, end up playing some role in the destiny of the universe itself—perhaps closing it or perhaps opening it or changing it, modifying it in some way that we can't conceive of yet? I don't expect human society to give me that answer, but I might expect the galactic club to tell me.

I find myself in a difficult situation or difficult mental frame, because on the one hand interstellar travel is enormously expensive in either time or energy, or both. Yet I can't deny to an extremely advanced civilization some technologies that might change that picture somehow, or enable one to influence matter on a cosmic scale. I just don't know whether those things happen. We're really not that smart yet.

Assuming that someday we find very convincing evidence that extraterrestrial life does exist, what would you as a scientist do with this evidence?

It's hard to answer this in advance because you don't know what the nature of the message is, but assuming that it was the kind of message that we expect, a tutorial type of thing that lets us determine how to communicate, then I would follow up on it. I would favor announcing the discovery to the whole world as soon as it occurred. I'd also attempt to learn whatever could be learned from the signal. If it's information bearing, then learn how to decipher it. If it's not, then get whatever information we can from it.

There are some people who fear that if we respond, we will be visited in a hostile way by extraterrestrials. I don't share that fear at all because I am acutely aware of the costs of interstellar travel. It's so expensive that I think it's an economic impossibility. If we were trying to go to the nearest star at chemical rocket speed, it would take on the order of forty thousand years, and another forty thousand to return.

Now you say we have very limited technology. So let's now assume we have overcome technological limitations and have a machine that will get there and back within a human lifetime, by traveling at a substantial fraction of the speed of light, like two-tenths the speed of light, and that we have a 100 percent efficient drive so that no energy is wasted whatever in bringing the spaceship up to speed and back down again twice. Then it turns out that the energy that would be expended on that single trip to the nearest star is enough to supply *all* the energy needs of the planet Earth for over a millenium.

I can't conceive of an appropriations committee like ours okaying a mission that would pay off in eighty thousand years or take enough energy to supply the world for a thousand years. I don't think it's in the cards; you don't get enough out of it.

I think that these people who say, "Where are they?" because we haven't been visited are being very unrealistic about facts of life. The energy I'm talking about is an absolute minimum; it doesn't assume *any* technological limitation whatever, only the limits imposed by natural law, so there's no way to get around it.

What do you guess might be the public's response to the announcement that ETI exists?

There would be initial excitement. It would be rather sensational to have proof positive that we aren't alone. Then, when the realities of the slowness of the data acquisition became evident and the round-trip light time was realized to be probably on the order of a century, I think the interest would fade, and people would go back to the National Football League and life would go on as usual.

But over the years an enormous amount of information would probably accumulate, and in the long run, another thousand or two thousand years, the impact would be tremendous because we would then probably have such information as all the natural history of the galaxy at our fingertips, and could see how convergent or divergent evolution is on different worlds. Are intelligent life forms more or less the same shape as we are, or is there a wide variety of shapes? Is DNA the universal building block for life—the universal blueprint, let's say,—or are there other bases for life?

We would also be in touch with those civilizations that have survived for a long time, because the others wouldn't be around, and we probably therefore would at least learn what their social forms are that confer longevity.

I have a kind of extraterrestrial viewpoint about this whole thing. I look at the galaxy as whole, with these islands of intelligence scattered through it like raisins, and each of these advanced civilizations can in principle last for billions of years. I find it incredible that they would go all through their entire histories in ignorance of one another.

In other words, I think it's almost certain that interstellar communication exists, because some of these races are much older than we, and therefore, when we get in contact with a single extraterrestrial intelligent civilization, we may be in contact with the whole network, and the first requirement imposed upon us might be to erect a beacon to continue the group process. The point is that I can't imagine these societies living in total isolation for their entire histories.

And since I can't think they do, I don't want my team to be a galactic recluse.

Melvin Calvin

CHEMIST

Born April 8, 1911
St. Paul, Minnesota

Professor of Chemistry at the University of California, Berkeley, Melvin Calvin conducted research that provided insights into the fundamental interactions by which extraterrestrial intelligence might subsequently be formed. He described in detail the chemical processes by which life probably evolves from inanimate matter. In 1958 he stated that "we can assert with some degree of scientific confidence that cellular life as we know it on the surface of the Earth does exist in some millions of other sites in the universe." He added that other, unfamiliar forms of life were possible, too.

While attending the 1961 Green Bank meeting on interstellar communication he learned that he had won the Nobel Prize for his work on photosynthesis. This gave the small group, and the fledgling SETI movement, some legitimacy and favorable publicity.

He received a B.S. degree from Michigan Technological University in 1931, and his Ph.D. in chemistry from the University of Minnesota in 1935. After postdoctoral study at the University of Manchester, England, he joined the UC Chemistry Department in 1937. From 1945 to 1980 he was director of the Chemical Biodynamics Division of Lawrence Berkeley Laboratory, and from 1967 he also served as director of the Lawrence Berkeley Laboratory. He directed UC's Laboratory of Chemical Dynamics from 1963 to 1980.

He has made discoveries over broad areas of physical, organic and biological chemistry. Recently his research has turned to renewable resources for energy.

His seven books include *The Theory of Organic Chemistry* (1940); *Photosynthe-*

sis of Carbon Compounds (1962); and Chemical Evolution (1969). He is the author of over six hundred technical publications, and holds a number of patents.

His many awards, in addition to the Nobel prize, include the Royal Society's Davy Medal; the Priestley Medal of the American Chemical Society; the Gold Medal of the American Institute of Chemists; honorary degrees from a dozen universities including Oxford, Paris and Columbia; and seventy lectureships.

He has served on many scientific boards for the U.S. government, including the science advisory committees for Presidents Kennedy and Johnson, and various NASA committees during the Apollo space program. He is currently on the Energy Research Advisory Board of the Department of Energy.

Active in many professional societies, he has been president of the American Society of Plant Physiologists and the American Chemical Society. Other memberships include the National Academy of Sciences, the Royal Society of London, the Japan Academy, the Royal Netherlands Academy of Sciences, and societies dealing with biology, geophysics, radiation research and photochemistry.

Interviewed June 1982 at Berkeley, California

Where did you live during childhood and when you were growing up?

Mostly in Detroit, Michigan.

What were your childhood interests?

I was in grade school and went through high school in Detroit. By the time I was in high school I already knew I was going to try and do science. In fact, I had already selected chemistry. I had selected it not for any intellectual reason at all, but for a purely practical reason. Keep in mind that I was in college during the Great Depression. I got out of high school in 1927, and 1929 was the Crash and so, after two years in college, I had to drop out of college and go to work.

I went to work in a brass factory in Detroit. By that time I had learned enough analytical chemistry to analyze for iron, zinc and copper in brass samples. I worked the night shift.

I picked chemistry because I figured it would always be in demand. I worked in a grocery store on Saturdays; every foodstuff had to have a chemist—even the labels on the cans had to have a chemist—so I figured I would always have a job. That's why I became a chemist. No grand questions about the universe.

I'm glad the choice was made that way. As it turned out, chemistry is right in the middle between biology on the one hand and physics on the other. I could go either way, and I did. I went both ways. I did some physics and then came back and now I'm doing some biology and physics as well. Having been primarily trained in chemistry, I felt I was in the right business.

From high school in Detroit I went to a mining school in northern Michigan, an engineering school, where I learned a lot of geology. In fact, the only biology I learned, I learned in paleontology, in a geology course. That was the only formal instruction in biology that I ever had. It really wasn't very much, but I learned what paleontology was from a very good geologist. I learned mineralogy there, which I found very interesting. It was mostly crystals and the chemistry of crystals. From geology, I learned that the earth had a long history, and that there was a paleontological record of it, of life on earth, and I learned a little bit about how it came to be. That's how I got interested.

The combination of things is what put it together because, as I said, the only biology I knew was paleontology. Therefore my interest in evolution was primary. That was the only thing I had ever learned about biology: the evolutionary record as exhibited paleontologically in the rocks. That's all I really knew.

When I started in chemistry much later, here, and began to sort out problems of how a green plant converts sunshine into energy, I realized that that must be a secondary effect. That can't be the primary way in which life got started on the earth; it's just too complicated. There must have been a simpler way. And the simpler way was the heterotrophic organisms which first originated in the primitive soup. Then the question was, "Where did the primitive soup come from?" And that was why the first experiments were done. It's as simple as that. It's a rational sequence, more or less.

Thinking back to your very earliest activities in science, what incidents involving science do you remember?

When I was about eight years old I did a kind of experiment in a friend's backyard. We'd catch grasshoppers and then we would drown them; hold them underwater till they stopped kicking. Then we found that if we took them out and laid them in the sun, they would revive and hop away. We did that many times, and I still haven't figured out how it works. Apparently, they breathe through the cuticle; they don't have a lung system. And keeping them underwater anesthetizes them; it doesn't kill them, it anesthetizes them. If you put them out in the sun they dry out, they get oxygen, and off they go again.

We did it over and over again. It wasn't just a trial. I don't remember if we extended it to other animals or to other insects like ants, but grasshoppers were so easy to catch and so lively that we knew whether they were alive or not; they got up and hopped away.

I recall another incident, when I was about twelve. I was walking home from school with my best friend, who later became a practicing physician. We were talking about chemistry things and I suddenly realized that the atoms that make up all the chemical world were all related to each other, and that they were built up, in some regular way, from simpler materials. I discovered the Bohr atom and the electronic structure of atoms by myself, almost.

Actually it wasn't new. That was in 1926. Bohr had made his discoveries ten years before, but I didn't know about them. I remember the shock when I realized that these atoms are all made of the same stuff, and they differ by one particle, each one next to the other. That was a discovery for me. It wasn't something taught to me, it was a discovery. And I felt that it was very important.

At that time all they knew about were electrons and protons. They didn't know anything else. I didn't know about neutrons, and I don't know when they were discovered. It may have been that year, 1926 or so. I don't remember how I made up for the different masses, and for the different isotopes, and also for the different mass numbers, which didn't correspond to the number of positive charges in the protons.

To me the discovery was the realization that atoms were related to each other, one after another, in a systematic way, and you could build one from another. That was the thing that impressed me: that there was a system to the chemistry of the atoms of the world. And as far as I was concerned that was a personal discovery.

That realization made it possible for me to make a commitment to chemistry, though as I said the choice was not on a grand basis at all but on a very simple, practical consideration. Chemistry was an essential component of modern life. The modern world couldn't get along without chemists, so I would always have a job. I could see people getting thrown out of work all around me. My father worked in the automobile factories in Detroit, and he was out of a job at that time. Things weren't going well. That was the major factor in my decision.

How was your schooling?

I found it interesting. I went to school and enjoyed it. It wasn't a pain to go to school. I don't remember much about pre-high-school days at all. I do remember the high-school days.

How would you assess its helpfulness to your scientific career?

Not much help. It didn't do much. I had a good math teacher, but that's all. I had a science teacher who was not very imaginative. All I can remember about him was a comment he made to me in physics class. I was awfully cocky. I would answer questions very quickly, without getting it all together. I was right nine times out of ten, but every once in a while I was wrong. And so he said, "You'll never make a scientist; you don't collect all the data."

I have since told my students that there's no trick to get the right answers when you have *all* the data; just the computer can do that. The trick is to get the right answers when you only have *half* the data, and half of the data you've got is wrong and you don't know which half is wrong. Then when you get the right answer you *are* doing something creative.

And I think the kids understand that after a few years. They realize that there's a creativeness in science which is not simply deducing from data. It's imagining something that *isn't* in the data, something that's not there. And then you go and find out if it is true or not. *That's* when you do creative science. The rest of it is computerized business.

This high-school physics teacher thought of science as that kind of thing. You collect the data and then you draw the conclusions. And there are still a lot of them like that. But the idea that there is a creative process, something that isn't obvious in the data, that a computer couldn't do, that your head *can* do, that *you* put together and you get new things out of it—*that* is creative science, and I get the kids to think that way after a while.

Turning now to the subject of extraterrestrial intelligence, when did you first think about the possibility that extraterrestrial intelligence might exist?

Well, my interest in that particular business is very easy to pinpoint. It came as a result of what ultimately was called the studies of chemical evolution. These studies began in 1951. The first paper that I published on that was about '50 or '51, in which we used the 60-inch cyclotron which used to be here, on this site. We used this 60-inch cyclotron as a source of intense energy to see if that kind of energy input could rearrange atoms into useful combinations. We started with carbon dioxide, hydrogen and water, and we got formaldehyde, formic acid, and glyconic acid.

I didn't put nitrogen in; that was one thing that was missing from that experiment. That was the first experiment in prebiotic evolution, 1951. Since I didn't use nitrogen, I didn't get any amino acids, naturally. That experiment has been redone by Cyril Ponnamperuna, one of my students who is at Maryland now. And, as I predicted, he has without any problem obtained amino acids; just the same experiment except he put nitrogen in.

Now we know you don't need a reducing atmosphere, but at that time there was a debate as to whether the first atmosphere of the earth was reducing or oxidizing. Since the most plentiful element in interstellar gas was hydrogen, the argument was that the Earth's atmosphere should have been a reducing atmosphere if it were an aggregation of that. At that time there was a big debate between Harold Urey and Wendell Latimer as to whether it was an oxidizing atmosphere. Urey said no, it must have been a reducing atmosphere.

I did my experiments with an oxidizing atmosphere, except I had hydrogen there. I didn't have methane, I didn't have ammonia. I had hydrogen, CO^2, and water. And I got everything which you would now expect.

Then about four years later, Urey did the experiment with ammonia and methane, and he got amino acids, and that's where things took off. Then I took to writing the book on molecular evolution, which I did in 1967.

Long before that, when we wrote the paper on CO^2 and hydrogen and water, I was thinking about it, and I came to the conclusion that there must have been evolutionary processes of the same sort elsewhere in the solar system, and perhaps elsewhere in the universe. That was in 1951. I thought about the astronomers' calculations as to how many different suns there were in the universe that were equivalent to our own. They come up with some very big numbers, and then they ask how many planets around those suns might have the same kind of temperature distribution that ours has and you come up with still big numbers, and so there must have been other places.

And that was the reason for my interest; there was no question in my mind about the Darwinian behavior of molecules. And if that were true, then that meant it must be true in the entire universe. Then there must be other places just like ours, and we've got to find them. That's the basis for this. And that started in 1951 when I began these first studies on chemical evolution of things.

Did you discuss the possibility of extraterrestrial intelligence much with other people?

Not really, except for an inner group. I didn't really discuss it outside, because all I was doing when I did that was to repeat what I had read—what Shapley had written, and what other astronomers had written about probabilities of planets with living things on them. Nothing beyond that, except within me I already had the conviction that Darwinian evolution acted on the simplest of atoms, which means it acts on hydrogen, and when hydrogen interacted with water or oxygen or nitrogen, then there was a gradual selection process. I discuss all that in *Chemical Evolution*.

If one accepts that, and accepts the astronomers' calculations as to how many earthlike planets there are, then you can't avoid the possibility of intelligent life. You *have* to have it. And that's where I was. Not because there was any evidence for extraterrestrial intelligence, but only on the basis, first of all, of accepting and demonstrating Darwinian evolution on molecules as well as on organisms, and accepting the astronomers' and astrophysicists' calculations as to how many stars there were, and how many of different ages, of different types, and how many had planets, and how many planets had conditions conducive to the evolution of biological molecules.

With all those numbers you can't avoid the possibility that there *are* other places where evolution has gone a lot further. And that's all there is to it.

When you did mention this to other people, what sort of response did you get?

Well, the only one I can remember that was not inside the community (inside the community is like preaching to the choir) was the first public lecture I gave on chemical evolution, which, believe it or not, took place in the chapel at Amherst College in Massachusetts.

I can give you the exact date. It was the Sigma XI Lecture, published in the *American Scientist* vol. 44, no. 3, July 1956, entitled "Chemical Evolution and the Origin of Life." There was an earlier publication but not a complete one; it was a very short one.

That was the first time I discussed the whole thing publicly in a large audience. It got a little of the Darwin-Huxley argument. After I was all through somebody in the back asked from the gallery, "Didn't you just get through proving there is no god?" I said, "Oh no. God told me to do this," and then there weren't any more questions.

In 1967–68 another course of lectures, the Oxford Press lectures, was given in the auditorium of the zoological museum in Oxford, the very same room in which the Huxley-Wilberforce debate took place. I didn't know it till afterwards; my tongue would have got tied. If I had known it I might have made something out of it.

What effect, if any, has your interest in extraterrestrial intelligence had upon your career as a scientist?

Not much except, as I said, there were a series of a hundred publications in the area of chemical evolution. That's different from extraterrestrial intelligence. It's something that has to do with molecular things and not with the question of intelligence. I never got into that.

Let's get a little more background information. You said that your father worked in the automobile industry. How much education did he have?

Very little in the United States. He was an immigrant. He came here when he was sixteen years old. He was very well educated in Europe, but after he came here he didn't make use of it very much.

He came from an urban family in eastern Europe. My mother came from a farm family in southern Europe. They were from different parts of Russia, and from totally different social strata: an urban intellectual in the north and she was a farm girl in the south. They never would have met there.

She immigrated to America at the age of ten to twelve. She was taking care of her little brothers and sisters, but she had no formal education.

After my father died, she developed a number of skills from which she earned her living. But they weren't things you went to school to get. She knew how to sew, knit and crochet. She knew how to do things like that. They were of commercial value.

My father was an intellectual, a European intellectual. Whatever he did, he did with precision. Even in his work in the automobile factory, he was a precision mechanic. He was proud of that. He worked for Cadillac, and I remember his sense of pride that he worked in the very finest motorcar factory. And he learned it all himself. He started out as a manual laborer.

How did your parents influence your eventual choice of science as a career?

Not any influence, except by their example of their hardships just before the Crash, while I still was in high school. And the fact that they tried a small business for a while: a small grocery store where I worked during my pre-high-school and high-school years. That was where I saw how important chemistry was. But otherwise they didn't influence me toward science, not directly at all. They didn't know what science was. There was no reason they should.

What sort of religious beliefs did they have?

They were Jews, both of them. Russian Jews. They didn't keep any religious practices. When I grew up I was without religion; a-religious, not anti-religious.

Are there other people that we haven't mentioned who influenced your ideas about science in those early times?

No, except, as I said, the mathematics teacher in high school was very good. She was very encouraging. But I can't remember anything she did or said; it was just her attitude.

The other thing that did happen that may have had an influence was that my father's brother came to America young enough to have a university education in Minnesota. At least he knew what science was and he was helpful to me, not by telling me what to do, but by helping me when I was in need, and in a way that my father couldn't help. So I suppose he had an effect, but it was a very nebulous kind of thing; it was just his presence and his understanding. Not that my father wasn't sympathetic. Quite the contrary: he was very sympathetic but he wasn't knowledgeable. There were no other influences except the circumstances—circumstances of the 1929 Crash.

How about brothers and sisters?

I have one sister, about ten years younger than I. She didn't have the opportunities of education that I did. I guess it wasn't customary, it was not part of the family tradition. The son, the first born, gets everything in education. The younger daughter gets what's left over.

What about your friends when you were growing up? Did they share your interest in science?

Only one, and that was the boy I went to grade school and high school with. It was through him that I made the discovery of the periodic table, or whatever it was. He became a physician. He went to medical school at Ann Arbor and has been practicing for the last fifty years. He's one of the leading internists in Detroit. I haven't seen him in the last three to four years, but I see him occasionally.

As kids we discussed many things, very many things. We did those grasshopper experiments in his back yard. He had a back yard and I didn't.

When you were in childhood and adolescence, what were you like?

I never had many friends; I only had one friend and that was Abe.

What other interests do you have?

We have a ranch about eighty miles north of here and my wife and I are growing oil plants on it. We are in the business of trying to grow oil because we don't want to run out. These trees in the picture are Brazilian trees. This tree here is from the Amazon forest in Manaus. There is a wooden bung in the tree, and you see black around it. The oil runs out through these little holes in the vertical pores in the tree, and drains into the funnel and then out into a bucket. This picture is the beginning of a plantation of that big tree.

Realizing it is only a guess, when do you think that contact with extraterrestrial intelligence might be made?

That could be today, or that might not be for one hundred years. I don't think we're ready today. If it is a matter of receiving signals and not receiving visitors, then I think it will take a little while, because the problem of reading those signals and understanding them is going to take some time. And then communication is going to take years because of the time it takes signals to move back and forth over the immense distances, if we ever get to that stage. By "ever" I mean within the life span of the human race as we know it. That means thousands of years from now.

What form might extraterrestrial intelligence take?

Well, as I say, it can be either communication by signals or it can be communication by visits.

What about the beings themselves?

There's no way to tell that, no way. The only thing you can say is that they'll have organs for sound, organs for sight—at least those two—and probably for touch, because the universe in which they live is a universe of light and of sound. Yes, sound, because if there are gases there, then there will be sound. And there's got to be gases there in order to have any kind of life we know. Gases mean sound, because solids and gases make sounds. Even in water, a liquid, there is sound transmitted.

There would also be light, a source of energy. It's difficult to imagine any evolution occuring in the absence of electromagnetic radiation. They have to have sensing organs somewhere for that kind of radiation. Whether the radiation is going to be in the visible region or not—that's uncertain. I would expect that to be the case, but that doesn't necessarily mean that it is. They may have sensing organs that can sense wavelengths that we can't sense. Our eyes are sensitive from 4000 to 6800 angstroms. Theirs may be from 2000 to 4000 or 6000 to 10,000. Infrared, or maybe in the microwave region. At any rate, some kind of sensors for electromagnetic radiation.

So, sensing organs for sound, sensing organs for electromagnetic radiation and probably for touch, so they don't bump into each other. I don't know what else they might have but these three would be essential. I can't imagine development of an intelligent arrangement of atoms that doesn't have those gifts, those capacities, aside from the information capacity which is equivalent to the brain.

Don't ask me what form *that's* going to take. Processing all the information that sensors receive—there's got to be some way to do that. But those are basic things that any kind of evolving system generates. And intelligence, by definition, means the ingestion of, the receipt of sensor information which is stored and is processed. So you have to have the sensors, and you have to have the storing and processing system, but I don't know what shape that's going to be.

How do you think most scientists view SETI?

Oh, I don't think it's considered not legitimate. It may not be looked upon as important, in the sense of being likely to yield useful data in the lifetime of one or two scientists, or one or two generations of scientists. And that's probably the main reason for an apparent present lack of interest: because it doesn't seem likely to have a consequence for one or two generations. On the other hand, if you don't start, you won't have it at all, so somebody's got to do it. So I wouldn't be adverse to seeing money spent on that.

Do I understand correctly that the amount of time you're spending now on SETI is close to zero?

Close to zero, that is correct. *Not* because I'm not interested, or that I don't think it's important. It *is* important. It's just I have no contributions to make, I have no way to help. I listen, and I periodically see if I can assist. At the moment, the only way I see it working is by detecting and understanding signals. I don't expect to have a visitor that I can analyze.

However, there is an interesting correlative subject just now arising. In the last few years they have discovered, in the south Pacific, warm water coming up through the bottom of the ocean, from the seabed. They've seen strange living organisms down there, organisms, that's what they are, not plants or animals, very deep. There's no light down there, no air, no oxygen, but they are living organisms. They move around. We have photographs of them and, in fact, we have been able to bring some of them up alive, and in the last few months we've learned how to grow them on the surface, in laboratories. They have to grow under 50,000 atmospheres in 500 degrees Fahrenheit or 150 degrees Centigrade. That is very hot, and 50,000 atmospheres is pressure! And they don't grow under any other conditions.

So far nobody has done the biochemistry of these organisms. They're too difficult to work with—they burst when they are not kept under pressure. We know they are made up of the same elements: carbon, hydrogen, oxygen, and nitrogen. But, what kind of molecules do they have? Do they have DNA or not? Do they have protein or not? We don't even know that yet.

Another big question is what is their source of energy? It looks like it's H_2S and CO_2. Not oxygen at all; there's no oxygen down there. But there's CO_2 coming up, and there's H_2S coming up. Now H_2S and CO_2, the thermodynamics of that is an energy-producing reaction. Maybe that's the source of the energy for these organisms.

The suggestion had been made that this is an evolutionary environment, quite different from the one we know up here on the earth's surface. These may be primal organisms, not simply degenerate surface organisms that had fallen down there and have made a Darwinian adaptation to that environment, but rather they originated down there. It's a very different thing. The question is, which is it?

It would be very important if these are not degenerate surface organisms. It would be another source of the origin of life. There are people in Oregon in the Oceanographic Institute who think that it may be really primal, different origins. They base this on oceanographic reasons, not chemical reasons. Until I see the comparative biochemistry, I can't be sure.

That's the kind of thing I could do something with. I can't do anything with signals, but I'm encouraging this. I said, "If you want me as a collaborator, I would be glad to be one. It's not extraterrestrial intelligence, but it's still a possible source of life fundamentally different from our own."

Look back at scientists of the past and ask yourself what would you like to know about Galileo, or Newton, or somebody that you admire, that we don't know today. What similar sort of information should be preserved about you?

I have been reading, for obvious reasons, a lot of Darwin in the last four or five years. I hadn't read the *Voyage of the Beagle* until four years ago. And I've been trying to read *Origin* for the past three years since then. I carry it with me constantly when I travel. The only time I can read it is on the way home, when I'm not too tired. On the way out I've got business in my head. And on the way back I'm usually too tired, so I don't get much chance to read it, except sometimes at night when I'm in a hotel room alone.

From what little I know of Darwin's life, it seems like it was a collection of accidents, but accidents which he made use of. That's what I think *my* life has been, a collection of accidents. Some of them, most of them, I used—not all, most of them. I put them together to make a useful combination. It's hard for people to realize how much of a life is such a collection of accidents and fortuitous juxtapositions of bits of information by which a person arrives at something new.

It would be interesting to know if Darwin had given any thought to extraterrestrial life. I have seen no evidence that he did think about it. Maybe when I get around to reading the *Descent of Man* there will be some sign in it of his thinking of other worlds: could there have been similar organisms someplace else? I have not heard anyone give quotes from Darwin on that, but the people who are doing this SETI stuff are not the kind of people who read Darwin. Certainly not the *Descent of Man*. They're mostly physicist and engineering types. They're signal readers, they're not biologists or biographers, nor even chemists. So they're not that kind of people.

I'm thinking of who they are. The only one that might possibly fit would be Josh Lederberg. He was tops in science, but he gave it up to become president of a university. You can't do both jobs.

I've never taken an administrative job in all my life, and never will. I had no hankering for it, ever. I founded the laboratory and administered it for thirty years, but I did it in such a way that it would support my work rather than hinder it. In other words, I got a lab going and got people to run it so that there would always be a substratum, what they call in industry an infrastructure, on which I could do my science, and I had somebody else run the lab. I was responsible, but I had various administrators actually doing it. And I did it largely to be sure that there was an infrastucture for my work, which was only a small part of the whole lab—10 percent or something like that. It's important to me to have that, so that I don't have to fight for a living every few months.

What was the most memorable moment or event in your life?

Oh, there's been lots of them, lots of them. Our marriage, seeing my children growing up on various occasions. Of the scientific events, I suppose the most important one was the moment of discovery which I recognized to be that moment, twenty years ago, when I really understood what the carbon cycle was, and I *knew* that it happened.

It happened rather quickly. I was at home. I had some data and I put it together, and finally I could see it working right. I had not written it all down but I had written the main essence, bits of it, on pieces of paper. I put it all together, and it worked! That was a big moment for me.

We don't have those very often, maybe two or three times in your life, but it does happen. When something new presents itself and you know it's right, you know it's right right away, then it's good. Then you do one or two experiments and it comes out right, then you know it's right for sure. Then it may take ten years to convince the rest of the world, but that's all right, that's not the problem.

What scientists or other people in the past do you admire?

Obviously Einstein had the biggest effect.

Who do you think are the outstanding figures in SETI?

The only ones I know are Frank Drake and Barney Oliver. There are others, but those are the only two serious ones.

How would you characterize the role of Frank Drake?

He is a very solid astronomer, and he made an effort.

How about Barney Oliver?

He's an electronics expert, and he provided a lot of the push behind the thinking of how to go about doing this.

How about the others? How would you characterize Carl Sagan's role in this?

He's a publicist.

How about Philip Morrison?

He's a first-class scientist, and his thinking about it has to be taken seriously.

What do you think the public thinks about the idea of extraterrestrial intelligence?

I don't think they're thinking about it right now. They have other things to worry about. I doubt there's much public discussion of it at all. In fact, the only related subject which has reached a high level of public visibility is the debate between the creationists and evolutionists. SETI hasn't even gotten to anything like that. The SETI debate is taking place at a much lower level.

So would you say the the public . . .

They are not even there. Look, my impression of the public would be that they are not even fully conscious of what the debate between the creationists and evolutionists really means. By "public" I mean those who are not directly involved in it at all, either as creationist or evolutionist or scientist. For the general public it's peripheral. They think the creationists and evolutionists are arguing about something unimportant. The don't realize how important that argument is. It's really the heart of the whole business.

The creationist-evolutionist argument is an earlier stage, much earlier than SETI. You've got to get *that* one settled before you can even talk about life on other worlds. The creationists don't believe in SETI. No, they wouldn't accept it, because there's only one place. God is a very limited person for them. God lives on earth, and made man and nobody else. He's a little guy, as far as they are concerned.

What stages has your thinking about ETI and SETI gone through?

As I mentioned, astronomers' calculations about the age distribution of the stars, and planetary distribution around the stars of different ages, suggested that there were planets with living organisms on them billions of years before us. That gives rise to the notion there are evolutionary sequences that have progressed beyond where we are, and others that haven't progressed as far, and there are others somewhere in the middle. That idea is not a new idea. You find it in Harlow Shapley's popular books on astronomy, written twenty years ago, which I read.

The book that really turned me on was *The Meaning of Evolution* by George Simpson, based on his Silliman lectures at Yale. I read it in 1949 and that prompted me to do the experiments in 1951, and that started the whole thing. Shortly after that, Shapley's book came out, *Calculations of the Planets*. There was a concatenation of factors, including a number of meetings.

About the only meeting I remember of any consequence was the one in November 1961 which took place at Green Bank, when Frank Drake was the host. I don't know if he was directing it but he had some official capacity there. There were people from all over, from NASA, NSF and the Navy, and the scientific people were Barney Oliver, Frank Drake and Carl Sagan, and maybe two or three others. We discussed at some length the question of how would you understand a message if you ever heard one.

Among the various presentations the one I most fully remember was the one by John Lilly, a dolphin man, as an example of how we learn to understand a completely alien language. We came to a general conclusion, at least that was my impression, that in order to make any sense out of an alien language you had to hear a conversation between *two* of them. You had to sit between them and hear a call and a response. You couldn't just hear one side of the conversation, you couldn't just receive.

That was one of the big problems. Frank Drake said, "Let's just put up receivers." But we finally decided that to understand the messages we had to put up receivers that would receive from two different directions at the same time, in order to see if there was any interaction between the two messages. I don't think that ever happened, but that sticks in my mind as one of the conclusions we came to in that November '61 meeting.

I think Frank Drake was the one person who had already *done* something, in the sense that he had a radio telescope in Puerto Rico that had been listening for some time. It's a big dish, built in a hollow in the Earth.

He also talked a little bit at that 1961 meeting about what later became known as pulsars. We didn't know what they were. They sounded as though they might be some kind of Morse Code signal system because they were periodic and had patterns. All kinds of imaginative things were said about them but that was before it was finally understood what they really were. At that time no one could prove anything about them. That was quite exciting, that period.

That 1961 meeting was the biggest of the series. There were others after that, but I don't think I was there. Or if I was, they didn't leave very much of an impression on me.

There were several meetings here on the West Coast hosted by Lederberg, and then the meetings in the '70s. I didn't participate in them. I wasn't active after 1970, not because I wasn't interested, but I just had nothing to contribute. My capacities have to do with chemical analysis and things like that, but we had nothing to analyze except signals, and that wasn't my business. So in the last ten or twelve years my main contact has been through correspondence, mostly with Barney Oliver. He came and saw me and we talked a couple of times about the possibilities. Several NASA people have discussed it with me, too.

My book *Chemical Evolution* gives the history of my involvement in this subject. It was a series of lectures I gave at Oxford in 1967 and was published a year or two later by Oxford University Press. In chapter six you can see the beginnings of my involvement in SETI. The first two references here, one is to Harold Urey: in 1952 he was discussing the planets. And then in 1953 there is a translation of Oparin's book, the Russian, on the origin of life. And then came my own paper from 1951, in which we did the first reduction of carbon dioxide in aqueous solution by ionizing radiation. And then the Miller experiment was done in 1953, where he used nitrogen; I didn't have nitrogen in mine.

It was this whole chapter that started us in the business. Then comes the selection and growth of molecules, the Darwinian selection of molecules on the molecular level. And once I was convinced of that, then in the last chapter comes a section on searching for extraterrestrial intelligence. Then, in the appendix I have the possibilities of interstellar communication.

To be sure we accurately record the stages through which your ideas developed, let's clarify the distinction between actually communicating with extraterrestrial intelligence, as opposed to the idea that it might exist. When did you first think of communicating with ETI?

In 1951 I came to the conclusion that evolutionary processes like those on earth must have existed elsewhere in the solar system, and perhaps elsewhere in the universe.

Going a step beyond that in your thinking: when did you first think of actually going out and searching for it?

That really didn't happen until it became possible to *do* it. That came through the NASA work.

Earlier I had said that if one accepts that molecules undergo a

Darwinian selection process, and accepts astronomers' calculations as to how many planets there are. . . . That was the Sigma Xi Lecture, 1956. That was when I first talked about it. That was before the NASA work.

At that stage were you thinking of trying to locate it, to find it, to communicate with it?

No, I didn't really start until I began working with NASA. NASA was created by Kennedy when he talked about going to the moon. That's when I got into it, after NASA started and began talking about it. Up until that time I was convinced that it existed but I had not arrived at any way to *do* it, to actually search for it. That didn't come until the space agency was created.

The Green Bank meeting in 1961 was the beginning of it. We went there to hear the various proposals for listening for ETI. It was not for sending a space vehicle but for listening from the Earth. By that time we were already thinking about it, and trying to get the agency to do something. That was what the meeting was for. Although the agency was not yet committed to it, there were these listening systems set up on terrestrial radio telescopes, but that was all.

Turning to the responses of other people, what did your colleagues think about your interest in searching for extraterrestrial intelligence?

Which ones? Those in the inner circle, so to speak, at the Green Bank meeting, were all doing the same thing.

What about people at your lab?

The lab wasn't very much involved in it, but I did have close contact with Professor Weaver of the Astronomy Department, and he was interested in the same thing, though from quite a different point of view.

What about the other scientists, who weren't involved in it?

If they weren't involved, we didn't talk about it. They were aware of my interest, but I didn't feel any great interest or great antipathy from those who were not involved.

I'm interested in what happens when a scientist goes into an unfamiliar field.

I can tell you that. Not for this study, but when I first began work on photosynthesis back in 1946 and I gradually had to learn biology, I was very much aware of the fact that the biologists didn't think much of me, because I was not a trained biologist.

That was *very* clear. That went on for many years, from 1945–46, when I first started it, until the late fifties, when I had begun to sort it out a little bit. It took years before they believed me. That was where I encountered that sort of thing—not antipathy but skepticism—not in the origins-of-life business.

I never encountered it there, at least not that I was aware of. It may have been under the surface, but nobody ever told me to my face.

I didn't *feel* it either, whereas in the earlier case, where I was moving from chemistry to biology, I *did* feel it. It was a very different thing. I knew it and they knew it, and they let me know.

What did your family think about your interest in extraterrestrial intelligence?

I've never heard anything about it, except my wife goes along with me. But my children were not aware of it; they were too young.

And your superiors, your bosses at the lab?

As long as I did my regular job, they didn't care. That is not their business, to worry about what I do in research.

And how about your friends?

I don't have any sense of rejection at all about that. Not in the same way that I got it from the scientific community ten or fifteen years earlier, when I moved from chemistry to biology. That was a real break. It took ten years to convince them that I knew what I was talking about.

What would you do differently if you were starting over again to study extraterrestrial intelligence?

I didn't really study it. I was a spokesman for it occasionally, because of my other interests, but I wasn't directly involved in any of that particular aspect of space exploration.

How would you characterize your role in SETI?

Well, only as a professional chemist, with some biochemical interests, and willing to try it, to find out if it could be done. I wasn't personally in a position to do anything about it, except I did a little meteorite analysis, but I found nothing there to encourage me. It was part of my organic geochemistry work.

What is the importance of this issue of extraterrestrial intelligence to you?

At the time that it was a hot subject for discussion, I felt that it was one of the most important things that the space agency could do, to determine it one way or the other, and they did make an effort on Mars. I never expected anything on the moon. We analyzed the moon rocks and there wasn't any indication there, but the idea that there might be something on Mars was still an acceptable idea, until our vehicles went there and did on-site experiments. They didn't bring anything back, but the on-site experiments were discouraging, to say the least.

There are those who say they were negative, but there is some ambiguity about it, so it still remains a question, and I think it is an important question, even today. If we could get an answer to it, and I can't say how, whether with radio telescopes or some other way, with landers and sample returns, something of this kind. I don't know what would be the best way to do it. I haven't been in touch with it for some time now.

What has been your biggest difficulty in the study or thinking about extraterrestrial intelligence?

I have never worked directly on that subject because I don't have the wherewithal to do it, either in my head or elsewhere. It isn't a direct subject of concern to me, although I feel, as any lay person would feel, that this is a very important question, which we should have an answer to. It would affect the way we behave on the earth. It's just a layman's view of it.

What do you think the public's attitude toward this issue is?

Now there doesn't seem to be an evident interest in it at all, anywhere in the scientific community as a whole, and probably none at all, or very little, in the lay community. However, ten or fifteen years ago, when the subject was still a matter of immediate concern, to at least a small group of scientists, it did make the newspapers because they thought the public was interested in it. That's probably the best way to judge the public's interest in it.

The public is interested in UFOs. What is your feeling about UFOs? What are they?

There is no evidence that they have been seen or that they exist.

Do you think they are peoples' imagination?

I don't know what they are. There are so many different ways in which that thing could happen that I won't even venture what they really are. Each sighting, if you want to call it that, has probably a different basis.

Do you think any of them might be connected in some way with extraterrestrial intelligence?

I don't think there is any evidence that they are, or that they even exist as such.

About a year ago you sent me an article about an intriguing formation on Mars that looked like a human face.

Yes, someone called my attention to that, so I sent it on to you. I don't think there is anything to it. There are many photos taken of Earth, from satellites, which look like that, too, and we know what they are: they are natural formations.

If scientists do find definite, verifiable signals or other evidence of extraterrestrial intelligence, what should they do with this information?

Broadcast it as widely as possible over the Earth.

What effect would this proof have upon us here on Earth?

It would have a marked effect. It's such a broad, major subject of concern to everyone, no matter where they are, that I think people would listen. It would have an effect on their behavior, but I can't predict what it would be. It's like introducing a new religion, I suppose, and having it picked up by a lot of people. That could be a major factor in the behavior of the peoples of the world as a whole. It would have that kind of an impact, if it were really established that intelligent life does exist elsewhere.

You feel that as soon as scientists check it out to be sure there is no mistake, it should be released immediately?

Of course! It has to be. There is no way to keep that a secret, and the more thoroughly the evidence is exposed, and the more honestly it is done, the better it will be, regardless of what it is.

Ronald N. Bracewell

ELECTRICAL ENGINEER

Born July 22, 1921

Sydney, Australia

Professor of Electrical Engineering and Computer Science at Stanford University, Ronald Bracewell has made many contributions to the theory of radio telescope antennas, and has stimulated discussion of extraterrestrial intelligence and methods of its detection. He originated the idea of using robot probes for the exploration of the interstellar neighborhood. He invented an infrared interferometer, suitable for space-shuttle launching, to use in searching for planets around stars other than the Sun. His book *The Galactic Club: Intelligent Life in Outer Space* (1974) has been translated into Dutch, Italian and Japanese.

He received a B.S. degree in mathematics and physics from the University of Sydney in 1941, and later a B.E. (1943) and M.E. (1948). During World War II he designed microwave radar equipment at the Commonwealth Scientific and Industrial Research Organization, Sydney. From 1946 to 1949 he engaged in ionospheric research at the Cavendish Laboratory, Cambridge University, England, where he received his Ph.D. in 1950.

Returning to the C.S.I.R.O in Sydney, he investigated very long wave propagation and radio astronomy. In 1954 he came to the United States and lectured in radio astronomy at the University of California, Berkeley. The following year he moved to Stanford, where his activities have included the construction of two radio telescopes, the automatic preparation of solar maps, and analysis of the tumbling of the first Russian and American satellites.

His current research is on two-dimensional imaging, especially medical imaging, a topic he became interested in when his algorithm for the construction of

Photograph by Chuck Painter. By courtesy of Stanford University.

astronomical images was universally adopted for use in x-ray brain scanners. A new algorithm for spectral analysis is described in *The Hartley Transform*, Oxford (1986).

He has also written books on the Fourier transform and radio astronomy and has contributed chapters to twenty-four books, on topics ranging from radar to indirect imaging. His papers have been published in over thirty scientific periodicals.

He has chaired panels for the National Science Foundation, the Advanced Research Projects Agency, and the National Astronomy and Ionospheric Center Advisory board, and has served on panels for the Office of Naval Research, National Academy of Sciences, National Radio Astronomy Observatory, and other organizations.

The Institution of Electrical Engineers, London, awarded him the Duddell Premium, and he was elected a Fellow of the Institute of Radio Engineers. His many memberships include astronomical societies in England, Australia and America, the International Scientific Radio Union and the International Academy of Astronautics.

Interviewed June 1983 at Palo Alto, California

Where did you live during childhood and when you were growing up?

Sydney.

What were your childhood interests?

Swimming, reading.

Tell me about your earliest activities in science.

Making explosives and smells. One of my earliest activities in science was to mix potassium chlorate and sulfur, and put it on the tram tracks. I mixed iron sulfide and hydrochloric acid to make bad smells. I used to squash pennies on the tracks and I also dropped a pound of potassium chlorate and sulfur on a concrete floor. On the way down, which took about half a second, I suppose, I had to do a lot of thinking. Fortunately it didn't blow up, or I could've destroyed myself. I was about thirteen or fourteen years old then, so chemistry was big for me, up to about the age of sixteen.

I was also very interested in those days in puzzles. I read all the books on puzzles that existed in the thirties. Though I regarded that as just entertainment at the time, I found that that experience has been very helpful to me in scientific work. The cracking of a puzzle, as contrasted with the working out of an exercise on something you've studied, is very much more related to real-life problems. In other words, when I teach you algebra, how to solve equations, then give you an example—that's less related to everyday life than being confronted with a puzzle where you've got to guess what it is that's really going to count.

So a lot of the puzzles, especially the mathematical ones, seem crazy or just amusing, but they do in fact teach you a certain way of thinking; how to get mental blocks out of the way and try to reason creatively and imaginatively and look for things that are not obvious. That's very much what scientists do, so I would recommend puzzles to teachers.

What about during adolescence? What activities do you recall from those years?

I also played chess and read extensively. I've totally given that up. I've totally given up doing puzzles; they don't entertain me anymore. But they're definitely formative.

And at school I got very interested in languages. I learned French and German at school and talked quite well with those. My Italian is very good now. I studied Russian, but never used it, so that's languished. So I was obsessed with that, and with English language for that matter, too.

I read dozens and dozens of books. I used to read very intensively when I got enthusiastic. So I read all the books in the Sydney Public Library on spiders; that took me a couple of months. Then I read everything on fungi, which is pretty hard reading, and not much of it stays in your mind. But that's the sort of thing I would do—read everything they had. I went around the Sydney Municipal Library, which was very formative for me, and read practically everything they had in it.

I would go to the section marked Psychology and there'd be four or five shelves going up to seven feet and I'd run through that, and economics. I'm just mentioning psychology and economics because they're two subjects that never managed to fascinate me; the books didn't look all that interesting. At one time I had heard that Freud was very good so I started reading everything that Freud wrote, and I can tell you that I found that very tedious, except for one book, *The Psychopathology of Everyday Life*, which I thought was marvelous. All his other stuff seemed to me to be hopeless. Of course, I wasn't in a position to judge, but it's the sort of thing that didn't stimulate me at that time.

How was your elementary schooling?

Now elementary school, I didn't find it dull or boring. It was on the positive side of neutral, but I don't recall it as being exciting. Although it was very repetitive stuff, I was always listening to something interesting.

Of course, I was interested in playing marbles then so I didn't pay too much attention to schoolwork. School proceeds very slowly; it's mainly to keep the kids off the streets, I suppose. It doesn't take all that long to teach you arithmetic or reading or spelling; you learn that very rapidly when you're in a learning mode. So school must be for the purpose of occupying your mind till three o'clock.

What about secondary school?

Now secondary school I found pretty stimulating. I enjoyed the chemistry and French and mathematics; I thought that was all very good. I didn't take geography or history, which I might've found dull. And I didn't even find English dull. I thought grammar was interesting; it's a sort of mathematics, of course, and I daresay it was taught for that purpose. It's a very technical subject which they don't teach much these days. It was broken down into parsing and analysis. A good deal of it was worked out by great minds, there's no question about that.

But I didn't much enjoy reading Shakespeare or literature in general. They made us read several plays and we went and saw performances. So I saw and read *Midsummer Night's Dream, Henry V, Richard II*, and I read one or two other plays because my aunt had given me a whole volume of Shakespeare in microscopic print; appeared to be everything he ever wrote. I had no trouble reading it at that time. But I didn't find that stimulating.

I never understood why Shakespeare appeared to be so great. I had the feeling that anyone could write a play, and I started to write a play. The fact that I didn't get more than a few pages into it seems now to indicate that there's more to doing it than struck me when I was about twelve years old. I've never had the urge to write a play again, so I've now come to conclude that there is something about playwriting that I don't contain, and if I were to force myself to do it, which technically I suppose I could do, it would not be considered a great play by critics.

I think I'm defective in music, too. I've noticed that other people get more pleasure from music than I do. And I'm slowly beginning to understand that I've got a dimension there that's not fully developed.

How would you assess the helpfulness of secondary school to your scientific career?

I think that secondary school was very helpful to me, that whole period.

When did you first think about the possibility that extraterrestrial intelligence might exist?

Age ten. I had experiences like Carl Sagan. I read Edgar Rice Burroughs. I read *The Chessmen of Mars,* and all those books. When I was about fifteen I was reading that sort of literature. When I was about ten I was already reading science fiction, which came in those days in pulp magazines. There were two magazines, *Amazing Stories* and *Fantastic Adventures.* These are well known. They were brought out by the famous magazine publisher who's name I've forgotten for the moment.

Hugo Gernsback?

Yes, that's the man, a great entrepreneur, who brought out a whole lot of pulp magazines. Some of the best authors were living from the fees he was paying, because they were writing stuff that was mostly read by youngsters, I suppose. Anyhow, those stories were pretty exciting. I found that if you took two of these used magazines back to a little store around the corner, they would give you another one. But to buy one you had to have cash. They were traded around pretty heavily among my friends. This went on until the magazines powdered away.

The stories dealt with fantastic visits of creatures from Mars and Venus, and usually some hideous fate befell the inhabitants of Earth in the meantime. Similar stories used to appear in comics that appeared weekly in Australia but which had come from England. All the Australian comics came from England, which was very curious since life in the two countries was quite different, and reading about this strange and different world is sort of odd.

Were there any particular comics that you recall?

Well, I'm trying to think, because in one of them there was a fabulous story about an invasion from Mars which ran on for some weeks. It was really marvelous. It may have been the *Boys' Own Paper*; I wouldn't swear to that. In any case, what I remember about that story was that these Martians had arrived in steel balls, or balls made of some impenetrable alloy, which could walk about on tripods. Efforts to attack them with artillery failed because of a force shield that they were able to develop and which emitted an eerie blue light. They brought out bigger and bigger guns and it never did any good.

That was pretty good stuff! Later on I realized that all that material had come quite directly from H. G. Wells. The stories were the same. They simply were reinterpreted week after week after week, with minor variations, on the same theory that pulp magazines work now: when you're on to a good formula, you stick to it. If you were to look through a volume of H. G. Wells' collected works, you'd find a number of science fiction stories there which very clearly were the basis for all that development in the pulp magazines. It was easier for youngsters to get hold of these magazines than it was to locate the hardcover books, which were circulating among a different clientele.

I didn't realize it at the time, but there was a demand for that sort of stuff, so these writers were taking H. G. Wells and Jules Verne and paraphrasing them in short, zippy stories. Anyhow I got onto these when I was about ten. I must've read everything for about two years and then dropped that and went on to something else.

So this is what suggested the possible existence of ETI to you?

Yes, I got the idea by bumping into the magazines; that's all I can say to that.

Did you discuss this possibility with other people?

I certainly discussed that with my friends, but I was very young then, so that would not have been discussed at a very high level, but it certainly sowed the seed in my mind. And then I didn't read any science fiction for years. In fact, I still don't read it. I stopped reading it because I felt I'd read it all, which is quite possible. So I got that out of my system, and after I left high school I doubt if I opened up that sort of material again.

What about communicating with, or at least searching for, ETI? When did you first think about this possibility?

Well, fortunately I have it all written down. I have a copy here of the manuscript of the *The Galactic Club* which I dredged out this morning. It took sóme finding, and I remembered that in it there was a section called the "Trail of Ideas," which subsequently was removed from the book by the editress, who was very fierce, because it didn't seem interesting to her. Finally I got it back in, partly. This is the original version, which I'll give to you. I was addressing this very question: What got me onto it?

For many years I thought that Cocconi and Morrison's article in *Nature* in 1959 had started my thoughts. That was the paper in which they suggested that you could make radio contact and that the wavelength to use would be 21 centimeters. So that stimulated my interest, and I think many others began thinking about it at that time.

But then when I was scratching my head I suddenly realized, with a bit of a shock really, that in 1954–55 I'd been at Berkeley in the Astronomy Department with Otto Struve, the distinguished department chairman who had invited me to come from Australia to lecture on radio astronomy. Struve had been thinking about extraterrestrial life as a result of a discovery he had made back in the thirties that certain stars like the sun had a slow speed of rotation, and he deduced from that that they probably had planets.

Also at Berkeley at the same time was a Chinese astronomer, Su Shu Huang, who subsequently went to work for NASA at Goddard. Su Shu wrote several papers that appeared in the *Astronomical Society of the Pacific Proceedings* discussing various subtopics. For instance, the habitable zone of stars is a topic that he took up and did some calculations on. There were a number of these brief papers with no exciting or stimulating introduction; just dry technical notes. As I look back upon it there's no doubt that Struve must have been talking to him about habitable zones around stars and caused those papers to be written.

The appearance of Huang's papers on this subject in 1959 and 1960 was independent of Cocconi and Morrison, so I believe that my association with Otto Struve and Su Shu Huang, both of whom I came to know quite well, must have prepared my mind for the subject of life in space. I was in a way, unknown to myself, in a hotbed of that type of activity at that time. But of course SETI was not an identifiable word or concept then, and I therefore forgot it. It was only when I came to write these notes down that I realized that that must've been lying dormant in my mind, and that when I read Cocconi and Morrison's article, it struck more sparks off me than it did off some of my friends. I had been prepared for that sort of concept.

Now another interesting thing is that Frank Drake heard a lecture by Struve in the early 1950s, when Drake was an undergraduate at Cornell and Struve was visiting there. I don't know precisely what was in the lecture, but it had to do with the possibility of other planets being inhabited. So Drake also traces back to that stimulating lecture from Struve. Drake and Sagan are interrelated through being colleagues at Cornell for a very long time now, so I have no doubt at all that Sagan's interest comes from Drake.

Another person, Sebastian von Hoerner, has written many articles on the subject, very original and stimulating. He must trace back to Drake, too, because he was at the National Radio Astronomy Observatory at the time that Drake was there. Von Hoerner came from Germany and worked there with him. I knew them both there from about 1960 on. That's how von Hoerner got interested, I would surmise.

At Green Bank?

Yes, you'd have to check this but I'm trying to track this back and I see that Drake is the seed for both Sagan and von Hoerner, and from then on for many others, because a meeting was held at Green Bank in 1961, at which all the people known to be interested were invited. I was away and unfortunately couldn't go. I think I was in Australia. They then got to know one another and it became a conscious activity, and then a little later they had a meeting in Armenia, which again I unfortunately missed, so the Soviet people made contact.

When I began thinking all these people were seeded by Drake and I was seeded by Struve, but so was Drake, I began to think that's fascinating because this whole thing stems from the fact that one man is thinking about it and talking about it.

Then I wondered, how could Struve have influenced Shklovskii, because Shklovskii wrote on this subject pretty early, too, but not before Cocconi and Morrison's article. So I think probably Shklovskii was influenced by Cocconi and Morrison. Now I don't know where Cocconi and Morrison got their stimulus from. You'd have to ask Morrison. But I began to wonder, was there some link that would bring the European and the North American sides together?

Then I thought of some very strange coincidences. Struve, after all, had come from Russia. He'd left there, I suppose, about 1919. He was in the White Russian army and had cleared out at the end of the revolution, or in the middle of it.

Of course all the modern speculation about life in outer space goes back to Tsiolkovsky.* So I wonder if Struve had been connected in some way, or aware in some way, with what Tsiolkovsky had been writing. He almost certainly knew that material, which was a mystery to all of us. None of us even knew about it. You wouldn't stumble across it in libraries because it was all written in Russian, but he had written books and there it was. Struve very probably knew that.

Now, I ran across a very strange coincidence. Tsiolkovsky died in 1935, but Alexis Tsvetikov, a Russian, a man with some interest in these things, had had correspondence with Tsiolkovsky while he was still alive, but this time now, Tsvetikov was living in the United States. The next thing I discover Tsvetikov was working here at Stanford. Incredible! Right down here where the linear accelerator used to be, in the Hansen Lab. He was working as a technician for some years, although I don't know the exact dates. Maybe that's attainable because he has a daughter still living somewhere in the area.

So here's a guy at Stanford writing to Tsiolkovsky about these mat-

*Konstantin Tsiolkovsky, 1857–1935, was the Russian who established the theoretical basis of spaceflight (Ridpath, p. 212).

ters in the same years that Struve is in Berkeley fifty miles away. It's very likely indeed that the Russian emigres had some kind of mafia going in the Bay Area. Struve very possibly knew Tsvetikov. Very possibly they had some casual acquaintance through some Russian connection.

Then there were other interesting things. Lederberg became interested. He was here at Stanford although now he's president of Rockefeller University. He was pushing very hard to have the American rocket sterilized before they went to the moon. He was afraid that the moon would be contaminated by organisms that could travel with the rocket. It seemed very odd to me, but I went to a few meetings with Lederberg and I very soon realized that if you have 10^{20} microbes under a rivet head and expose it to ultraviolet light for 4 days on the way from here to the moon there'd still be 10^9 left. The mortality would be enormous, but the number that survive will still be enormous; microorganisms are very hard to kill. So he was interested in that.

Now, when I throw my memory back, I realize that Carl Sagan spent six months or so here at Stanford, working in the genetics department with Lederberg. The three of us used to talk on occasion. We drove down to Monterey one time for a meeting on this sterilization of rockets while they tried to convince NASA to put some kind of a drape over the rocket as they launched it. So there's another thread weaving all these individuals together at a very early time, before any of the recent developments took place.

Well, my thoughts on ETI lay dormant until I read Cocconi and Morrison in 1959. I read that in *Nature*, within weeks of it coming out, and then I began to think, and within a year I had my own letter in *Nature*, which has a number of ideas that I've often rewritten at greater length. But I haven't added very much to what's in that first paper. So those ideas came all in a sudden rush.

For a few weeks I thought about that and talked to everybody about Cocconi and Morrison's paper, and I gave a lot of talks in the Physics Department and here and there around the campus, and also to alumni groups, around about 1960. I got very enthusiastic and soon found that the talk was very popular. After a couple of years of that, I had to decide not to do it anymore because I get tired of telling the same story over and over again, but it's a great talk.

After you completed your schooling and began working as a professional scientist, who or what has influenced your thinking about ETI and SETI?

As I say, it was Cocconi and Morrison who stimulated my next burst of interest.

What effect, if any, has your interest in ETI had upon your career as a scientist?

I don't think it's had any effect on my career as a scientist. I'd have to say very little; it's being conducted in a separated compartment.

What do your colleagues think about your interest in SETI?

My colleagues have never exhibited any negative reactions; I'd say they were neutral. I can't say they ever thought that my time spent on such things was important or trivial; they just accepted it as one of the things that someone might get an interest in.

Did they ever discuss it with you or ask you about it?

I certainly had discussions with people; if I go to the faculty club, someone is bound to make a comment. If they've seen something in the paper or have a question, they're bound to say, "Ask Bracewell." I often find people remembering that I'm associated with the subject.

What about your family?

What my family thinks about it, I've got no idea. I'm sure it doesn't bother them. I don't take it all that seriously, you see. I would have to say they're neutral.

What about your superiors, bosses, etc?

As to superiors and bosses, I think that the department chairman regards it as negative, or perhaps it would be better to say he regards it as not positive; it's not part of electrical engineering, which is his responsibility to foster, I daresay.

Is his "not positive" feeling because of the specific nature of SETI or simply because it's not part of electrical engineering?

Because it's not part of electrical engineering.

So it's not because it's SETI; it could be Shakespeare or spiders?

Oh yes. Astronomy to him is not part of electrical engineering. But of course it *is*, just as ETI is. That just represents the parochial view that an individual might have. I think it's quite clear that ETI has, and is going to, involve electrical engineering considerations. You only have to read the reports that Barney Oliver wrote on Cyclops to see that it's an exercise in antennas and communication theory from beginning to end—a great way to get students enthusiastic about dry technical aspects of electrical engineering. Radio propagation, electromagnetic wave propagation is involved, a whole many aspects of electrical engineering. Fourier analysis of the signals.

So it's very much there, but it doesn't look at first sight like the sort of design of integrated circuits that one does on the bench.

Nevertheless, it's certainly a part of electrical engineering, and of course without electrical engineering we wouldn't have had radio astronomy, which has come to be an important part of the whole of astronomy, and which now furnishes the best images that are available, both as to resolution and dynamic range. The best images now available on Earth are made by microwave interferometry, which is a radio technique coming straight out of electrical engineering. So astronomy has been the driving force there. You just have to say the two subjects are inextricably connected.

What about your friends? What do they think about your ETI interests?

My friends just seem to think that it's interesting that I do these things. They read my name in the paper occasionally and it doesn't bother them. If anything, I suppose it amuses them.

Before going on to discuss your ideas about SETI, let's get a little more background information. What was your father's occupation?

He was in insurance.

How much education did he have?

He was not very educated. He had to leave high school to take a job in a newspaper office when he was probably about fifteen. But he was very bright. His father was a schoolteacher and so was his mother. So he had a lot of academic background from his childhood. He always read extensively and was interested in dictionaries and grammar and things like that. He had a copy of Roget's *Thesaurus*, which he used to consult frequently.

What about your mother?

My mother had even less formal education and never went to high school. Their experience wouldn't be in any way unusual. She grew up on a farm in poor circumstances. She was good at needlework and sewing machines and cookery and laundry, and I'd say that she was probably typical of the whole wide range of relations and friends that I had at that time.

How would you describe them?

My father was outgoing and had a manner suited to selling insurance. He was certainly gregarious; not reserved, not quiet. He was serious. He also had a sense of humor. My mother was an underdeveloped personality, as was forced by society on women at that time, kept at home to do home duties, and so she was slow in developing because of that restricted outside contact over the years of childraising. So I wouldn't describe her as quiet. She would be outgoing, but not in the degree that I'm describing for my father, who dominated the household.

How did they influence your eventual choice of science as a career?

My mother would've had no influence on the choice of my career, but my father was very interested in finding out what opportunities were available for someone like myself. About the time I was finishing high school, he made many contacts with high school teachers, at my school and at other schools. The people that he met in the course of his occupation were from industry.

He also hired a psychologist to question me and write a report on what I should do. I still have these reports; there were two of them, and they are rather interesting. The fellow said I should learn Japanese and go into wool buying or go into industrial chemistry or be a court interpreter. He worked on these things he saw that I was good at in school. And he was employed at screening applicants for jobs at Woolworth's and things like that. So he obviously met a lot of people, and was a pretty shrewd judge of character, and he made a living from it.

So my father paid a lot of attention to that, and I ultimately went to the university to study chemistry. I lost interest in chemistry the first year and moved to physics and mathematics. So it all worked out quite well.

And what about ETI? How did they influence you in this respect?

My parents had no influence on me with respect to ETI. They probably knew I read science fiction, but didn't think that was unusual.

Were there other people we haven't mentioned who influenced your ideas about science and ETI?

I didn't have any influences about ETI from teachers or relatives, but I certainly got influences about science from teachers, both from high school and the university.

What religious beliefs did your parents have?

My mother had been brought up as an Anglican of no great strength of persuasion. She'd been baptized, but she had never been confirmed. Now my father, on the other hand, had a strict orthodox Anglican upbringing. He had been confirmed and so he'd had the full religious instruction, but he didn't go to church. By the time he was forty, maybe earlier, he was an unbeliever, consciously a nonpracticing, areligious person. But he acquiesced in my receipt of the full-strength religious instruction. And on my insistence, actually, he came to communion in the morning on one occasion. He was quite neutral in expression of religious belief.

My own development was that I was a firm believer until about the age of twenty. And then, over the course of five years, it evaporated.

How many brothers and sisters did you have?

I have one brother, six years younger than myself. He's a water-supply engineer.

What about your friends when you were growing up? Did they share your interests in science and ETI?

Yes, I always had one or two friends that I was very close to, with whom we'd discuss everything, so I had someone to share my interests with. I've always had someone to talk to about the latest story I'd be reading. In fact, there would be a group of us who always read the same comics and magazines and traded them around, so we must've had casual conversations. I wasn't isolated in forming opinions about such things.

What were you like during childhood and adolescence?

When I was growing up, I was not outgoing, not gregarious; I was very timid, very introspective. I can't say I had *few* friends—I've always had friends, so I'd have to say I was average in that respect but I was certainly very shy and withdrawn. The reason was, apart from any genetic contribution there might have been, that I was always smaller than my school friends and so they used to beat me up. This may have nothing to do with ETI, but it certainly formed my personality. Being beaten up regularly means you have to defend yourself someway or other.

What other professional interests do you have, apart from SETI?

As to the professional interests, I'm interested in everything in astronomy and electrical engineering, which is a pretty wide field.

What recreations do you enjoy?

I'm interested in growing trees, and in wood, and small tools, working with things and carving things and painting them. So working in the garden; that sort of thing.

What was the most memorable moment or event in your life?

I seem to lead a fairly even life. The most memorable moment in my life this year is I had my heart operated on; that was memorable. I've almost forgotten that now.

Turning now to SETI, about what fraction or part of your working time do you spend on it?

Well, let's say 1 percent. That's not to say that it's a small amount, but I do a lot of other things, too.

What stages has your thinking about ETI and SETI gone through?

My ideas haven't changed very much, but I've thought of one or two extra things as time went along. So if I have to write a chapter for a book now, which I guess I have to do every couple of years, you'll find that my ideas are basically the same, but I've hit on one or two little extra ones.

For instance, I thought that establishing contact (as distinct from communicating), establishing contact by using a space probe made a lot of sense. Although it takes a lot of time, the time is in proportion to the longevity of society. It has nothing to do with the longevity of an individual. And so I thought that space probes were a good proposition, but generally speaking, that hasn't received very wide approval. The majority of the people, whose names are familiar to you, comment adversely on it. But a few have taken the opposite view.

When I first thought about that, my idea was that the space probe would go to the prospective star and then put itself into orbit at the distance where a planet could be expected to be. But now I think that it's not necessary to do that. It would be sufficient if that probe just went *through* that solar system. It would take more than a year, and that would be ample time to answer the question, "Is there any intelligent life on one of the planets?" After that, the probe would merely expend itself, continue on into space. And that would be the whole of its mission.

That is preferable because the cost of the probe is then much lower; it doesn't have to carry with it the rocket that would be necessary to get it into orbit. Or, if you didn't use a rocket, there might be some very elaborate solar sail, but it would be a technically elaborate thing. But if the probe merely has to fly through, and send out a few signals while it's doing that, it can be relatively small, like about the size of a human head.

People making critical remarks about the probe idea usually talk about something the size of a Volkswagen. But I have in mind something about the size of a football. The virtue of small size is that the reliability will be better if it's less complicated, and it will cost less to launch. So you will be able to launch more of them in a given time, in the direction of more stars. That was an idea that just came to me, maybe two or three years ago, that you could cheapen the scheme.

I also had another idea that I didn't have originally which I think is important, and that's the concept of working out the ratio between the time taken for a civilization to evolve, and the time taken to travel through the galaxy at some small fraction of the speed of light. Now when you do that calculation, you find that we can travel from here to the center of the galaxy in much less time than it took for us to evolve.

Consequently, the first traveling civilization to evolve is going to colonize the galaxy and beat out any opposition from parallel evolution that may be taking place elsewhere. That's not to say there may not be two or three places where life has evolved to a technological level; but the first one that begins to spread is likely to dominate. And then after a time the galaxy will be occupied by the descendants of that first race to succeed.

I believe I am the first person in print with that idea, and in fact it's in my book, but it's hidden on one or two pages because the editor of the book felt that it was better to plug the idea of widespread intelligence than to plug uniqueness and colonization. But the idea is there in full, though it was trimmed down a little bit. And I believe I published the idea even earlier than that in an obscure place.

Since then there's been a number of unpopular people who've adopted the idea. For instance, Hart and others who have had symposia and written many papers, like Tipler, on why we haven't observed other civilizations. They are pushing that idea on their own very strongly.

Their thoughts have received just as much hostile attention as I experienced. It is not something that people, the SETI community in general, like to contemplate. There's a psychological block, I believe, that causes them to resent the notion that we may be unique. The psychology of it seems to be that it undercuts their active efforts to get to work finding these other beings. If they, extraterrestrials, are not there, then the whole rationale for SETI activities is eroded somewhat.

Consequently, I myself have experienced very spurious counterarguments. People will say probes are too expensive, and they will do the numerical calculations and it comes out an astronomical sum. Then you'll find they've thrown in a factor of a thousand, which they introduced themselves. When you point this out to them, it's not very long before they come out with another irrational objection. All the best-known names have done this.

For instance, Phil Morrison called space probes "tendentious" and "drones." What to me is an automatic space probe with the latest high technology, with intelligence able to emulate a human packed into something the size of a human head—to him that's a drone, some kind of inanimate object. And I don't know what "tendentious" means except that it's slightly negative. But that was the substance of his argument.

Shklovskii refers to this in his book. He devotes a whole chapter to my paper. It says that because of the capitalist system that we operate on in the United States, it won't go that way. Well, what he had in mind was something I'd implied about civilization being terminated by a bomb explosion. He's arguing that capitalists might do that but we communists would not.

It's a totally illogical argument because if we capitalists bomb them out of existence, the fact that they wouldn't is irrelevent. So you find that he doesn't agree with the idea but his argument against it is just crazy.

All the other authors who have criticized the probe have done so on similar strikingly emotional and irrational grounds. I don't think they resent me personally, but it's rather amusing to see the way they jump on both the probe idea and the idea of uniqueness with very strong, very fierce feelings.

After I began to realize this, it suddenly struck me what is the true analysis of this situation. These criticisms are all being written by action-oriented people who would like to do something. And when it's suggested that we don't have to do anything except wait for the other party to contact us, they have a subconscious negative reaction. So they immediately write against it, and it doesn't occur to them to coldly analyze what their argument is. They write down the first thing that comes into their head that denigrates it and pushes it off to the side.

I've been vaguely surprised that my friends, at least five of them, have done this, and I think over the course of twenty years it's surprised them that I haven't packed up this idea and thrown it away. I'm still giving it out because I want to hear an honest argument about it. As soon as they talk me out of it, I'll admit it. But the arguments are all of this crazy variety. Each time I shoot that down—I try to do this from time to time; I try to rebut these things, and generally you can use a little humor—it produces more comments, and always will as long as it's viewed as interfering with some project for action, like building huge antennas or building huge data processors to look for transmissions, or whatever the current ideas for activity may be.

So I've now got off onto another tack that I think may help. I'm now talking about an equi-partition principle. It is not original with me, but the idea is that, even though the other party may be much more advanced than we are, both parties would put in an amount of effort that's in proportion to their capability.

If you think about it, that idea was already built into what Cocconi and Morrison said. They said the more advanced civilization is going to beam a radio transmissiom towards us but we will have to contribute, (a) by having invented radio and (b) by having put up a suitable antenna and receiving system and paying attention. So we've done something, they've done something.

With an alien space probe as I originally described it, we did nothing. The space probe arrived in our solar system, went into orbit around the sun at the right distance, listened for our radio communications and, if it heard any, then radioed to us on one of the frequencies that it could hear was in use. We'd have to do nothing. We'd just notice this as interference and they'd make their contact. It would require no input from us.

So I began to think, what input could we make to facilitate their problem, which is to locate an intelligent community?

Well, the answer to that comes out in terms of economics. You put yourself on this other civilized planet where they have a man in charge of making contact with newly developing intelligent communities, and his job is to get as many as possible but he has a finite budget. They may be very rich and very able but they still have finite resources which they can parcel out onto this project or that project, and no doubt they have many demands on their budget just as we do. So he gets a small slice of the pie. But with this small slice of the pie that he is administering, his job is to find as many intelligent communities as possible.

If he can send out ten cheap space probes in place of one expensive one, and each one of these has as good a chance of finding a new community as any other, then the cheap one is better because he can send out more, and he has more chance of success. So if you can think of a cheap probe that would be as effective as an expensive probe, that is the way he will go.

So we should ask ourselves, what could we do to make their probe a cheaper one, because they will be banking on us to have thought that out, just as they would be banking on us to have invented radio. If we hadn't invented radio, the whole thing would be hopeless. They might as well not worry.

The answer seems to me to be this. As I suggested earlier, a cheap probe would not go into orbit around a star—or our sun. It would just be a fly-by. It would go *through* the solar system. A very tiny jet would be able to make the necessary mid-course corrections to assure that it went through the solar system and didn't miss by a long distance, but it would fly through.

So what we could contribute would be a space watch system that watched for objects arriving in the solar system and, if one was detected approaching, then we would attempt to rendezvous with it or at the very least communicate with it during the decade or so that it was here, and then, when we drain it of all its information in that length of time, it would pass on out the other side. So that would be our contribution, to maintain a space watch.

I've got two comments. I think that's a valid proposal. Secondly, it has a psychological component which is likely to make it more acceptable to the official world because it does not call for a large space activity at this time.

It calls for setting up a space-monitoring system, with space probes that we would send out, which would lurk out in the outskirts of the solar system; or prepare launching facilities which would make it possible to do that should a warning come in. This would be combined with some kind of telescope search of the outer reaches of the solar system—something like what the people looking for comets do now, but on a bigger and more automated basis. I think NASA would be delighted to have a telescopic space watch. It would be a big operation, lots of budget, keep lots of engineers employed, and would be prepared to launch a rocket out into the vicinity of such a thing.

Not only would it have those benefits but it would be free from the criticism by the astronomy community, which very often criticizes NASA for spending very much larger sums than astronomers spend, and for sometimes claiming that the benefits will be astronomical. Whereas the astronomers say, "If we had that money we would spend it for astronomy, but not *that* way."

But in this case a ground-based space watch, with telescopes which looked for approaching space probes, would also pick up not only comets but meteors, asteroids, things that may be close to the earth and of intrinsic interest but may collide with us. So there's a practical justification there. And, of course, astronomers would be delighted if a rendezvous mission could be arranged to visit Halley's comet, or any other comet.

So I think I've got a formula here which is valid in its own right from the SETI standpoint, and which may overcome the cool reception that the probe idea has previously received.

Would it be correct to say there might be four steps in the process by which you came to be thinking about the search for extraterrestrial intelligence? One would be space flight by humans, then the existence of other life, not necessarily intelligent. Third, the existence of intelligent life, and then the final step would be the communicating or searching for this life. Is this sequence the way it happened for you?

No. I've been aware of the idea of space flight by humans since about 1930, by reading *Amazing Stories* and those magazines, but it never occurred to me that that was anything other than fiction. I didn't regard it seriously, and that attitude has remained with me right up to this time.

I've consistently been surprised by every success of the space agency. It really astonished me when they landed men on the moon, and I was astonished when they fired Shephard down the Atlantic range and got him out of the water. Every time they've fished people out of the sea after coming back from the moon I've honestly been surprised, and I'm only now getting accustomed to the notion that when they fly off in that space shuttle it always seems to make a successful landing. I really found that very difficult to believe.

I'm very cautious as an engineer and would probably not be the right person to design a spaceship which had to be secure for the people on board. I'm too much aware of the difficulties.

For a long time I didn't know how NASA actually succeeded in doing it. The government bungles so many other things that I didn't really understand how they did it, but now I'm beginning to appreciate this massive redundancy they have so that there's not a single element that can go wrong which will wreck the mission. Carried to extremes that really paid off, and they do it very, very well.*

The existence of other life, I'm quite open on that. It doesn't bother me or have any effect on me at all. I just would be interested to know if there is other life, but there's all sorts of living creatures on the earth that haven't even been discovered yet. Within my lifetime the number of known insects rose from 500 thousand to a million species, so obviously there are lots more to be discovered. If such things were discovered somewhere in space I wouldn't be surprised at all.

It wouldn't surprise me if there were no other intelligent beings, but no one will ever be able to prove there are no others, so I don't need to contemplate that. And intelligent life, if it showed up, I'd think was really exciting, though I don't have any prior belief in it. Of course I would wish to communicate with them and find out whatever they have to say. That would be tremendous.

As for beginning to wonder about these different subjects, did they all start to emerge in your thinking around the same time, when you began to read science fiction, Edgar Rice Burroughs? . . .

*Three years later, space shuttle Challenger exploded.

Oh yes. I went through that phase when I was in my teens, but my recent interest all started with Cocconi and Morrison's paper about 1960, and I've had a connection with this subject ever since.

The reason I'm looking into this is that John Billingham had spent a great deal of time thinking about space flight by humans but it wasn't until he came to NASA that he began to think about these other things; first about the existence of other life forms, period, and then about intelligent life.

Well, John Billingham's example shows that he was influenced by what other people were thinking. It's quite clear that these ideas have diffused around and haven't emerged spontaneously in different peoples' minds. I think we just pick up these ideas mostly from other people, and that may be generally the case.

If you went back and tried to find out where Tsiolkovsky got his ideas, you might find that he didn't originate them either. We certainly know from history that Giordano Bruno was thinking about it, and so were other people throughout the centuries, and while you might say their writings or ideas disappeared for the time being, still you never can be sure. Books have been lying around in libraries, and ideas get handed on by word of mouth, so it would be very difficult to prove that all these thoughts don't go back in a continuous way to antiquity.

I was more interested in the case of each individual scientist; when he started thinking about this.

Well, where did Edgar Rice Burroughs get his ideas? It would be very surprising to me if he generated all that out of his own mind. He was probably aware of earlier authors who are not read now. He might have got it all out of Jules Verne, and where did Verne get his ideas? What about Swift? He had people on Mars, and he knew there were two moons around Mars before they were discovered, which has always excited people.

What would you do differently if you were starting over again to study ETI?

If I were starting over again, I can't imagine I'd do anything differently. I have just had occasional ideas. I've been assiduous at getting them into print. I haven't been just sitting talking. I think everything that I thought of I have got into print one way or another.

Why is ETI important to you?

It's not very important to me; I found it just an intellectual interest. And I didn't select that as something to study, in contrast to other things. I do dozens of things all at the same time.

What has been the most difficult aspect of your ETI work?

I haven't had any difficult aspects.

How do you think most scientists today view SETI?

I think it's clear that scientists view SETI as legitimate. The fact that the International Astronomical Union has now set up a commission to engage in it will help to legitimate the subject for scientists who haven't thought about it yet. That will, of course, help.

We ought to bear in mind that the people who got the IAU to do it are the self-same people who have been involved in it all this time. It is not as though you could think of the IAU as some independent external organization which put some catchet of approval on it; it's not that way at all. The main figures in SETI have always been well-connected in the scientific hierarchy, and they just decided to make themselves even more legitimate by this move. The fact that they've been successful and have not been opposed shows the general assent among scientists to the view that it's a worthwhile and respectable activity.

Of course, the scientists involved have all been very responsible. We haven't had crackpots or crazies associated with it. Despite what some people might say about Carl Sagan, you know he's not really a crackpot.

What about public attitudes toward SETI?

The public is enthusiastic but they don't know what they are enthusiastic about. I met a man the other day who didn't believe me when I told him no work was going on on Project Cyclops. He didn't know that it was only a summer study that began and ended in a few months, at most. He had seen the photographs of it and thought it was in Arizona or somewhere, and he really didn't believe me when I told him not only that it hadn't been built but it wasn't even being worked on; no one was designing it, and furthermore Senator Proxmire had managed to cut off the authority of NASA scientists to *think* about SETI for a year or so.

So the public is enthusiastic and every three months there's big bursts in the newspapers. They enjoy reading about it, but *what* they are reading about is always a little bit vague, and the writers in the newspapers are not very good either. The usual way of the papers is slightly sensational, with not too much respect for what they wrote a few months ago or what is actually true.

Realizing that it is only a guess, when do you think contact with ETI will be made?

I've got an objection to answering that question. My response to it is an orthodox scientific attitude. I think we should make every effort to contact ETI, and the only way to find out when it will happen is to notice that it *has* happened. Otherwise it might never happen, which means that it would take infinite time, or it might happen tomorrow, so the whole range from zero to infinity is perfectly acceptable, and the only way to find out is to try, and even then we may never find out. It's possible to say, "Yes, ETI exists," but it is not possible to say, "No, it doesn't."

What might ETI be like?

The variety of forms that intelligent organisms might take is of course very extensive. They could be in the form of intelligent scum. You could imagine some organism that lives in colonies as a scum, like coral, which is a colony all interconnected. But it might be a scum on water, and in the course of time, as it gets ability to do things that colonies can do, such as bee colonies, it might be that it needed some mechanical protection. This scum might manage to fold itself up and get itself into some kind of bony exoskeleton and then begin to re-semble what would be a brain in a skull, and yet by origin it might have been a colony. Now that is as far from individual intelligence as you can get.

It may not be as far as if you go the other way. You might think of Fred Hoyle's black cloud, which was an interconnected cloud of gas and solid particles communicating by radio. That's quite unlike our organization, although I suppose that you could say, as he envisaged the black cloud, it was an "individual."

Now to talk about six-legged horses and things like that, which we've seen lots of illustrations of, I think that shows a lack of imagi-nation as to the variety we can find. If you look at all the insects and spiders and plants and slime molds that we have on Earth, that's a tremendous variety already, but that doesn't encompass the full range of things that *could* happen.

So I don't think we've got any basis for judging what ETI could be like; it just could be anything. All it has to do is be able to think and have a capacity to observe the universe the way we do, and to form observations, to learn about physics of the universe. It could be a black box but if it could do all that and communicate, you'd have to say it was an intelligent life. What was in the box would be almost irrelevant; it would be a great fascination, but as far as our interaction is concerned, it wouldn't matter much.

What scientists or other people in the past do you admire?

Archimedes was a great man, and then there was an early Greek astronomer, Aristarchus, a very clever fellow. I thought Ptolemy was clever, but he's been criticized for plagiarism recently. And there were some great Greek geometers. Euclid was their spokesman but there were many others. I think that Newton was a great man, no question, and all the great physicists and chemists—I have an admiration for them. Darwin I would put way up near the top of my hierarchy. He's an interesting example of a nonmathematical scientific thinker. Faraday would be another person of similar description, who did extremely well without the power of mathematics. And Maxwell and a few Frenchmen and Germans, a lot of Russian mathematicians.

So, you name them, I've got admiration for them. If you've heard of them, I think they're pretty good. Of course, they're all dead, the ones I've mentioned.

The only role they've played in SETI was to provide us with a background of knowledge about our universe in physics, chemistry, astronomy, which has been essential to modern progress. Earlier scientists speculated about SETI, but they had no constraints placed on their imagination by what we now know of physics and astronomy. So you could reason freely about voyages to the moon.

What changed the game in recent years is that speculation since Cocconi and Morrison has had to meet the boundary conditions of not conflicting with what we now know of physics, chemistry, astronomy and biology. And that's made it much tougher, but nevertheless, what's come out is much sounder.

Who are the outstanding people in SETI?

Those people would be Frank Drake, Carl Sagan, Barney Oliver. No one in England or France or Germany comes to my mind immediately. In the Soviet Union I would mention Shklovskii. I would give a lesser rating to Kardashev and Troitskii. Of course Tsiolkovsky was a very great and influential Russian. The name appears to be Polish, but there's no question that he was a Russian.

Other names you might not have heard so often from asking this question would be Otto Struve and Su Shu Huang. As I mentioned earlier, the two of them worked together for a while at Berkeley.

Phil Morrison. Other people around the United States. But as to outstanding—that would go pretty close to covering it, I think.

Barney Oliver took the idea of trying to make contact by radio very seriously on a large scale. He foresaw that a very large installation could be funded if enthusiasm could be developed. His role has been to try to raise that enthusiasm by various techniques. He's been as successful as you might hope. He hasn't managed to get a great Cyclops project funded, but I doubt whether anyone else could have, either. Without his drive, we wouldn't have had what we do. We'd have the situation that we see in the Soviet Union, where they never made much progress at all. That's how I see his role.

Frank Drake and Carl Sagan have been very effective at getting public interest; as advertisers they've done extremely well. Sagan's name is now known to everyone. Frank Drake may be less known, but he contributed popular articles and popular books and many talks. So I'd put those two fellows together in the publicity department.

Phil Morrison has also contributed on the publicity, but his function is generally being a kind of decorative elder statesman who adds adornment to serious meetings and is influential with other scientists. He certainly speaks well to the general public and has done a fair share of that, appearing on television shows, more than once. He speaks with a tone of concern and seriousness that's very effective in influencing fellow scientists.

Shklovskii has written popular articles and popular books and things that appear in the Soviet press. So he's been pretty good on the popularization side, too. And he has brilliant ideas now and again.

Kardashev? Well, his main claim to fame was that he thought of Type One, Type Two, and Type Three civilizations. I never thought that was very important, but I couldn't help noticing that everyone else *did*. So I thought the answer must be if you invent a classification system, everyone will adopt it. So I then wrote a paper of my own in which I did the same thing, which I think I called Case One, Case Two, Case Three. All I can say is it went over like a lead balloon, so I now no longer hold the opinion that a good classification system helps. I think Kardashev's idea was just an appealing idea. And in addition to that, it had the convenience of pigeon-holing and classifying these ideas so you could think about them.

Another example of this would be Drake's famous equation. Now Drake's equation is obviously wrong, but it's regarded like motherhood, and is very, very frequently reproduced; it's hard to pick up a book or a paper that doesn't start with Drake's equation. So there's something appealing about it. The idea that you can write an equation in mathematical symbols and define the symbols—even that itself has an appeal.

So I thought about that, too, and I converted quantitative ideas that I had in my 1960 paper in *Nature* which were, however, expressed in words. It's true they were expressed graphically with numbers, but there was no equation. Then I began to think that an equation has more prestige in getting attention, so I converted the graph to a pair of equations. But I haven't noticed that they've ever been used.

I'm still in favor of an equation if you're trying to impress certain sorts of people. But the idea itself had better be appealing, too. And the idea that if there's any life a thousand light years away, then it won't live long enough for us to get in contact with it, is not an appealing idea.

The people who think about radio beacons would like to be able to pick up beacons a hundred light years away, a thousand light years, ten thousand light years, wherever. They don't want to think about the prospect that at a thousand light years there won't be time for a round-trip message before the civilization is extinct. So I've got a nonappealing idea, and expressing it in equations is not enough to offset it.

Why do you think the Drake Equation is wrong?

It's trying to estimate how many civilizations there are, something that we don't know and that we will have to arrive at by observation. His method of calculating it, fundamentally, is this: We multiply together two other quantities that we don't know. One is the rate of formation of intelligent communities; the other is the longevity of a community. This might be disguised in the form of five or six or seven different factors. But it boils down to just these two in the end.

Now the fact is, we don't know the rate of formation of intelligent communities any better than we know how many intelligent communities there are. And we don't know the longevity any better than we know how many there are. So what we've done is to replace the guessing of one quantity we don't know by the guessing of two quantities that we don't know, and then we multiply them together. The fact that it looks like an equation carries a certain conviction with some people; it's undoubtedly influential, as is obvious to me from the fact that it's reproduced so frequently.

Now suppose that I give you a question that you're more familiar with from your own experience: I ask you to estimate how many cats there are in Palo Alto? There are various ways of doing this. You could go out at night, walk all the streets, and write down how many cats you heard. Then you could do an experiment on one block and find out how many cats there are by counting how many you see, and then walk down that block at night and count how many you hear. Then you would say, "Well, I saw a hundred cats there, but I only heard ten." So if you heard a thousand in all of Palo Alto, you would say ten thousand cats. That would be a rational sampling method.

Unfortunately, we can't do that in space because we've only got ourselves, and we don't know how many blocks to assign to the Earth. So we can't do it that rational way. Now here is, by analogy, Drake's equation for the number of cats in Palo Alto.

He says we don't know how many cats there are, but we will take the birthrate of cats in Palo Alto; we will say there are a hundred cats a week born in Palo Alto. Then we multiply that by the longevity of a cat. Let's say a cat lives for three years; that's about a hundred and fifty weeks. So we multiply a hundred by a hundred and fifty and we get fifteen thousand cats in Palo Alto. That's how he did it. He guessed the birthrate and he guessed the longevity.

Now here's the problem: How do we know if cats live three years on the average? Well, we've all seen three-year-old cats and we've seen cats that are ten years old. But cats are usually born in litters of five or six, and it very often happens that only one of those survives, if any. Their owners frequently destroy them or mysteriously arrange for them to not survive.

Then, there's more than one category of cat. There's the kind of cat that's born into a loving household; there's the cat that is born under the buildings at Stanford where its mother was fed by secretaries; and there are cats born in the streets. They are different categories of cats, and their survival rates are quite different.

And the birthrate of cats is very dubious, too. In order to know the birthrate of cats, you'd have to do a study that'd be very complicated. I don't know how you'd do it. We have a cat and it gave birth to five kittens about three years ago. As soon as it had those kittens we arranged that it never have any more. So there is a category of cats that are interfered with, and there are others that run wild and aren't interfered with. How we would find out how many of these categories there are, then average them, is beyond our ability. So we don't know the birthrate; we would have to guess.

The longevity of a cat is a guess, too. The average lifetime of a cat in Palo Alto might only be week. There must be the possibility that enormous numbers are born that are killed; we really don't know.

Both of these two guesses are just as wild as if you asked directly, "How many cats are there in Palo Alto?" I might say a hundred thousand or I might say ten thousand; it would just be a guess. But to guess the other things and multiply them together is not going to do any better.

That's my criticism of Drake's equation. I don't think Drake likes the equation much, but as the owner of the equation, he can't help being proud of the fact that it is frequently reproduced. I don't know whether I've published my criticism of it, and I don't know whether it's been cited in the literature. I bet it would be pretty unpopular.

How do you characterize your own role in SETI?

Well, I've had a fair amount of publicity. I guess I've contributed to influencing public opinion, together with many other authors. I doubt whether the public remembers names, but I've certainly contributed a noticeable fraction of all the popular material, through a book and articles and talks to alumni and rotary clubs and senior-citizen clubs and physics seminars; I've been pretty vocal. So I've played a contributory role.

On the official side, people certainly have read what I've written. They don't always like it, but they certainly invite me to the next meeting. So I've just been a minor tile in a big mosaic; a noticeable tile, but not out of proportion.

I'm a careful writer, and when history comes to be written I don't think you'll find internal inconsistencies in what I've been saying. I don't say something unless I put a lot of care into expressing it, so I haven't written anything that I wish to repeal.

What do you think UFOs are?

I think they are reports.

Do you think there's anything to them?

I believe that there are such things as UFO reports. That's as far as I can go. The reports themselves are not hoaxes or misconceptions. I don't know what lies behind these reports. It's kind of mysterious.

There are genuine astronomical and atmospheric objects behind a lot of reports. We know that because of newspaper accounts where something has been seen one day and then seen again the next day by crowds in the street, and then the third day it turns out to be Venus. Then you go back and look at what people were saying: that it was colored, it was flashing, it moved up at tremendous speed. Many people saw it. Yet, by golly, it was just Venus.

So we know that a lot of reports are generated by physical objects. Of course, they get crossed off the list, and the question then goes back to the residuals that are not yet explained. I'm not very helpful about UFOs. I'm just an agnostic, you might say.

I admired the action of the NASA administrator who said he'd be delighted to put the whole resources of NASA into the analysis of any piece of a UFO that was brought to him. That's a very clever statement. It can't dissatisfy proponents of UFOs because he's offered to deal with the stuff when they bring it, and it doesn't dissatisfy those who are suspicious about whether they exist, because they may believe that no pieces will ever be produced. On the other hand, if a piece is produced, you can't have any objection to analyzing it.

If scientists do find definite verifiable evidence of extraterrestrial intelligence, what should they do with this information?

It will be in the hands of the people who make the discovery to decide what to do next. I recall what happened in Fred Hoyle's novel *The Black Cloud.* The President of the United States immediately provided scientific facilities for the discoverers of the black cloud, and then built an impenetrable fence around them, so they could not get out or communicate, so the decisions as to what to do next were made at a political level. I don't see any reason to think that would not happen in this case also.

That's an interesting idea as to what will *happen. What do you think* should *happen when scientists get this information?*

I don't know what they *should* do. I can guess what they *would* do. I suppose that, if the discovery was made at Goldstone on the 210-foot dish, that the scientist who made that discovery would immediately communicate on the NASA network to scientists with comparable equipment in other parts of the world. He might do that, but I'm not sure.

He might *not* do that, and instead go to the chief of the Jet Propulsion Lab and inform him privately. And JPL might then contact the administrator of NASA, who might contact someone in the White House for advice as to how to proceed.

If that happened, I believe that a political decision to keep the discovery quiet would be most probable. In fact, I think myself that it would be correct.

So I don't know whether the scientist who makes the discovery would immediately let the cat out of the bag or whether he would be more cautious, having perhaps thought about this in advance, and proceed through official channels. For all I know there may already be some secret directive, or maybe not a secret directive; there may be an instruction at government labs or labs closely connected with the government, such as JPL, to proceed cautiously in the event of such a discovery. I don't know that there is such an order out, but one could inquire.

The same may be true at the National Radio Astronomy Observatory, which is less connected with the government, except that it does get its money from the National Science Foundation.

So it's conceivable to me that through government contacts the large radio observatories may have some standing orders for behavior by the employees in the event of the discovery of extraterrestrial intelligence.

More likely, however, I would guess that the government hasn't got its act together on this, and such things might on occasion have been discussed, but not implemented. I could understand that because many people would judge the likelihood of this discovery being made in the next year or two to be rather low, and not feel any sense of urgency about having rules in place, so probably nothing has been done about it.

And in that case, I think that whoever made the discovery would immediately tell everybody. Within hours it would rattle around the world. The Europeans would know essentially overnight.

You don't get agreement between people on what *should* be done, so there is no answer to what should be done, on almost any general question. To ask me what a government employee *should* do, I might answer in terms of religious convictions that I have, or in terms of political convictions that I have, or I might say a government employee should do something which might be a crime for him. A Russian's answer would be different from mine. His answer would be influenced by *his* background. So I think you'd get closer to agreement when you question people if you ask them what *will* happen, or what *might* happen. You're likely to get considerable spread but more consensus.

What would you do if you made the discovery?

If I made the discovery I would sit and think about it very carefully, although I have already thought about these questions. I can't say in advance because this discovery could take many different forms. It could be that you intercepted a long message that could take years to receive, let alone to decode. In that case there would be no particular urgency.

Or, if you were to make contact with an extraterrestrial probe that was in the solar system, that would be kind of urgent, because there would be the possibility of interference by some other party: some government might wish to shoot it down. That would be urgent, and I'd think very carefully about the wisdom of widely spreading news of that contact immediately. I would obtain some advice. I would sit down and consider this very carefully before acting.

It might be very important. There might be dangers involved for the human race. There might be national dangers involved because the information could be valuable, and it might be deemed desirable by the Soviet Union to deny access to that information. It could bring about a serious military situation.

What effect would information of such a discovery have upon us here on earth?

I suspect that it would be a nine-day wonder in the press. If, as I imagine, we just break into the middle of a transmission—we didn't have the beginning and we wouldn't have the end, and it might run on for years—the news that some extraterrestrial message had been intercepted would be very exciting, but only for a few days, because it wouldn't continue to produce day-by-day developments.

It would be like finding the Rosetta Stone. I don't even know whether *that* got into the newspapers. It would only have been reported for a day or so because it took many years to decipher, and that's not the nature of something that produces wild reactions in the newspapers. So I suppose there would be lot of excitement in the press, there would be lots of articles, opinions by people.

That would taper off over a relatively short period, and then the main effect would be on some kind of organizational structure to continue receiving the message and to consider how to disentangle it, to find out what it meant. It might not be instantly intelligible, although it might be instantly obvious that it was extraterrestrial in origin. So there would be a long period under that sort of scenario while people worked on it.

It would be rather like digging in a building construction site and finding some archaeological remains. A lot of excitement in the newspaper the day of the discovery, and then there might be discussion about whether there were some remains that should be handled by some ethnic group, but then it would be two or three years before you read an article in the *Scientific American* in which it had been reduced to some coherent story. There would be a long interval in which a few people were working, but the day-to-day results would not affect others much.

Suppose we got the whole Encyclopedia Britannica written in some strange language. It would take a long time to transcribe that. Then to interpret it and figure out what it meant would take even more time, maybe decades. So there would be a burst of excitement followed by who-knows-what.

What will be the general public's response?

I suppose the general public would go on, regardless. Einstein made his great discoveries, and it didn't affect the public at all for many decades. When the atom bombs went off was the first impact on the general public, and that was more or less unpredictable from the nature of the original discoveries of Einstein. It was not obvious that those would lead to a bomb. In fact, some thirty years went by before that occurred to anybody.

So I don't suppose the effects of receiving information from outside the earth will be obvious when we've got it, and since I haven't even got that information, it's even less obvious to me what the impact would be.

An identified star, say twenty-five light years away, I don't think would cause any panic. There would be articles in the press saying that it would take hundreds of years for anybody to cross that distance, and that we'd simply receive a communication, so I don't believe people would worry about that.

If, on the other hand, some object were discovered inside the solar system which was of extraterrestrial origin, and maybe even heading our way, then I don't see how you can avoid having a lot of apprehension, and there would be doomsayers in the press, and that would whip up enthusiasm, and there are some people who like to predict the end of the world.

They would undoubtedly raise their voices, and they'd have more to go on than they usually do, so I think there would be a lot of apprehension and maybe panic in some parts of the world, particularly if the extraterrestrial object was giving out messages. Whether you could interpret them or not, it would be very alarming.

There would certainly be pressure to attack the thing, and there would be some discussion as to whether it was safe to shoot at it. But any one country could shoot it, so it would be a rather interesting case if maybe the UN Security Council could get their act together and debate the pros and cons. But in the end the sovereign states have the authority to do what they like, so maybe the French or the Chinese would shoot at it, and if it was defenseless, there would be no further risk to us. But it might be dangerous to do that, because I don't believe that we would find any spaceship that had taken the trouble to come all this way and was not armed. It doesn't make sense.

So it would be a shame if we were to go into a panic and destroy this artifact before there was an opportunity to extract all the information it might contain.

Iosef S. Shklovskii

ASTROPHYSICIST

Born July 1, 1916, Glukhov, USSR

Died March 5, 1985, Moscow

I. S. Shklovskii, one of the world's foremost astrophysicists, was a leader in raising the search for extraterrestrial intelligence to a level of scientific respectability in the Soviet Union. His 1962 book, *Universe, Life, Intelligence* was influential in the USSR, and his subsequent collaboration with Carl Sagan, *Intelligent Life in the Universe*, (1966) had a similar impact in the West.

Originally optimistic about the existence of extraterrestrial intelligence, he became discouraged as time passed and no sign of it was detected, and eventually considered the possibility that we are alone. Nevertheless, he signed the international petition urging a concerted search for extraterrestrial intelligence.

He was born in the Ukraine, the son of a rabbi, became a railroad worker in Siberia, and studied at Vladivostok's Far Eastern University for two years. He graduated from Moscow State University in 1938, and began graduate work at the Sternberg State Astronomical Institute, but, when the Nazis invaded, he was moved to Central Asia along with a number of other students, including Andrei Sakharov.

Returning to Sternberg after the war, he earned a doctorate in mathematical-physical sciences in 1950, and became head of the Department of Radio Astronomy. In 1968 he was also appointed to the Institute for Space Research of the Academy of Science.

He performed early studies on the evolution of planetary atmospheres, and later proposed that nearby supernova explosions may have played an important role in the evolution of life on Earth, perhaps triggering the extinction of the dinosaurs.

His major work dealt with radio astronomy, x-ray astronomy and theories of the solar corona. In 1948 he investigated the 21-cm line in the radiation of neutral interstellar hydrogen. In 1953 he explained the properties of the Crab nebula in terms of the radiation of electrons in a magnetic field. In 1965 he proposed a new concept of planetary nebulas, linking them to the end of the evolution of red giant stars and the birth of white dwarfs. During the 1970s he studied galactic and extra-galactic x-ray sources.

Many of his students became leaders in Soviet planetary exploration or radio astronomy, including N. S. Kardashev.

His articles in Soviet journals include "The Solar Corona" (1951), "Cosmic Radio Frequency Emission" (1956), and "On the Distant Planet of Venus" (1960).

He was awarded the Lenin Prize and the Order of the Badge of Honor, and became a Corresponding Member of the Soviet Academy of Sciences in 1966. He was also a member of Britain's Royal Astronomical Society, an honorary member of the American Astronomical Society, a recipient of the Bruce Gold Medal of the Astronomical Society of the Pacific, and a member of the U.S. National Academy of Sciences.

Interviewed in December 1981 by Bernard Oliver at Tallin, Estonia

When were you born?

I was born in 1916. I am 65.

When is your birthday?

First of July.

You are a little younger than I am. I'm May 27th. Where were you born?

The place is a very small city in the Ukraine, named Glukhov. A very small city, twenty thousand persons.

So it was a sort of small town environment. How long did you live there?

All my childhood was in this same village.

What were you interested in as a child?

In my childhood I was a painter.

House painter or artistic?

Artistic. In the Ukraine at this time, right after the Civil War, the first year of establishment of Soviet power, after famine, after war there were no pencils, no real paper, so I employed a piece of coal. To draw on the walls without paper. When I was three or four years old I could not write but I could draw very well. Principally, living animals: mice, horses, dogs. I continued this interest up to twenty, when I was a student at Moscow University. I achieved more or less good results. I was quite good in portrait drawing. For example, in a half hour I can make a more or less realistic picture.

Was this in charcoal still or was this in oils?

No, never in oils.

You should illustrate your books with your drawings.

Ah, yes. Maybe if I live in the United States.

With that sort of start, how did you become interested in science and astronomy?

In science, fantastically. I did not complete middle school. I was finished at only the third term of middle school. Only seven classes.

I was finished with my middle school education when I was fourteen. I was a small boy and I had to go to work. I worked in railway construction in Siberia. My parents worked in this field, naturally.

They brought you from the Ukraine to Siberia?

I journeyed many places. I lived in Khazakhstan, middle Asia and Siberia. In Siberia I began my independent life as a worker in railway construction.

Did your father work for the railway?

Yes, also. And when I was 16 by chance I went to a very small library in the small station where I worked. I read a little booklet, a Soviet literature magazine entitled *Novi Mir*: New World. I read a popular article devoted to the discovery of the neutron. It was very astonishing because I knew, from my very low middle-school education, that matter consists of electrons and protons. The neutron—that was new! It was very interesting. I concluded from this that, for me, it is necessary to study physics.

Everything hadn't been discovered yet?

Yes. I still remember it quite well. The stimulus was that article in *New World*. During the next year I prepared myself to be a candidate. By self education I prepared for examination for the university. My first university was the Far East University in Vladivostok. I studied there two years. That was when I was very young. I was only seventeen. Because the university was very poor and the quality of the professors teaching was fantastically low, there was no possibility for education at a high level. The head officer at Far East University decided to introduce astronomical specialization for my course.

You mean he recommended that you study astronomy?

No, not just me. All boys simultaneous with me, maybe twenty or twenty-five persons, should study astronomy.

I was very much against such an idea because my dream from Siberia time was studying neutrons. Astronomy was not for me.

You wanted to study physics and he was urging you to study astronomy?

I changed my plan. I forced my education and in one year I prepared to take examinations for two steps. I made a jump from the first course to the third, only to avoid astronomical education.

You studied hard to avoid your subject.

Yes, that's it. Life drew me back to astronomy. After the third course I went from Vladivostok University to Moscow University, and I finished with the physics department of Moscow University in 1938, when I was twenty-two. By this time I was married and life was very hard, not comfortable. I had no apartment, I was very poor. In that situation, chance brought my attention to astronomy because there was a possibility to do postgraduate education in astronomy only. I considered my change only as temporary.

Because you needed a job.

It was necessary.

Fate was just shoving you right in there.

Oh yes. Maybe by chance, maybe not, but I'm absolutely satisfied and today I cannot imagine any other kind of work than astronomy.

After this I finished the postgraduate degree at Sternberg Astronomical Institute of Moscow University. The time was World War II. I was not in the war because my eyes were very feeble. Maybe that was the reason why I'm living.

And you?

I was working for the Bell Telephone Laboratories when the war came, and so I continued to work on automatic tracking radar. That was my specialty.

Somewhere along the line you discovered that there might be extraterrestrial intelligence, or you began to think that there might be. When did that first happen?

A very interesting question. At first for several years I worked in radio and solar astronomy and solar physics, when I finished the postgraduate course in Sternberg Astronomical Institute. I had very good results with solar physics problems because at this time there was an absolutely new idea about solar physics. Maybe you remember the hot corona? Fantastic new idea because all people before World War II believed that the temperature of the sun was six thousand degrees, and it was absolutely unexpected that the corona is one million degrees.

Of course. How can that be?

The physical condition was absolutely different from what had previously been thought; the absence of thermodynamical equilibrium, the absolutely unexpected state of ionization, and radio emission from the sun.

I was a pioneer in solar radio astronomy also. Together with Dr. Ginzburg, who is an academician today, I explained the meter radio radiation from the Sun, which had recently been discovered. I divided these radio emissions into two components. One was a thermal emission of the hot plasma connected with the solar corona, and the second was connected with special kinds of nonequilibrium phenomena connected with plasma oscillation, nonthermal processes.

But in 1948–49 I very quickly changed my scientific interest in the direction of galaxies, because new and very important radio astronomy was giving me a very strong push in the direction of galaxies and meta-galaxies. For example, just after Vanderhulst had his brilliant idea about the 21-centimeter line, I calculated in 1949 the probability of transition in this 21-centimeter line and predicted the possibility of observation of this line. And besides the 21-centimeter line I was the first to calculate the OH line, the hydroxyl line, the 18-centimeter line, and many others.

I was the first to apply the new idea about synchrotron radiation to radio astronomy, real radio astronomy. I was the first to explain, for example, the optical radiation of the Crab nebula as a synchrotron process; the optical radiation with a continuous spectrum in addition to the line emission from the filament and gas structures of the Crab nebula.

In extragalactic astronomy I was the first one to predict the secular diminutions of flux in the radio emissions from Cas A. It's a real strong radio source. And I am very proud of my book devoted to cosmic radio waves translated into English in the United States. It is very interesting to remark on that moment because it was very new and interesting. There began to be many radio astronomers in America after the discovery by Karl Jansky.

Yes, Jansky started it.

By the way, you knew Jansky?

Yes, I knew Jansky very well.

This is the only example of a giant science developing from only one person. I don't know of any other example.

It was kept alive by Grote Reber during the war. Can we get back to the question of what got you first interested in ETI and SETI?

I describe for you all the climate and spirit of my work. In 1957 there was the launch of the first Soviet Sputnik. I very enthusiastically took part in this investigation and made several projects. In 1961 the President of the Soviet Academy of Science, Mr. Keldysh, proposed the writing of several scientific and semi-scientific books devoted to a memorial five years after launch of the first Soviet Sputnik. For that occasion I declared my aim to write a book devoted to interstellar intelligence.

Maybe by chance the work of Cocconi and Morrison made a very strong impression on me. I read this paper soon after it was published in 1959, but several months delayed, naturally, in arriving in the USSR. I was very interested, and I thought that a first book in this direction may be interesting for the public, a popular book. I declared my proposal and people supported me and I worked very quickly. I spent only four, maybe five months writing this book. I go each day to my job, all day, and only in the evening time can I write such a book.

Besides the compilation of old ideas, the historical part, the general description of the situation, was purely astronomical side: the new idea by Cocconi and Morrison, which I knew, and Dyson and Bracewell's ideas. At the same time, I developed my own ideas.

For example, I drew attention to the power of TV. Our TV system is quite enough for transformation of a very small and very modest planet into the second most brilliant radio source in our solar system after the sun. TV waves go freely and fast; and from other parts of the solar system, maybe other stars, our earth is a bright point source.

As another example, I drew attention to the technological aspect of the problem. The transition from the First to a Second Type of civilization in the famous division was proposed by Dr. Kardshev, maybe two years after my book. In my book I drew attention to the possibility of very highly developed technological civilizations, but not to the kind of divisions proposed by Kardashev. But many times I emphasized the rich possibilities of fantastic power that can be achieved by such kinds of civilizations.

I think Cocconi and Morrison sowed the seeds here because they pointed out that we actually had the possibility of communicating with ETI, but there had been people before them speculating about the existence of extraterrestrial intelligence.

Before this I was not actively working in this field but my book was well received by the public.

It was sold first here in Russia, wasn't it?

In Russia in 1962, and after this, five editions. The last edition was in 1980. The Sagan-Shklovskii book is very fine story because Dr. Sagan asked me about the possibility of translating it into English. I sent him the manuscript of my book and asked him to make translations and some additions. Because Carl was well educated in biology and I am not, I asked him to bring it up-to-date and add a section on biology.

Sagan spent three years on such preparation, not only from the biological side. After this, the book was published by two authors. The text which was written by Carl is divided by preamble. It's very useful and important for me because of the political situation. Because I am Soviet author I must show respect to Marxism and work "according to Marx," but after preamble Sagan can write "according to Kant. . . ." This is fine because I can show Soviet officials, who say this book has a mistake, that it is not *my* mistake, it's Carl's [chuckles].

I want to tell you that one or two months ago the famous New York publishing house John Wiley and Sons made an agreement with Soviet special firm about the translation of my own book, fifth edition, in the United States.

This is yours and not the collaborative book?

I think it's necessary to make some additions because it is a very quick developing field.

Are you married?

Yes, I am married. I have two children, daughter and boy. I am grandfather. I have two granddaughters.

You're ahead of me. I have three children but no grandchildren.

They are in your future.

I hope so. You've told me a great deal about yourself. Could you tell me some deeper things, some personal things? What do you think has been the greatest happiness in your life?

To behave like a humorist, for me it is to be with Dr. Oliver, vice president of an internationally famous corporation. But, my main happiness has been the realization of predictions. I felt such happiness two times only in my life. The first time when I predicted the polarization of optical radiation in the Crab nebula. It is the key problem for all models of development in the studies of the synchrotron theory, the theory of nonthermal radiation.

The situation was that the first observations were performed by very small telescope in Caucasus observatory in Soviet Union, only a sixteen-inch telescope, with no possibility of detailed investigation of the optical polarization of the Crab nebula. Always the proper polarization was averaged or smeared by the disk of the nebula, but the measured value was 12 percent. Fantastically high. For example, the percent polarization usually starts 1 percent, 2 percent only. Next investigation showed that in some parts of the nebula the degree of polarization is up to 50 percent, 60 percent, 70 percent. But only because of the ability to resolve different locations. When you employ very bad, very small telescopes, you average, and the result is much smaller. This made me very happy.

Second similar situation was when I predicted in 1960 the continuous diminution of the radio flux from Cas A. Cas A is the strongest radio source in the sky, and I came to conclude from a purely theoretical investigation that this source must diminish flux maybe two percent per year. For example in ten years, 20 percent. Fantastic! It was the first attempt to consider a cosmic radio wave source as a variable source.

The method I used for such a kind of exploration was the comparison of one radio source with another very strong radio source. And with Cas A the ratio of the two sources was secularly increasing, and it became clear, because Cygnus A was an external galaxy source and constant in flux, that Cas A was diminishing in flux 2 percent per year.

I wrote to England and Dr. Struve in National R.A.O. Why I wrote to England? Because Cas A was discovered at Cambridge Radio Observatory.

Martin Ryle?

Yes. And it was important to use the same telescope, same apparatus with different side effect to repeat this obervation with the same method. Very quickly he established that since 1948 when this source was discovered the flux had decreased by about 30 percent. That's very strong.

That's a thrill. Have you suffered any real sorrow or disappointment or just plain real tragedy in your life?

Tragedy? No, absolute 100 percent satisfied. Life is blue dream.

I see. You've been sailing along on moonlight bay.

The drudge work is satellite, part of everyone's life.

Now, you were concerned with the hydrogen line about the time Cocconi and Morrison pointed out the significance of it for SETI.

Their work is stimulating mine. If you remember, Dr. Kardashev said the same thing. It's very stimulating, this work. Very important.

I didn't know Cocconi. I never met him. But I know Dr. Morrison quite well. For example, last time I met him four months ago in Albuquerque, New Mexico. Very fine man.

Kardashev was a student of yours, right?

He was my student. He is one of the most brilliant Soviet astronomers, I think. He's an absolutely independent man. The best kind of relation exists between me and Dr. Kardashev. He has an absolutely different position about the SETI program. But he is my best friend. He's a very beautiful man.

He's in science very deep, very deep. For example, the scientific achievement of Dr. Kardashev. You know what a radio recombination line is? It was an invention of Dr. Kardashev. He was one step only from the discovery of pulsars. In 1964 he came to the conclusion that a new form of neutron star was rotating very fast with very strong magnetization.

That's the luck of scientists.

It's very difficult to make the last step.

Would you say that he's your foremost student, or do you have other brilliant students, too?

I've got many students but he's my best student. He is a member of the Soviet Academy of Science. He was elected five years ago. He's not so young. He's mid-fifty but his face is very childlike. I remember very strongly in 1967 at the Prague astronomical meeting, when I introduced Dr. Kardashev to many Americans they were very astonished. He looked so young.

Have your ideas about SETI changed since you first started to think about it?

There is a common opinion about the change of my mind but it is not correct. For example, my book devoted to intelligent life in the universe had five editions published. Naturally each edition was different from the previous edition. Sometimes additions and improvements, and sometimes changes. The most recent edition was published in 1980; before that, 1976. In 1976 and in 1965 also I made short remarks that it's quite possible we are alone in the universe, that intelligent life is an extremely rare phenomena. (I show you small footnote from my third edition.)

With advent of large radio telescopes—in the United States, German system, Soviet system—in last ten years sensitivity increased a hundred times, several hundred times. The sensitivity is the level of a millijansky. One jansky is fantastically small value for flux ten years ago, and today we speak about millijansky.

The main question I would like to emphasize in this edition is very important work of Michael Hart in Texas. Why can't we observe signals from them at the level of our radio astronomical instruments? It's not clear. Dr. Hart emphasized the problem of absence of signs of colonization of our planet by foreign visitors. My main emphasis is connected with the absence of artificial signals.

The problem connected with attempts to establish contact with Tau Ceti, or maybe other star, is because in the vicinity of several hundred light years the total number of stars more or less similar to the sun is several tens of millions. Their address is very undefined. In such situations the best strategy is to make a signal addressed simultaneously to *many* star systems.

Let us suppose there is a super civilization in Andromeda nebula. Andromeda is galaxy two or three times larger than our galaxy. One of their civilizations achieves fantastically high technological level. In such a situation this civilization makes a big radio signal in the direction of our galaxy. This beam must cover maybe one degree, corresponding to an angle subtended by the inner part of our galaxy viewed from the Andromeda nebula. In such a situation the signal is maybe simultaneously observed by many hundred million possible receivers.

It is not an isotropic signal. The gain on this system may be ten to the fifth. The angle is only one degree; less than one, maybe half a degree. Quite enough! And the power, for example, is not as well determined because there are free parameters connected with the width of the band. But I can believe ten to the seventh. A dish with 100-meter diameter—with power, for example, one megawatt per dish—is quite possible technologically, and it is also possible to have many dishes working simultaneously.

One ought to organize such a search and very slowly to make a scan of this area, the inner part of our galaxy. We would then observe in middle part of the Andromeda nebula a very strange pulsed source, an artifical pulsar for example. A very strange and fantastic narrowband signal; absolutely artificial!

It's very important that as yet we don't see such signals. Naturally, similar investigations must be continued in the future. But from such kinds of pieces of evidence, it follows that such a high level of technology has not been achieved in the vicinity of our galaxy, maybe in our local group of galaxies. This is my main argument.

Is there anything you would like to have people in the future know about your life?

I am *not* satisfied completely with my life.

What would you like people in the future to know about your feelings?

.... [pause] A little more degree of freedom?

Yes, of course.

That is common.

Nikolai S. Kardashev

ASTROPHYSICIST

Born April 25, 1932
Moscow, USSR

Dr. Kardashev is deputy director of the Space Research Institute of the Academy of Sciences in Moscow.

After graduating from Moscow University in 1955, he studied under Shklovskii at the Sternberg Astronomical Institute, receiving his doctorate in 1962. The following year he conducted the first Soviet search for extraterrestrial signals, examining a quasar. He worked at the institute until 1967, when he joined the newly formed Space Research Institute. In 1972 he began a search in all directions, looking for sporadic pulsed signals.

A leader in launching SETI in the Soviet Union, in 1963 he organized a SETI research group at Sternberg and initiated arrangements for the first All-Union Meeting on extraterrestrial civilizations, which was held at Byurakan the next year.

He was also an early advocate of the idea that supercivilizations have evolved and may be billions of years more advanced than ours. He believes there are no limits on their activities, and he proposed a classification of civilizations according to the amount of energy they could harness. A Type I civilization is capable of using all the energy falling on its planet from its sun. A Type II civilization captures for itself all the energy emitted by its sun. A Type III civilization utilizes the total energy output of its galaxy. Since we have not yet reached the first stage, we could be called a Type O civilization.

Advanced civilizations would be capable of immense astroengineering projects, including constructions around nuclei of galaxies and quasars. Such structures could be detected by their thermal radiation or by the way they screen or reflect the cosmic background radiation. These civilizations could also put enormous energy into deliberate signals. To detect structures or signals, Kardashev recommends searching in the infrared and millimeter ranges, in contrast to the centimeter approach of most SETI scientists.

The most promising region in which to search is toward the center of our galaxy, because the density of the stellar population is greatest there along the line of sight. The closest galaxies should also be observed, and the cores of exploding galaxies might reveal evidence of meddling by extraterrestrials.

A Corresponding Fellow of the Soviet Academy of Sciences, he has also been active internationally in SETI, including co-chairing sessions of COSPAR. His current projects include design and construction of a 70-meter radio telescope in Central Asia, and planning for an international space-ground interferometer. He is also interested in cosmology.

His reports appear in Soviet journals and international publications including *Nature* and the proceedings of the International Astronomical Union and the International Astronautical Federation.

Interviewed December 1981 at Tallin, Estonia, by Bernard Oliver

I'm sitting here a moment after the afternoon session of the SETI conference in Tallin, Estonia in December of 1981. I have with me Nikolai Kardashev. Dr. Kardashev is unquestionably one of the outstanding pioneers of SETI in the USSR. We're very fortunate to have him here to ask a few questions.

I'd like to ask about your personal history, some things about your background that may have been important in bringing you into the SETI picture, causing you to choose SETI as part of your career.

Where did you live during your childhood?

In Moscow. I stayed there all during the war. In the first days of the war I saw German aircraft over Moscow.

Tell me something about the things you liked as a child.

I remember that my mother took me to the Moscow planetarium when I was five years old. The planetarium was very old and had a very strong influence on me. It was unusual because it gave not only lectures on astronomy but was organized like a theater: dramatized situations connected with Galileo, Copernicus . . . living, those people.

So they made astronomy live.

I asked my mother how many points were on stars in the sky, compared with the red stars of our flag. Her answer was "five also."

My interest in astronomy started very early in my school days. Each week we convened a special small group of small boys interested in astronomy. A scientist from the Sternberg Institute organized this group at the Moscow Planetarium. We went there once a week, for observing celestial phenomena or for discussions. At night we looked through telescopes. We had many lectures by volunteers about different celestial phenomena.

Was this in primary school?

Yes. I was only about ten years old when I started astronomy education in that special small circle at the time of the war.

Were there other activities in your childhood that we should note here?

This was during the war. Military destruction of Moscow by planes. Near the planetarium were buildings that were completely destroyed by bombs. This was a very strong influence on childhood, of course.

When you were young did you read any science fiction?

Yes, of course. Science fiction books connected with astronomy. One was an interesting, imaginative story about a scientist preparing a metal that was exactly the same as the metal inside a white dwarf star, and a small piece of this metal was inside his suitcase. I forget what happened but it was an intriguing science fiction story.

Did you read comic books?

No, it was science fiction books. Also many people read a book about history of astronomy. The writer was Berre; probably British. It was translated into Russian.

What happened then?

I finished school in Moscow and I finished at the Moscow University also. At Moscow University my first teacher was Professor Shklovskii. That gave me a very nice situation connected with astrophysics.

So Shklovskii was an influence in your early life?

Yes, Shklovskii, of course. During my education at Moscow University I directed my attention to experimental radio astronomy.

You chose that as a major subject?

Yes, because in 1953 the first courses, lectures were given by Shklovskii, especially in the direction of radio astronomy.

Those were the first courses he taught.

Those lectures after some time were published in his book about radio astronomy and translated into English, and became very popular.

After I finished Moscow University I worked with Shklovskii's group. Shklovskii was interested in many new directions in astronomy. He was the person who first understood that it was necessary to use all the electromagnetic ranges to understand each astronomical object. And he had many young people working in new branches of astronomy—radio astronomy, infrared astronomy, X-ray astronomy, gamma-ray astronomy—and it was connected with a group of people working at Sternberg Astronomical Institute.

From 1967 most of the group changed its location and worked together with Shklovskii at the Space Research Institute, a new institute founded in Moscow. This was the main institute connected with science using space programs. Shklovskii stayed as chief of the department at the Sternberg Institute, and simultaneously he was chief of the new astrophysical department at the Space Research Institute.

Two positions?

Two. And the people working together with Shklovskii at the Sternberg Institute divided into two parts. One part, together with me, changed location and now worked at the Space Research Institute; a small part stayed at the Sternberg Institute. The two groups are working together even now, in different areas of astronomy.

Let's be sure we have those dates when you finished your secondary education.

We have in our country normal, basic school that continues for ten years. I started in 1940, exactly one year before the war, and finished in 1950. After that I studied at Moscow University. I started in 1950 at the mechanical mathematical faculty, and in 1953 the branch of astronomy at Moscow University changed its position, separating from the mechanical mathematical faculty and joining the physics faculty, because the decision was that astronomy was now more connected with physics than with mathematics.

I finished the physical faculty at Moscow University in 1955—five years of education. After that I joined Shklovskii, working with the Sternberg Astronomical Institute. I prepared the first stage of my candidate thesis in 1960, and second thesis, as doctor of science, in 1962. After that I continued to work with Shklovskii and his group.

In preparing my first thesis I was connected with a program observing the 21 cm line of the galactic emission. It was exactly in the same direction described in the Cocconi-Morrison paper.

But you started before that?

Yes, yes, yes.

You started because Professor Shklovskii suggested that you do that?

Yes, yes. One of the many interests of Shklovskii was exactly 21 cm that is connected with the emission of hydrogen atoms, the most common atoms in the universe. He gave strong pressure to many experimenters in our country to start examining at 21 cm the different objects in our galaxy and other galaxies also. This observation was in Crimea on the new radio telescope. I started to prepare such kinds of observations in 1954 when I was just a simple student, and this telescope was not completed yet.

In 1956 the first observation on this telescope was completed, and after the appearance of the Cocconi-Morrison paper we had many discussions about the possibility of finding something from our observation, and discussed what kind of shapes we detected from our records. If it was some kind of unusual record, we had a science fiction discussion about these problems, and it improved our interest.

The Crimea radio telescope was connected with the Crimea station of the Moscow Radio Physics Institute. It's not the Sternberg Institute, but Sternberg Institute with Shklovskii was a strong collaborator with the Radio Physics Institute, and each year Shklovskii was in Crimea, too. He liked Crimea very much; he liked nature in Crimea, and very much liked to swim. And there were many discussions about SETI problems on the beach in the Crimea, near the small village Sireez, a Greek name.

Did your colleagues and friends also get interested in SETI?

Yes, yes, of course.

So there were many discussions between you?

Yes. After publication of the Cocconi-Morrison paper we had all our group involved with this. Many discussions were centered on this topic.

I can see it's a wonderful subject for Russian philosophers to talk about.

Yes. Interest in SETI was very strong after the Cocconi and Morrison paper, but before it I had some interest in the same problem, but connected with the probability of life on Mars, because we had in the country a unique laboratory operated by Professor Tihov. He was, if I remember, professor of the Pulkova observatory, but after the war he stayed at Kazahkstan. At Almata, the capital city of Kazahkstan, he organized a very unusual lab named Extraterrestrial Botanics, and he compared spectrum of the different places on Mars with spectrum of different trees in his garden, and selected the trees nearest to the spectrum of Mars.

When did you first think about the possibility that extraterrestrial beings might exist? Was that earlier, in your childhood?

No. It was only after publication of the famous article by Cocconi and Morrison.

That was about the thought that we might communicate with them or detect them. But did you speculate earlier that they might exist? Did you entertain that possibility in your discussions?

Yes, of course, because before this time I met with Professor Tihov, who was searching the surface of Mars, and he had many publications comparing the different colors of Mars with colors of the different kinds of vegetation on Earth.

He was the Percival Lowell of Siberia.

Yes, yes, exactly. All discussion was connected with life on Mars only. All people thought that maybe it's possible to detect life on Mars or maybe on Venus, but other stars were thought to be too distant from us. "It's impossible today" was the general position of all people at that time.

But by the time the Cocconi-Morrison paper appeared, there had been the launching of the first Sputnik and the first negative results about life on Mars. Before that, nobody thought about life more distant than Mars. It was discussed but not strongly, because it was expected that the very near planets may be populated, too. But after negative results of the first space searches, there was increased interest in searching other planets and more distant objects.

The attitude about extraterrestrial intelligence differs widely in the scientific community. Some scientists are enthusiastic about it, some are negative. How do you think most scientists in the Soviet Union view SETI? Do they think it's a legitimate thing or do they think it's foolish?

It differs very much because we have, for the upper level, for the level near the government, two very different kinds of people. For example, Academician Kotelnikov has a very positive attitude toward this problem.

I met him in 1965 when I was here with the IEEE and I know he was enthusiastic even then. But you may have very negative people, too. Is that right?

Yes, yes. It's right.

So you run the gamut. Do you think there's been a change with time? Do you think that SETI has become more popular or less popular?

I think it's more popular. It's a very strange correlation. Political relations now [1981] are very bad, but many people search for other possibilities for positive influence on their spirits, and want to find something unusual.

Has your interest in SETI had any effect on your career as a scientist?

Different people have different relations to this field. Skeptical people think it is crazy activity. We have a very favorable situation in the Academy of Sciences on this problem. In the Soviet Academy of Sciences there are not too many people who are skeptical about this problem, but in different directions sometimes it happens that I meet people who think that scientists working in this field are crazy.

What group, what kind of people would think you are crazy?

It's not a group. It's usually people not connected with some philosophical direction of thinking but centered on a small, concrete practical problem. Also a small number of astronomers are very skeptical, but in my opinion these astronomers have a very narrow direction for their research. These people do not have a broad look on the universe.

What does your boss, your director, think of your interest in SETI?

I am the director; I'm deputy director of the Space Research Institute.

What about your family? Were they interested in science? What kind of background did they have, your mother and your father?

Both my father and mother are now dead.

They are gone.

It's a difficult question for me. My father died before the war, and it was connected with the Stalin regime. My mother died about two years ago. She was a medical expert.

What was your father's occupation?

I don't have information about it. I only know that at the end of April 1937 he was arrested. At the end of the same year he was executed. I don't have any information about his business exactly. It's very strange but it is very difficult to collect this data today. I know only that he was working as an engineer in construction of a channel near Moscow.

He was an educated person?

Yes, he finished education but I don't know exactly what kind of profession he was officially.

You mentioned that your mother was a medical expert. What kind of medical work did she do?

I don't know exactly. She was working at the last as a doctor in the field of epidemiology. She completed her medical education in Moscow, at the time of the revolution.

Did you have any brothers or sisters?

I had a younger sister. My early childhood was during a very bad time. My sister was only about three months old when she died. She was in a camp with my mother.

I just met your wife, so I don't have to ask if you're married. How many children do you have?

I have one daughter. She is nineteen. She's studying computer techniques. Second courses, not at university but a special institute that specializes in computer technique.

Do you live in Moscow now?

Yes. All the time.

Have you had great satisfactions in your career?

Yes, yes.

What is the greatest of those that comes to mind?

It's very strongly correlated with many discoveries that were realized exactly at the time of my initial stage of working in radio astronomy. All kinds of galactic and extragalactic sources were suddenly discovered: pulsar, quasar, at the same time of the discovery of background radiation and molecular lines and other sources.

I've had business connected firstly with thermal radio emission from different sources. Secondly from work connected with 21 cm survey that was prepared in the Crimea observatory. After that I calculated the possibilities of detecting recombination lines, and this line was detected. After that my work was slightly connected with SETI problems because at Crimea there was a special deep space communication center. It has a very big antenna system.

The second step to the SETI problem was observation of very unusual source, CTA 102, and it was surprising that we detected that this source varied with time. It was very strange.

That was before pulsars.

Yes. Yes. Yes. Before detecting quasar radiation.

And on a time scale that meant they were very small.

It is interesting that this problem is not resolved up to date, because we cannot satisfactorily explain why, at the decimeter wavelength, we have radiation flux.

So you were involved in that discovery?

Yes.

So what you are saying is that those years were very happy years because there was a sequence of discoveries—not one particular thing but many things that happened together. Well, that's very interesting.
Let's review your other interests, apart from SETI.

My interests started with searching the HII region thermal radio emission from the ionizing from the galactic plasma clouds, and after that I observed 21 cm emission from the neutral hydrogen clouds of the galaxy.

After that I had some interest to observe extragalactic sources, and at this time was found unusual variability of extragalactic sources, and I proposed that maybe some unusual variability may be connected with SETI activity of the very distant objects. After that I was connected with the program searching for pulsars, shortly after pulsars from the universe were detected. They were explained as neutron stars that have very unusual conditions of emission.

Now I mainly am interested in application of astronomy to cosmology. I also have interest in some technical problems of constructing new telescopes, and techniques connected with very long base interferometers. We've started to construct a new observatory near Samarkand in middle Asia that will be connected with our department of our Space Research Institute, and we hope to finish this program in 1990. It will have a big telescope, seventy meters in diameter.

After much adjustment of the surface, that will give us possibility to use this mirror up to the shortest millimeter wavelengths, because we've excluded many errors of the surface. Also I'm connected with construction of a space-ground radio interferometer that will be launched after a few years. Now I'm preparing technical devices for the first space radio telescopes that will be used together with ground telescopes as radio interferometers.

What other interests do you enjoy?

In Moscow we have a very highly theoretical group who have interest in understanding more general questions connected with astronomy. We have very bright person such as Shklovskii. We have very bright person such as Zeldovitch, who is high level theoretician physicist, and a third person is Ginzberg.

The main interest of these people, and the people included in their circles, is cosmology: understanding how all kinds of physical processes were in the early universe, and how they generated the universe that we observe today. My interest also changed to this. Now I think about experiments in this.

Aside from these scientific interests, do you have other interests like music, playing cards, walking, watching birds? Professor Shklovskii liked to draw.

No, I have no such interests. I very much like nature, and I like travelling. Yes, I prefer to look at new places.

Does extraterrestrial intelligence exist? Are there intelligent beings out there somewhere on other planets?

My position is very optimistic but many people around me are skeptical.

Even Shklovskii.

Yes. Even Shklovskii now has changed his position.

I called him a "defector" this morning.

But it seems to me that Shklovskii's position may be connected with his life situation. He was many times ill. Very strong correlation.

Why is extraterrestrial intelligence important to you?

Two sides. One side, I think it is very important because the most interesting answers to the famous scientific questions will be received very soon if we have contact with extraterrestrials. Many basic scientific questions we don't expect to answer ourselves by experiments because of limitations of funding for real scientific experiments. We don't expect, for example, to construct telescopes above one kilometer in diameter. But from contact we may find, I hope, all answers to the basic questions.

The other side, of course, is to understand the future for ourselves, our civilization, the social questions. I am also interested in this. Now in Moscow for the first time was published Toffler's book, *The Third Wave*. It is an interesting question. I want to buy this book.

What fraction of your time have you been spending on SETI?

It is very difficult to estimate. Most of my time is connected with constructing the new observatory near Samarkand. The main instrument of this observatory will be a dish, a very steerable mirror, 70 meters in diameter. It will be used for searching for extraterrestrials. Now it's an engineering problem. About 50 percent of my time is spent organizing this new observatory and completing all the facets of it.

What about five years or so ago? Were you spending about the same fraction of your time then on SETI?

Yes.

Have your ideas about searching for extraterrestrial intelligence changed since you first began thinking about it, or are they about the same?

The same. We have a more detailed idea of the kind of electronic system needed for searching for extraterrestrials, because it is nearer the time when our new telescope will be working. We are starting to develop all the electronic systems for the search; not only the telescope itself but also many additional devices exactly for the problem of searching.

If you were starting over now to search for extraterrestrial intelligence, is there anything you would do differently?

Yes, of course, because we have many new details of unusual phenomena in the sky that may possibly be objects formed by extraterrestrials. One example is a very unusual source in the galactic center, or surrounding the galactic center; most details were found in the last few years only. Another example is the many sources we may find in the infrared ranges from the all-sky survey that will be undertaken by the American and western European infrared satellite IRAS. My opinion is that some of these sources may be connected with extraterrestrials.

It is interesting, surprising that from science discussions with different physicists we have no exact limits for the upper level for extraterrestrials, no physical laws that limit the power of extraterrestrials. It is difficult to predict by natural physical laws the limits of the size, the power, the mass of their activity.

What might extraterrestrials look like?

I think it must be a very big structure in space. It's impossible to predict how it is constructed, but I hope this construction can be detected by its thermal or special emissions. It's impossible to predict what kind of activity or position.

A more simple question—how to predict changes in *our* civilization for the next hundred years—is also without answer. It is impossible to predict.

My position is to search for new objects in the universe, that are difficult to explain by natural causes. I think 21 cm is one place for such searching but it's not the only one. Another possibility is to search for emissions from the solid-state matter of astronomical objects, and analyze what these objects are: dust clouds around normal stars, or generation by planets, or maybe thermal emission of very big constructions?

It is possible to discriminate between the cloud of small particle emission and a very regular, big artificial construction by analyzing the spectrum of the emission. Usually the cloud of dust grains does not give radio emission but gives only emissions in infrared, because size of the grains is very small. It is simple to discriminate between emissions from an ETI structure and a cloud of dust.

What about beings? Opinions vary widely about what extraterrestrial beings might be like. Some people think only in terms of hominids or humanoids and some people even think about life on such exotic things as neutron stars and so on. What are your ideas on the subject? Do you expect to find carbon-based life or do you expect to find other kinds?

I think there may be strongly increasing computer techniques in the future. After a hundred years, for example, each person may be changed like a robot system. There might be better model young people.

Do you think that carbon-based life may only be a precursor to a silicon based life?

It seems that electronic life is better.

I expressed that thought back in 1963 in connection with an article I wrote on electronics, but it hasn't come to pass yet.

It is difficult to predict, of course. It's like fantasy. I think in principle it's possible to prepare such devices that would be much better than the human being, with the same emotions. This construction must be immortal, and if you put in the program all positive emotions connected with the human being, this position is exactly the position of Dr. Minsky.

I completely agree with the position of Dr. Minsky. He was on the Byurakan symposium that talked about development of modern processor systems, and predicted that combination of the very developed processor system with automatic system-like robots may be the future of life on earth—very strong possibility of how it will change.

So you agree with Minsky?

Yes, I agree. I think it's our future, after a hundred or maybe two hundred years. It may be, for emotional reason or another reason, possible to conserve all differences connected with human beings, and, using the new techniques, to have multiple forms.

Will these artificial creatures, robots, androids exist along with living creatures or will the machines make living creatures obsolete?

I think it will be mixed, both; because in physics the process is un-limited. Both kinds could be improved in such situations. It is difficult to predict but I hope it will be possible to put in such beings all that is *positive*.

Toffler is thinking the same, not about SETI, but differences be-tween capitalism and socialism and communism now start to decrease because they are connected with another new wave of distribution. The third wave, if I understand, is the wave of information. The first wave was distribution of land, the second wave was distribution of the manufacturing plants, and the third wave is distribution of informa-tion. Information has the main role in development.

I have great interest because in my publication in Kaplan's book I also give some general description of the development of civilization. The development of civilization used mainly new information. The highest prize in civilization is new information, not management of land, because in the universe there is too much matter, too much space, too much time, but new information is the highest prize. Tof-fler's position is very convenient for my position.

When was your book published?

It's not my book. It's a translated book containing many articles. The editor is Kaplan. It was published in the Soviet Union, but an English version was prepared, and I saw this English version in many libraries. It was translated in Israel.

What is the title?

Extraterrestrial Intelligence.

In the years that you've been thinking about SETI I suspect that your ideas have gone through some changes from the early Cocconi and Morrision paper to the present. Could you say a few words about the changes in your thinking about SETI and how you look upon it differently now than then? There's nothing much happened in the way of actual discovery but could you describe maybe how your thoughts have changed?

Our group held many discussions about possibilities of SETI. Especially there was much discussion about what was the highest level of the proposed extraterrestrial intelligence. It was connected with discussions with Shklovskii and other groups. Approximately one year after publication of the Cocconi-Morrison article a special section was organized in Moscow on this topic. This section was connected with the radio astronomical council that was organized about five years before. Mainly this discussion involved the problem of what is the maximum power that may be emitted from extraterrestrial intelligence.

Is this when you came up with the idea of your Type I, II and III civilizations?

Yes. It exactly correlated with observations of CTA 102. After some time it was determined that this quasar, with very big red shift, is very distant.

So, having been studying things that were so very powerful in nature, it didn't seem unusual to speculate that maybe man or intelligent races could achieve this power. Do you actually think they do? Do you think there are Type II and Type III civilizations?

Yes, I think it's possible. There are no limitations for the activity if there is a very long time for improving the intelligence. Very possible. There is no limitation for very big transmitters, for example. Troitskii's more skeptical than I, of course.

What has been the most difficult aspect of your work in this field, or has there been any problem with it?

It is generally difficult. We need much time on a big aerial system.

Better observing time.

Yes, observing time is the main problem. In our country we used only the 22-meter diameter dishes for the problem. If you want to do observations with point sources but on a survey of the whole sky, you must use Ratan 600 system.

Realizing it's only a guess, do you have any thoughts as to when you might expect contact to be made? Would it be next year, next ten years, a hundred years?

I think the next ten years.

You think the next decade might bring it?

Yes. In astronomy in general we are in an interesting time because all wavelength ranges are possible to use for examining new kinds of sources. Soon we will have a total map of full sky for each wavelength range. For example, it is extremely interesting to observe the whole sky in infrared wavelength because exactly in this range we must observe emissions from rigid body constructions, and find in what position of the galaxy you have maximum concentration of these rigid bodies.

Who do you consider the outstanding people in SETI today?

In our country or in general?

In general.

In general for me the first two are Cocconi and Morrison, but I don't know where Cocconi is living now.

He's living in Switzerland, I believe. But he is not pursuing SETI.

And next person is Dyson, because his ideas are strongly correlated with our discussion about highest levels of civilization. And, of course, our friends Drake and you and all the people who were at the Byurakan symposium.

And in your own country?

In our country, Shklovskii of course.

Even though he is pessimistic?

Yes, but in initial stage he was very optimistic. Simultaneously, in our group there are younger people: Dr. Zeldovich, Dr. Slysh, Dr. Panofkin.

What would you say your role in SETI has been?

My position is very near to the position of Dyson, because he completely agrees we need to search for unusual phenomena in space. His papers also I think are excellent. Freeman was in our country, in Moscow, and I met with him.

What scientists or other people in the past do you admire?

Of course I was strongly connected with Professor Shklovskii; very close to him for many years. I was also very close to Professor Pekalmer who was also working at the Sternberg Institute.

He was a high level expert in different fields of astrophysics; he made some of the best plasma physics applications to astrophysics. I was also connected with Professor Kaplan at Gorky Radiophysical Institute. He was an expert in astrophysics and worked together with our group and with Shklovskii.

How about in previous times, long ago?

Of course, this was only from books, but a strong influence in my early education was Flammarion. Also the ideas of Schiaparelli on Mars channels.*

One of the things that most interests me about SETI is that it answers the question of our uniqueness or else our existence in a universe filled with life. But that is really only part of the deeper question. The deeper question is whether life is an accidental phenomenon in the universe or whether it exists in some way to assume an important role with respect to the evolution of the universe.

Have you had thoughts like that? Are these philosophical speculations that have occurred to you, and what are your reactions?

It seems to me that generation of life may be by chance, or maybe it's a regular process; it's difficult to say. But very probably we have in space a very high level of organization, much higher than our own level of organization. The problem is to find the location of this level; the direction of this civilization.

And after this problem is approached, after the first contact, how will it influence civilization on Earth? It probably must first be a shock that we meet a very high level.

It seems to me that the first reaction would be to look further. I mean, you will hold on to the source you discovered but you will probably also search with much more effort then, to determine how much—not just "if" but "how much?"

My position involves size: we really have only one civilization in the Galaxy. Many civilizations have been generated in our galaxy. But, immediately after contact, two civilizations combine, and as a result we have only one civilization, because there is big profit from combining the civilizations. For me it makes possible speculation that after contact we will be combined also with, for example, galactic center.

We must think about a very big ship that must start to the galactic center.

---

*Camille Flammarion. French astronomer who wrote in 1892, "The present inhabitation of Mars by a race superior to ours is very probable" (McDonough, p. 27).

Giovanni Schiaparelli. Nineteenth-century Italian astronomer who observed straight-line markings he called "CANALI" on Mars (McDonough, 25–26).

You'd like to explore that?

Yes. It is one of the nearest very drastic objects. From the astrophysics perspective, the galactic center may be a very big mass, a black hole, but it has not been detected yet. We observe in the center of the Galaxy a very strange radio source, a point radio source, not explained yet. So I'm thinking to keep in mind two explanations.

One is a natural explanation: that it is a black hole that gives off the phenomena observed. But maybe, because it is a very favorable position in the galaxy, this object is connected with some kind of intelligent extraterrestrial activity.

We have not any observational data that favors one or the other of these hypotheses. If we detect something that seems of intelligent origin, we must be thinking about some kind of expedition to the center of the Galaxy. But this is only one example.

Another example is maybe we find some construction in another direction, not in the center of the galaxy. It must be the object of future explorations.

But the galactic center is the most interesting region of our galaxy, a distance of about 30,000 light years.

Well, that's certainly an ambitious undertaking. I've said that if there are many, many oases of life in the Galaxy, it is unthinkable to me that they will all evolve throughout their entire existence without coming into contact. And I have speculated about the proximity of galactic civilizations and why there might be easy contact between pairs of them, and a network might be established this way.

Do you think this is a possibility? Do you think that there's a galactic club that might have grown up?

It seems to me that for the highest level a compact society is needed because, with a very big delay between different points, it is impossible to give, for example, new discoveries immediately to the opposite side.

It takes a thousand years to get across.

Yes. But if it is compact, it's possible immediately.

This is why you want to go to the galactic center?

Yes. And if life is also connected with *extra*galactic sources, it is difficult to construct a system for receiving signals between galaxies. But if all civilizations are combined inside one galaxy, it's possible to combine technology, and that gives you possibility of communicating.

Often in talking about extraterrestrial intelligence UFOs are mentioned. The public is interested in this. What do you think UFOs are?

I'm very skeptical. It seems to me that in principle it is possible for other beings to come to the Earth, but really all situations described in the newspapers and special magazines are usually connected with unprofessional searching with different kinds of techniques. It is difficult for unprofessional types of people to explain what they see in the sky, because there are many new military technologies that are unpublished. I think the situations described in the newspapers are mainly these military technologies which I know, but may also be geophysical phenomena.

What should a scientist do if he does receive evidence of extraterrestrial signals?

It depends on the psychology of the scientist. I have two examples. When we found variability of the source CTA 102—very unusual. We didn't forbid publication of this information by the newspapers. It was published immediately in the newspapers and was announced by radio from Moscow, and the next day I received a telegram from Frank Drake with questions about the position of this source, and next day there was a press conference at the Sternberg Institute with many journalists and it was open to all people. We told what we really observed at the observatory: the frequency, the location, all that. It was not secret from the public.

Another situation was in England when the first pulsar was detected. It was a secret for half a year. The hypothesis about green men was a secret hypothesis. There was a possibility of proposing a false situation, but after half a year or a year it must be published because it is impossible to cover up such a secret.

What would you do?

My philosophy would be to publish it immediately, because it is possible to use many different kinds of telescopes, different technical devices that I don't use personally, but are at different institutes, different countries, that may be used for this. But it depends on the psychology of each scientist, and there are strong differences.

What do you guess would be the public's response to the news that you have found absolute proof of extraterrestrial intelligence?

It will have a very strong influence on people, of course, and it is difficult to predict what will happen. But I hope that, in the time between the first detection of some unusual object and the final decision about it, there will be prepared a general opinion so more people can understand what happens after discovery.

Also an interesting question: what happens if we found something unusual and we detect that the extraterrestrial is more developed than our earth civilization? Our position now is that we are unique, we are the most developed, but if something more developed is detected, it will be a shock to general opinion. We will need to understand, if we are not the most developed, how we must develop afterward.

What will be the long-term effect on humans—a hundred years, a thousand years after contact?

My opinion is that it is probably impossible to develop our Earth civilization up to the highest levels in the universe. Maybe there is only one decision, or maybe two decisions. We know that we are not the best in the universe, not the most developed, but we might stay at our level and search for information from other beings. Another possibility is to combine with more developed civilizations. This means that we disappear, because we conserve only our historical past as we combine with more developed beings.

These two decisions are possible. It's a very interesting question.

Do you have any guesses as to which one it will be?

I don't know. My feeling is that the distance between very developed extraterrestrials must be very large, and so there is small probability for visitors; small chance that each star system will have guests from the very distant objects. But it's very difficult to predict all this.

I think the best process is searching for new astronomical objects, and I hope that some will be detected, because powerful astronomical telescopes are being constructed. The results may not be for cosmology, but maybe connected with SETI is the problem of the 90 or 99 percent of the hidden mass of the universe. It is unexplained. What is it?

Finally, is there anything else that you would like future historians to know about you that otherwise would not be printed in regular journals or biographies?

I want to say additionally that it is possible to discuss, in the frame of the SETI problem, *all* problems connected with the best organization for living on the Earth.

That is a very positive result.

You make better people. That's an interesting observation.

Yes.

The hour is getting late. I thank you very much, and I must say I'm delighted to learn that Shklovskii is the first pioneer to have another pioneer as a student. He must be very proud of you. Thank you very much.

Thank *you* very much.

Vsevolod Sergeevich Troitskii

PHYSICIST

Born March 25, 1913

Mikhailovskoe, USSR

Physicist V.S. Troitskii, Vice-Director of the Radio Physics Institute at Gorkii, takes a middle-of-the-road approach to SETI.

In contrast to Kardashev's assumption of supercivilizations, or Shklovskii's belief that we are alone, Troitskii thinks that there are civilizations at our technical level, or slightly higher. Such civilizations, if they were trying to contact others, would be sending out highly directed signals toward the star systems they believe might be inhabited. These signals would be narrowband, both to maximize the distance obtainable from limited energy, and also as a recognizable sign of artificiality.

Troitskii advocated this view at the first All-Union Meeting on Extraterrestrial Civilizations at Byurakan, in 1964, and when the Academy of Sciences of the USSR established a section to search for extraterrestrial signals, he was appointed chairman.

In 1968 he searched the twelve nearest stars for signals, and in 1970 he began a radio survey of the entire sky, using a network of coordinated sites stretching 8000 kilometers across Asia. The most ambitious SETI project to that time, it logged over seven hundred hours. Additional observations, from a research ship in the Atlantic, led to detection of previously unknown sporadic radio emissions generated in the upper atmosphere and magnetosphere by solar radiation.

He completed technical school in 1932 and worked in the Central Radio Laboratory at Nishnij Novgorod until 1936. Then he entered the University of Gorkii, graduating in 1941. During the war he worked in a factory. He joined the staff of the Gorkii Radiophysical Institute in 1948.

From his primary interest in radio astronomy, he has applied radiometric techniques to a very broad spectrum of studies, ranging from distant stars and quasars, the Sun and Moon, the Earth's atmosphere, to the internal organs of the human body.

A pioneer in the measurement by VLBI of quasars at meter wavelengths, he has also developed diagnostic equipment for studying the body's physiological condition and temperature. Recently he has become occupied with cosmology and has developed a new cosmological model.

His publications range from proceedings of the 1964 Byurakan meetings, to "Search for Monochromatic 927-MHz Radio Emission from Nearby Stars," *Soviet Astronomy A.J.* (1971) to "Physical Constants and Evolution of the Universe," *Astrophysical Journal* (1987).

He was awarded the A.S. Popov Prize of the Academy of Sciences of the USSR for his investigations of radio emissions from the Moon. Other awards include the Order of the Red Banner of Labour and various medals. He has been a member of the Communist Party of the Soviet Union since 1944, and is a Corresponding Member of the Academy of Sciences of the USSR.

Interviewed December 1981 at Tallin, Estonia, by Bernard Oliver

Where were you born?

In the middle of Russia, in Tul'skaya gubernia, in Bogoroditskij county, in the village called Mikhailovskoe.

I lived there until 1918 and then afterwards with my parents we moved to Nizhnij Novgorod, now called Gorkii, where my father was director of a factory. He was an engineer. He died in 1920.

Was that a city or a town or a rural area?

It was a big city.

What were your childhood interests?

I was interested in many things. In the first place, airplanes. I later found a book on the history of philosophy. It was my uncle's book. I read it and began to dream of studying philosophy. I enjoyed it very much.

Then I became interested in radio techniques, detecting radio waves with crystals. I made my own receiver and played with it very much. It was a very pleasant time. That was 1925, when I was twelve years old.

That was the time when radio amateurs began to develop. It was a very interesting time. I attended many popular lectures on techniques of radio reception, which were delivered at Nizhegorodskaya Radio Laboratory.

How was your schooling?

The first year I studied at home; the second year, in a village school not far from Nizhnij Novgorod. The last two years of the elementary school, 1924–1925, I studied in a town school and lived with my aunt.

Education in the elementary school was not interesting. At that time I began to be interested in amateur radio and that predetermined the direction of my scientific activity.

In school I studied much physics. There was a very big physics textbook by Michelson, which contained five volumes, and I read all five volumes.

In seventh grade I performed some simple experiments and played with wireless and Morse code. My only tools were scissors and a hammer. It was not yet scientific activity; rather it was childhood amusement.

In the eighth grade of middle school I began higher mathematics, and after finishing middle school I entered a higher technical school, a trade school of radio communication. It was very good.

Education in the secondary school, 1926–1930, from the fifth form up to the ninth, top form, was interesting and pleasant, which stimulated aspiration for knowledge. At that time, together with my friends, I studied independently the basis of differential calculations, took part in technical classes, and read books and popular scientific literature on the classics of modern physics. Especially attractive for me were miracles of electrodynamics and special and general relativistic theory, though I had but little idea about this literature.

This activity and interest defined my future scientific direction.

I graduated from the technical school in 1932 and began my work in a research institute as a laboratory assistant. In four years I became an engineer and wrote my first scientific work. The research was on the propagation of radio waves of decimeter wavelength. I successfully explained what, in my view, had not yet been explained by scholars.

After middle school I became interested in astronomy. In the winter I studied the sky and read remarkable astronomy books by Flammarion on constellations, and sometimes would run outside into the street without a coat to look at the stars and the various constellations, and then come back in and look at the books again.

Then I continued at the research institute in the field of radio. At that time I studied by correspondence, but the study went very slowly and I didn't get any results, so four years later I applied for admission to the university. I was admitted, and became a student in the physics and mathematics department, in order to understand what I could not yet understand, and to seriously read the literature.

Before that, in school I started thinking about the stars. I read Einstein in the original, but of course I couldn't understand much. Then I also read Lebedyev's *Pressure of Light*, and this awed me.

When did you first think about the possibility that extraterrestrial intelligence might exist?

It seems to me that up to 1963 I did not meditate about the problem of extraterrestrial civilizations. Probably the possibility of life near other stars was considered to be evident.

What about communicating with, or at least searching for extraterrestrial intelligence?

I first began to think about the possibility of searching for ETI and communicating with it in 1963, when organizers of the first All-Union Symposium on CETI offered the possibility of making a report on this problem. The symposium was organized by the Astrophysical Council and by I. S. Shklovskii.

As I began to think of the idea, the question naturally arose as to how to send artificial signals. At the symposium, which was in Byurakan, I delivered a report on the basic problems: like questions of contact, of artificial signals. My conclusion was that a monochromatical signal was the most logical one to expect from outer space because it would be very different from natural signals. I still believe this is the most likely kind to be expected: that signals would be monochromatical signals.

During your college years, did anything stimulate you to think about ETI and the possibility of searching for or communicating with it?

I studied at the university from 1936 to 1941, when I graduated, and then the war broke out. I had to work four years in a factory during the war. After the war I became a graduate student. I defended my dissertation. It was apparently the first dissertation in the field of radio astronomy.

What effect, if any, has your interest in ETI had upon your career as a scientist?

It is hard to say, because at the same time I had many other interests; for example, radio astronomy, the study of the moon. This was finished around 1967 when I received all the data about the moon and finished the measurements on the radio emission of the moon. For measurements with great accuracy we used the method of artificial moons, satellites. All this is now known and is being written about.

When I see that a problem is about solved, I leave it, because it isn't interesting anymore. Such was the case with the moon. I studied it for maybe fifteen or twenty years, and we obtained all the data on the characteristics of the moon even before the landing on the moon, so my data coincided with the direct measurements made by the Americans.

From 1965 on I began to study VLBI. Here were physical problems and the problems of proximity to the earth, like the rotation of the earth, the movement of the poles—all these were related.

Parallel with this we began to search for emissions of artificial origin from the cosmos. We understood that we had to pursue this observation through many channels or frequencies, so we constructed twenty-five channels with a bandwidth of 13 hertz each.

In 1968 the first observations were made using the 15 meter telescope. We observed the twelve nearest stars. We worked several nights. We were greatly interested.

In 1970 we employed a new method of observation at various locations in the Soviet Union, searching for coinciding signals from several telescopes. We have continued this experiment to this time. As every research finds something new, so we have found sporadic radio emissions of the magnetosphere. That certainly is of great interest.

What do your colleagues think about your interest in SETI?

Most of them joke about it.

Do they ever discuss it with you or ask you about it?

They're simply joking, ironic questions. Yeah, it's a joke, a funny subject. But these are my colleagues. Some begin to be serious about it, but only a few.

What about your family? What do they think about your ETI interests?

My wife is a specialist in astronomy. We work together in the same institute for radio astronomy. My older son is interested in it, but my younger son is especially seriously interested. He even did a paper at the university on extraterrestrial civilization. He's also studying related fields. Both of them are here for this conference. Both graduated from the same university where I graduated.

Their specialty is radio physics. The older one defended his candidate's dissertation last year. The younger one is now going for his diploma on the subject of applications of radiometry for measuring temperatures of internal organs in human beings—an application of radio astronomical equipment.

What do your superiors think of your interest in ETI?

My superiors in the Institute are neutral to this question.

What about your friends?

I can speak about three groups of friends. The first group is interested in the theoretical investigations or takes part in experiments. The second one accepts it as crankiness. The third group considers it to be useless effort and a waste of money and time, though I must say that not much time and money are spent for this problem, and they are justified by the secondary results. Much more expenditure would be justified because of obtaining the accompanying physical information.

Before going on to discuss your ideas about SETI, let's get a little more background information. What was your father's occupation?

My father was an engineer in the food industry. In 1913 he was graduated from the Moscow State Higher Technical Institute. After the First World War in 1918 he was appointed the director of the food industry factory in the village Zeletsino, located twenty kilometers from Nizhnij Novgorod.

What about your mother?

My mother was a teacher. After Father's death in 1920 she worked at the same factory as a foreman, after graduating from special courses. Then from 1925 she worked in Nizhnij Novgorod in different establishments as a bookkeeper. She died in 1975 at the age of ninety-two.

How would you describe them?

My mother was an even-tempered, calm, reasonable person. She never punished her children and never thrust her opinion on her grown-up children, but softly gave her own view on our actions.

How did she influence your eventual choice of science as a career?

Her influence on the children's future was enormous. She gave a higher education to all her children. She stimulated the interest in reading books, working at technique. She never scolded us for ruining scissors or other things necessary for housekeeping.

Were there other people we haven't mentioned who influenced your ideas about science and ETI?

Two persons influenced me greatly: the teacher of physics in middle school who supported my interest in physics, and in the technical school, the teacher of radio technique, Peter Ivanovich Kondrat'ev, who became my friend for many years. I worked in the Central Radio Laboratory from 1932 to 1936 with many intelligent and excellent specialists who played an important role in my life and my scientific choice.

How many brothers and sisters did you have?

There were three children in our family. I was the oldest child. My brother was a year younger than I. He was a surgeon and died in 1977. My sister is still alive. She is a housekeeper. She had graduated from technical school of communication, but all the time she was a housekeeper.

What about your friends when you were growing up? Did they share your interests in science and ETI?

In school, in technical school and especially in the period of working at the Central Radio Laboratory in Nizhnij Novgorod, after graduating from the technical school and before entering the university (1932–1936), there were friends who longed for knowledge, who were interested in achievements of scientific fields and technique, but at that time the SETI problem had not been introduced.

What were you like during childhood and adolescence?

I was very shy, a silent boy and youth. I had many friends among those who longed for knowledge. Now I have overcome the shyness but I remain a man of few words. That is why I hate loquacity in other people.

What other professional interests do you have, apart from SETI?

Partly I have answered this question. Radio astronomy, radio sources, quasars, radiometry of faraway astronomical objects. We were the first to measure the size of some quasars at meter wavelengths. In America they did it in one-centimeter wavelengths. I asked once, I believe it was Kellerman, why didn't they start with meter waves; this would be simpler. He answered, "It didn't work out for us," but for us it did work out.

So, first, research of far stars, the sun, moon; then the earth's atmosphere by methods of radiometry; and now internal organs of the human body, their physiological condition and temperature. To study the internal organs we made some equipment which is now installed in a clinic in Gorkii doing diagnostic research. A group of physicians work on this equipment. Also Allan Barrett, an astronomer, is working in this field in the U.S.A.

I am beginning investigations in the field of cosmology. I suggest a new explanation for the red shift of galaxy radiation, on the basis of which a new model is being developed and published for the evolution of the universe without the singularity.

What recreations do you enjoy?

My hobby is science. My relaxation is active. I don't believe in passive relaxation. I like mountaineering and hiking. I like to go in a small boat on clean rivers where you can drink water from the river.

What has been your greatest satisfaction in life?

The greatest satisfaction for me was the result of my research. I got very much satisfaction from my research on the moon. Any kind of solution of scientific problems gives me satisfaction, especially now in cosmology.

What has been the greatest sorrow in life?

It's hard to say. I cannot remember sad and sorrowful things.

What was the most memorable moment or event in your life?

I remember my first attempts to receive a station on my self-made radio receiver. I would sit for hours, listening to the station with earphones, waiting and waiting.

Turning now to SETI, about what fraction of your working time do you spend on it?

About 10 percent for sure is spent on this, and I suppose the time will increase.

What stages has your thinking about SETI gone through?

The main ideas remain the same, but new ideas came. The main idea was that I had to conduct experimental work, and so I paid little attention to theoretical considerations and contemplations. I had reacted to and answered practical questions, observing technique, and so I was occupied only by practical questions, but in 1978 and 1979, when we received negative results, when the crisis came that some scientists think that we are alone in the universe, there is nobody out there, when we would have to close the program, then the inner protest arose and we began to think "Why?" and "How?"

Then came the question of theoretical speculation. Now I am attracted by theory, and now new ideas emerge and they will continue to emerge. I now have to prove these ideas by numbers. I don't care only to speculate on how to move the stars. We need numbers, figures.

We could do much more if we could attract young people, but this is hard.

What alternatives, what other possibilities about ETI have you considered?

I do not see another way of the ETI problem development.

What would you do differently if you were starting over again to study ETI?

I would begin the same way.

Why is ETI important to you?

The ETI problem is important for me because of the fact that it unites all mankind's sciences together, permitting one to understand better the Earth's civilization and the possible way of its development. The principal thing is to induce interest in the problem, "Is there a civilization in our galaxy?" I am sure that the broad investigations which are now scheduled may give an answer to this problem.

What has been the most difficult aspect of your SETI work?

The psychological resistance of the "surrounding environment": people around me.

How do you think most scientists today view SETI?

I think the majority of the scientists I come into contact with consider this not to be serious. Although they don't bring arguments against SETI, even the great scholars don't regard it as serious. They are indifferent at best.

What about public attitudes toward SETI?

The public is very enthusiastic but interested only in the most outstanding aspects, and wants to hear definite and probable results. They do not understand the colossal difficulty that is involved in that kind of research. We are just beginning on the path of the SETI program.

Realizing that it is only a guess, when do you think contact with ETI will be made?

Absolutely not soon, not soon. I don't want to think about it. Man can only guess, but what is the use of guessing?

What form might ETI take?

I think that forms will be natural. As in biology, nothing new can be imagined, I cannot think of something else. Forms may be different, but since the conditions and the laws of nature are the same, logic and thinking must be similar. Although the forms may vary, the essence remains the same, and will probably be easy to contact.

What are UFOs?

I am sure that UFOs exist but yet there is no satisfactory hypothesis for explaining this phenomena.

What should a scientist do upon receiving convincing evidence of extraterrestrial intelligence?

Look for a method for information exchange and mutual relations.

What would be the public's response to this information about the existence of ETI?

The response of modern society to the information on ETI contact will differ: stormy, enthusiastic, joyful and depressed.

What would be the long-term effect upon humanity of evidence that ETI exists?

The discovery of ETI existence will be a long-acting force on thinking, making people conscious of the universe.

What scientists or other people in the past do you admire?

I admire every physicist and biologist. I am now reading a book on biology. There are many outstanding people who created the fundamentals of science. They are worthy of great respect.

Who are the outstanding people in SETI today?

I believe that these are the people who started in the United States and with us: Shklovskii, Kardashev. In the U.S., Drake, Morrison, Sagan. Oliver has done much: he directed Project Cyclops, a remarkable thing.

What role has each played in SETI?

Cocconi and Morrison gave the first shock. They are the pioneers of the search and communication problems. Drake was the first who carried out experiments and observations. I.S. Shklovskii was the first scientist in our country who understood the value of this problem and made a number of suggestions on ETI. N.S. Kardashev made several assumptions on regularities of ETI. He suggested an energy principle of evaluation of power and possibilities of ETI.

P.V. Makovetskij founded the choice of the search direction and the beginning of observations of signals which are similar to the wave of communication suggested by Cocconi and Morrison.

Academician V.A. Ambartzumyan was one of the first prominent scientists who understood the actuality of this problem and organized the first All-Union Symposium on ETI in Byurakan in 1964 and then in 1972, the international conference. Academician V. A. Kotelnikov, the vice-president of the Academy of Sciences of the USSR, lends a constant support to the investigations of ETI problems.

How would you characterize your own role in SETI?

I began to carry out observations of nearby stars after Drake, in 1968 at the wavelength of 21 cm, and from 1970 I organized observations of the whole sky at 21, 30, 100 cm which have been carried out up to the present. Some new ideas explaining the regularities of ETI development and the cosmos silence have been suggested and published.

The silence of the cosmos, the absence of signals, is explained by the hypothesis of almost simultaneous origin of life in the universe. Due to this fact, the Earth's civilization may be the first in the galaxy to possess radio communication. The absence of strong radio and optical signals, as well as of any unusual phenomena in the sky, "miracles" that might be expected from powerful civilizations, I explain by the physical limitations of energy generated. These limitations are dictated by the conservation of the necessary parameters for life, including cosmic life.

Finally, what else should future historians know about you that would not otherwise be preserved?

In my opinion there is nothing interesting so far.

Carl Sagan

ASTRONOMER

Born November 9, 1934

Brooklyn, New York

Professor of Astronomy and Space Sciences and director of the Laboratory for Planetary Studies at Cornell University, Carl Sagan is one of science's most ardent expositors. He works to increase the public's appreciation and support of astronomy through activities such as the *Cosmos* television series and the science fiction novel *Contact*. He is cofounder and president of the 100,000 member Planetary Society, the largest space-interest group in the world. He initiated a petition in which seventy-three scientists from many nations urged the organization of a coordinated worldwide search for extraterrestrial intelligence.

He attended the University of Chicago, receiving an A.B. in 1954, S.B. in 1955, and S.M. in 1956 in physics, and a Ph.D. in astronomy and astrophysics in 1960. Before going to Cornell in 1968, he served on the faculties of Harvard, the University of California at Berkeley, and the Stanford University Medical School.

Along with SETI his interests encompass the physics and chemistry of planetary atmospheres and surfaces, planetary exploration, and the origin—and preservation—of life on Earth. An active proponent of arms control, he testifies to Congress, appears on television, and writes scientific reports and popular articles about the devastating effects of nuclear war.

His involvement in NASA planetary exploration ranges from planning experiments to designing the Pioneer 10 and 11 interstellar plaques, and the Voyager 1 and 2 recordings.

The scientific advisory groups he has chaired include the Viking Lander im-

By courtesy of The Planetary Society.

aging team at Jet Propulsion Laboratory, and NASA Headquarters Study Group on Machine Intelligence and Robotics.

He also serves on advisory boards concerned with social issues, including those of Mothers Embracing Nuclear Disarmament, Educators for Social Responsibility, and the Center on the Long-term Consequences of Nuclear War.

His numerous awards include NASA medals, the John F. Kennedy Astronautics Award, the Honda Prize, the Joseph Priestley Award, and public service awards of the Federation of American Scientists and of Physicians for Social Responsibility. He holds honorary doctor's degrees from a dozen universities, and an asteroid has been named after him.

In addition to professional societies related to astronomy, he also belongs to the Genetics Society of America and the World Association for International Relations.

Besides hundreds of scientific and popular articles, he has published a dozen books. *The Dragons of Eden* won the 1978 Pulitzer Prize for nonfiction. *Cosmos* (1980) is the best-selling science book ever published in English, and has been translated into many other languages, including Arabic, Korean and Serbo-Croatian.

Interviewed February 1983 at Ithaca, New York, by telephone

Where did you live during childhood and when you were growing up?

I lived in Brooklyn until 1948, when we moved to Rahway, New Jersey. I graduated from high school in 1951 and then went off to the University of Chicago.

What were your childhood interests?

What you want to know is where SETI came from?

Yes. We're looking back into your childhood interests generally and your earliest activities in science, what you first remember. For example, a passage from Cosmos *recalls some of your early experiences:*

Even with an early bedtime, in winter you could see the stars. I would look at them, twinkling and remote, and wonder what they were. I would ask older children and adults, who would only reply, "They're lights in the sky, kid." I could *see* they were lights in the sky. But what were they? Just small hovering lamps? Whatever for? I felt a kind of sorrow for them: a commonplace whose strangeness remained somehow hidden from my incurious fellows. There had to be some deeper answer.

As soon as I was old enough, my parents gave me my first library card. I think the library was on 85th Street, an alien land.

Immediately, I asked the librarian for something on stars. She returned with a picture book displaying portraits of men and women with names like Clark Gable and Jean Harlow. I complained, and for some reason then obscure to me, she smiled and found another book—the right kind of book. I opened it breathlessly and read until I found it. The book said something astonishing, a very big thought. It said that the stars were suns, only very far away. The Sun was a star, but close up (Sagan 19:168).

Also in the book that I wrote called *The Cosmic Connection* there's a discussion on my fascination with Edgar Rice Burroughs who, following Percival Lowell, populated Mars with a variety of intelligent beings. That was certainly the first specific contact that I can remember in the writings of others with the idea of intelligent beings on other worlds. As I said in *Cosmos*, on my own, as soon as I realized that the sun was a star, that the other stars were suns, it seemed likely that they would have planets, and that some of those planets ought to be populated. But that was an intuition; it was hardly a scientific conclusion.

How was your elementary schooling?

Fabulously dull. In fact, the high school principal's name was Conway, and we called the school CCC: Conway's Concentration Camp.

Did you feel you got any helpful preparation for your scientific career out of either elementary or high school?

Elementary school, almost nothing. High school, in mathematics I had a very good mathematics teacher named Joseph Persons, and at least adequate science training on the high school level in other courses.

When did you first think about the possibility that extraterrestrial intelligence might exist?

Well, somewhere in the six, seven, eight-year-old age, when I was contemplating this amazing statement I had read in the book that the Sun was a star; and then shortly after that, when I started reading the John Carter novels by Burroughs.

Did you discuss the possibility of extraterrestrial intelligence with other people?

Yes, but I soon discovered that this was considered a crazy subject and people looked at me with . . . I guess the most positive response was bemusement. There were certainly contemporaries who thought it was completely stupid.

What about communicating with or at least searching for extraterrestrial intelligence? When did you first think about this possibility?

At very early stages, of course, I knew nothing about radio astronomy. My schoolboy fantasies were connected with space flight and the idea of travel to Mars. Mars was foremost among those worlds advertised in popular fiction to have intelligent life and it seemed to me perfectly reasonable that we should travel there. Why not? We've traveled everywhere on the Earth. Again, it was an intuition almost completely separated from any knowledge of the relevent technology. But nevertheless I felt it strongly.

And in the first year or two of college, in the very early 1950s, I can remember being invited over to girlfriends' houses for dinner and when the subject came up it was always a kind of social gaffe. It confirmed the fears of protective uncles that the young woman was seeing somebody crazy. So after a while I learned to some extent to moderate my enthusiasm publicly.

During these college years was there any stimulation to think about ETI and the possibility of searching for or communicating with it?

Yes. I would say the critical event was finding a set of first-rate scientists who did not think it was crazy. I argued with them. We didn't see eye to eye on every aspect of it but they considered it a perfectly respectable subject to talk about. That encouraged me, because I certainly got a lot of feeling that many people thought the subject was entirely crazy. I can mention three of them in particular who were very important for my involvement, in part because of specific content ideas, but mostly because they convinced me that this was a perfectly respectable subject, and I shouldn't feel ashamed by my enthusiasm for it.

The first of these was H. J. Muller, who was professor of zoology at Indiana University, Nobel Laureate in physiology or medicine for 1946. He was the man who discovered that radiation makes mutations. Through a set of coincidences I met him and worked for him for several summers, beginning in the summer of 1952, in Drosophila genetics and other things. I was mainly concerned about the issues of the origins of life, in turn connected with the question of life elsewhere. And we had many long discussions on the subject. He was just tremendously supportive and encouraging.

I would say that was the critical event. If not for meeting Muller I might possibly have bowed under the weight of conventional opinion that *all* these subjects were nonsense, not just interest in extraterrestrial intelligence, but interest in extraterrestrial life, interest in other planets, interest in space flight.

All of that was considered quite disreputable, both in the minds of non-scientific adults and especially in the minds of a number of scientists whom I talked to, who seriously advised me to go into something more promising, or looked at me with some bemusement. There wasn't anything especially about *me*, I don't think. That was the general sense people had about the subject, in part because of the legacy of Percival Lowell, who had in a way discredited the whole idea in the minds of the scientific community.

Beyond Muller there was G. P. Kuiper, under whom I did my doctoral work at the University of Chicago. He even wrote about the possibility of life on Mars, albeit not highly advanced, not intelligent forms of life; but he was open to the possibility and discussed it seriously at the time when virtually no professional scientists were doing so.

The third critical person was Joshua Lederberg, in the middle 1950s, whose interests were moving very much in this direction. We had wonderful discussions on these subjects. I remember them with the greatest of pleasure. Again, not so much on the issue of extraterrestrial intelligence, about which he was and remains rather skeptical. But still, it was worth it to talk about it, and the idea of extraterrestrial life on planets in the solar system or in the interstellar medium was something that he was very interested in. We talked far into the night on this.

He became the chairman of the Exobiology Committee of the Space Science Board of the National Academy of Sciences, and asked me to join that. So we continued these discussions in an official context, advising, among other things, NASA about how to go about searching for extraterrestrial life. This was just after NASA was formed in the late fifties. Nineteen fifty-nine was when the committee was established and 1958 was when NASA was established. So I sort of glided effortlessly between some kind of bull sessions late at night to advising the government on the issue.

There were many other people, and all three of these people, of course, precede the SETI pioneers. There's no question that I got enormous stimulation from Frank Drake, Philip Morrison, and others, but the key people in my own development are those that I just mentioned.

By 1960 or so I was already dedicated to the subject and writing things on it. I guess another major thing that happened was the invitation from I. S. Shklovskii in 1963 to collaborate with him on the English language revision of a book on extraterrestrial intelligence, which put me into some intimate intellectual contact by mail with one of the most brilliant thinkers in the field.

What effect, if any, has your interest in ETI had upon your career as a scientist?

I think it has been a minor impediment. There are colleagues who have worried that this is such a thoroughly disreputable subject that no serious scientist could really be interested in it. I think there have been colleagues who wondered if my interest in this must reveal some deep-seated lust for publicity. Certainly there are people who think of it as an issue that will excite the public but has no scientific underpinnings.

But a remarkable thing has happened between the late fifties and now. That is, the subject has become very much mainstream. The recent SETI petition that I circulated is an excellent example of that. Seventy-three absolute first-rate scientists, seven Nobel Laureates, fourteen countries, lots of Soviets. It's by no means disreputable anymore.

I should also say that even planetary astronomy was considered disreputable in the fifties. In fact, Kuiper was the only full-time planetary astrophysicist in the world at that time. To the best of my knowledge, there was not a single other person who was working full-time on that subject: just studying the planets with techniques of modern physics. At the most there was a handful of others.

So this kind of stodginess was not restricted to SETI. It included all the topics I mentioned before, the search for extraterrestrial life (never mind intelligence), and even the study of the physical environment of the other planets—all of that was considered nonsense. You remember that Woolley, the Astronomer Royal, less than a year before Sputnik, described the idea of space flight as "utter bilge."

What about your colleagues now? Do they ask you about it or discuss it with you? That's aside from your SETI associates.

Oh, certainly. The topic comes up all the time and people talk about it in the most serious way. The thing that delights me is how many distinguished astronomers of the old school who have made no public pronouncement on SETI are excited and interested in it. A few very distinguished such astronomers have signed the SETI petition. But there are a whole lot more who for various reasons I did not ask or did not want to sign but who nevertheless have expressed very serious interest in the subject. It's not something that just young whippersnappers are in. It's permeated the, for want of a better phrase, entire astronomical establishment. The International Astronomical Union has established a committee on extraterrestrial life. It's here. It's respectable.

It doesn't mean that everybody thinks that there is extraterrestrial intelligence. The ontological or epistemological aspect of this subject ought to be very clear. When someone's interested in SETI it doesn't mean that he or she has made a commitment that extraterrestrial intelligence exists; for, merely a commitment that it is such an important question and the tools to test it are so powerful that we ought to go ahead and search.

What do your administrators, bosses, and so forth, think about your interest in SETI?

As far as I can tell it's just considered one additional subject, certainly a speculative one, that one of the faculty members is interested in. More than one—Frank is here and there are others here who are interested in it. And there's certainly no sense that it's some source of great embarrassment or some source of unique pride. It's just another subject.

What about your friends?

Well, my old friends have known about this interest as long as they've known me. So they've seen it grow along with me. I guess new friends, when they meet me, discover it's an interest of mine. It's not absolutely all consuming; I have a large amount of interests. But I think it's just considered part of the landscape, part of my characterological landscape.

How about your family?

My wife, Ann Druyan, is a real source of support on this subject. She herself is personally very interested in it. I have great pleasure talking to her about some new wrinkle. I've gotten some very good ideas talking to her about it. She has on occasion made some significant suggestions. As far as my children, I have a twelve-year-old son who, growing up in this culture today, is exposed to a whole lot of pop interest in SETI, because of movies and so on. I think it's a source of some pleasure for him that we can talk about it in a very serious way.

How about your parents?

Both my parents are dead but they died only a year ago, two and a half years ago. I think they were delighted with this, as they were with any other things I did. I don't think they made a distinction that the planetary exploration part of what he does is terrific but the SETI part is a little far out. They made no such distinctions. Certainly I got no opposition from them at any stage. In the very early stages they were marvelously encouraging: "Whatever you're interested in, that's terrific."

Did they give you any particular, specific push towards a career in science?

No, not at all. Their attitude was, "If that's what you're interested in, that's fine." There were a couple of times when they were concerned about whether it's possible to earn a living. I guess once or twice I was asked if it might be better to be a doctor. But nothing beyond that.

What was your father's line of work?

At the end of his life my father was the foreman or manager of a large factory that manufactured women's and children's coats and suits. Much earlier he was a cutter in that same industry. He manipulated very large saws, in effect, to cut huge bolts of cloth according to predetermined patterns.

What formal education did he have?

My father had two years of Columbia University before dropping out in the Depression. My mother went to night school after high school but also did not graduate from college. They both graduated from high school.

Did your mother have occupational skills or work experience other than that of the typical wife and mother of that time?

Yes she did, but in sales and so on. She also had some writing experience. She had published some things before marriage.

How would you describe them?

Difficult question. It's hard to do in just a few words. They were both very rich and full personalities. I would say it was very interesting growing up with them.

What was the general religious orientation of your family when you were growing up?

Both my parents were Jewish but of reformed persuasion, with a fair amount of assimilation into American culture. While I certainly spent some time in formal religious education, my skepticism was not discouraged. I think that's the way to put it.

Did you have any brothers or sisters?

Yes, I have a sister seven years younger.

What does she think of your general orientation towards science?

I think she's pleased by it. I suspect better than being a specialist in some subject which is impossible to describe.

Has she gone into a science field?

No, although her husband is an expert in computers for a large chemical corporation.

Is there anyone else from your earlier years, before you graduated from college, whom we haven't touched on who might've had an influence one way or another on your interest either in science generally or in SETI?

Well, science generally, the whole University of Chicago system at that time was tremendously good for anyone interested in science. This would take me into a long diversion on the nature of the University of Chicago educational system under Robert Hutchins, the chancellor. Not just science, but learning in general. It was a superb educational experience at the University of Chicago at that time.

That was in contrast to your earlier experience in school?

Exactly. It was like moving from a desert to the Garden of Eden.

How about your friends in school—elementary school or high school? Were there other children there who were interested in some of the same things that you were, or were you on your own?

Not a one, as far as I can recall. I certainly had friends who didn't discourage me, but I can't recall anyone before college who had a genuine, self-initiated interest in the subject.

What other professional interests do you have apart from SETI and apart from what I can get from your curriculum vita? Is there anything that would not be on it that we ought to put down here?

No, the vita will give you a sense to date of what my professional interests are. Lately one subject which has been occupying a lot of my time is the nuclear war issue. I suppose that's a professional interest.

What recreations do you enjoy?

I guess not many. My work is for me often a recreation. I do it because I have to. And I would say beyond that, spending time with my wife and children. Reading books, which I occasionally get to do, books that are not connected with the work I'm doing. I like to scuba dive.

Turning now to extraterrestrial intelligence, about what fraction of your working time do you spend on it?

Less than 10 percent. It depends very much what year we consider. For example, there was a period when Frank Drake and I were doing observations of external galaxies for SETI from Arecibo in which I

spent considerably more than that amount of time. I wanted to understand the instrumentation, I wanted to understand the significance of the results, if we got any results, so then I was spending a lot
of time, or when I was writing the book with Shklovskii or editing
the SETI book that was the proceedings of the Byurakan meeting in
Soviet Armenia. Those years, the amount of time I spent on it probably went well above 10 percent. I think on the average it's below
10 percent.

What stages has your thinking about ETI and SETI gone through?

I like to think my understanding of the subjects has become more
thorough, deep, or to use a word I don't much like, "sophisticated."
But my general approach has been the same for the longest period
of time, namely, that there's a lot of places out there, there's nothing
unique about the general course of events that happened here, so
therefore, there ought to be a lot of places in which beings more intelligent than us have evolved. The question is, "What's the best way
to find some sign of them?" The electromagnetic spectrum is obviously a lot easier than space flight, and radio seems to be the best way
of all parts of the electromagnetic spectrum. So that is a very crude
summary of my frame of mind, since the early sixties anyway.

*What other possibilities about extraterrestrial intelligence have you considered
along the way?*

The one which I think anybody seriously interested in ETI has to
spend a little time on is the possibility that they've come here or that
they are here right now. And so I've spent a fair amount of time on
the UFO issue. I was on the Air Force Scientific Advisory Board committee that looked into the Project Blue Book, the Air Force project
that handled UFO reports. As a result of that study, the Condon study
at the University of Colorado was set up, and I've spent a lot of time
with some of the leading legitimate UFO researchers.

Likewise the kind of work of Eric von Daniken, saying that there's
evidence of past visitations to the Earth by alien beings in the historical or archaeological record—I spent some time on that as well, and
in fact Shklovskii and I anticipated von Daniken. We raised this kind
of issue in our *Intelligent Life in the Universe.*

But my conclusion from all that is that there isn't a smidgen of evidence of us being visited now or having been visited in the past. It's
out of the question; there's just no evidence for it. So those are some
other approaches; they're sort of fun. You learn something about history or archaeology or human nature, psychopathology from them,
but I don't see any evidence that they are the most effective or promising way to go.

I've also spent some time concerned about interstellar space flight. In general, the question is how do you do it and how do you make it work? Is it something feasible for contact among civilizations? I wrote a 1963 paper on that and I'm still writing papers on that.

If you were starting all over again to study extraterrestrial intelligence, is there anything you would do differently?

Knowing what I know now?

Yes, from today.

That's the hard part. Well, there's a certain sense in which I wish that the pace had been faster. It's only right now that we are mustering some really serious systematic searches for extraterrestrial intelligence. And because this kind of thing is likely to take decades, just for personal reasons I wish that it had started earlier so I'd be sure of being around when it finishes, and we would know one way or another that this approach either works or doesn't work.

But even that is not very realistic because the reason that things are doing so well right now is because of advances in electronics, silicon chip technology and so on. I think the searches had to really wait till those technological advances occurred. So I don't think it could've been hurried much. Maybe a few years, but not much more than that.

Why is extraterrestrial intelligence important to you?

It's not the only subject I've chosen. I'm very heavily involved with unmanned planetary exploration, and physics and chemistry of planetary atmospheres and surfaces—lots of topics that have nothing whatever to do with extraterrestrial life or extraterrestrial intelligence. So it's not as if I've put all my eggs in that one basket. But having said that, I'll go on and try to answer the question.

You find out who you are. It's a basic question. Are there beings in some sense like you, elsewhere in the universe, or are we the only ones around? It touches deeply into myth, folklore, religion, mythology; and every human culture in some way or another has wondered about that kind of question. It's one of the most basic questions there is. We are fortunate enough to live in the first time in human history when the subject can be examined in some depth. In my opinion what has to be explained is not that some people are interested in the subject, but that some people profess *not* to be interested in it. *That* is the amazing thing to me.

What has been the most difficult aspect of your ETI work?

I think it's getting up the speed, getting a significant fraction of the scientific community and people in Washington convinced that this is something worth doing. That takes a lot of work but the progression has been very steady and systematic, and I think we've now reached the point where we can go ahead.

Difficulty; there are lots of kinds of difficulty. There can be a difficulty in a specific search program, there can be difficulty in figuring something out, but because all this has been so sputtery—a small surge on for a few months, then off, and so on—I don't think the technical aspect of this can be said to be the most difficult, yet. We're just reaching the point where the technical aspect will be the limiting step but up to now our limiting step has been that we haven't had a program.

How do you think most scientists today view SETI?

Well, "most" is a bit of a problem. There's a range of opinions. There are people who are enthusiastic about it, see it as one of the most significant and important subjects that science can approach today. And at the opposite end of the spectrum there are people who are absolutely convinced that there is no extraterrestrial intelligence, that we can be quite sure of that, and that it's a waste of money to begin. There's a wide number of people who have intermediate positions.

I would say the average opinion is that most scientists are happy that someone is dealing with this but they consider it such a long shot that they would not themselves want to dedicate a significant fraction of their scientific lives to what is probably, in their view, a will-o-the-wisp.

What about public attitudes towards SETI?

It depends again on what you mean by SETI. For example, starting with Lt. Arnold's invention of the name in 1947, flying saucers have attracted enormous public interest. Now what is that interest? Is that scientific interest in SETI? Is that a fear of Soviet invasion, in the same sense that the panic of the Orson Welles broadcast of 1938 was a fear of German invasion? What is this UFO public appeal?

There's been a continuous public interest in things of this sort through this century. Starting with Edgar Rice Burroughs and Percival Lowell and going on to Flash Gordon, Buck Rogers, and various comic-book adventurers of the forties and fifties, then to UFOs, motion pictures about extraterrestrial intelligence, up to the present with Star Wars and E.T. and all that, there's been a continuous public interest. Never has it dipped down so low that nothing in pop culture is dealing with the idea of extraterrestrial intelligence. It's always been with us; certainly this whole century.

But at the same time I think it's fair to say that there has recently been a significant increase in public attention. If you look at the highest grossing motion pictures of all time you'll find an astonishingly high number of them are concerned with this issue. That can't be a coincidence.

Also take a look at the ten highest best-selling books on the fiction list right now. You'll find something like five or six of them have to do with extraterrestrial intelligence one way or another. That's really quite striking. Just in case you want to pursue it, it's the Michener book which talks about it at the end, Asimov, the Clarke book, the Hitchhiker's Guide to the Galaxy book, the E.T. book and I may be forgetting one.

Realizing that it's only a guess, when do you think contact with ETI might be made?

There's no way to tell. If I could answer that question then I would be delighted. But that's simply one of those questions that if I gave you an answer and you asked, "Why do you think that's the answer?" I couldn't answer the question. It's silly for me to give an answer to that. But one thing that's sure is that if we don't look we're unlikely to find.

What might ETI be like?

Imagine yourself on Mars, let us say, and imagine that you're something totally different from a human being. Suppose you're a 20 meter diameter, squamous pink blob, trailing fifty tentacles and floating a meter off the ground. Through your telescope you look at the Earth, and you can determine something of its physical environment. Now tell us, what do the dominant organisms on the planet look like?

How could you tell? There's absolutely no way you can tell, because humans look the way we do because of a very long series of random genetic accidents, coincidences, the selection of one genotype and not another. The question of why do we look the way we do is deeply tied to our evolutionary history. But even then, there are major random events, such as when a cosmic ray caused a particular mutation. There's simply no way for us to know all the relevant factors influencing extraterrestrial forms.

One thing to me that seems rather clear is that extraterrestrials won't look like humans (not closely like humans), but what they *will* look like is a matter of speculation. It's one of the reasons why it's so interesting to try to find them and answer such questions.

If a scientist receives evidence of extraterrestrial intelligence, what should he or she do about this evidence?

Suppose there is compelling scientific evidence that we are receiving a signal from an extraterrestrial civilization; what do we do with it?

I suspect that before we do anything with it, the public will know. It will be very hard to keep it secret for more than a short period of time, as I tried to indicate in my novel *Contact*, because it is too important. It is impossible to imagine that astronomers will keep quiet, and not tell colleagues and friends and wives and children. It will get out, and that is a good thing, because the information is too important to be kept in any one institution or, indeed, in any one nation.

So we won't have to worry about that?

I think the problem will take care of itself. Beyond that, it does make sense to try not to release the data until confirmation has been achieved, because what a large embarrassment it would be to make a public statement that you've found extraterrestrial intelligence, and then find that it's some astronomical radio source you've misunderstood or, even worse, a hoax.

So naturally there will be some tendency to be very careful about releasing it until it is confirmed. That's, again, as it should be. But a confirmed signal should be announced immediately.

What will be the impact on the general public of such an announcement?

I've written a whole novel about this. I think it will be very complex. Some people will be absolutely delighted and others will be extremely fearful and threatened by it. You'll have involvement of every sector of opinion on the planet. Certainly religious people will be heavily involved in the discussion of what this means and what to do about it. I believe it will have political, social, academic, intellectual implications. Intellectuals will start reexamining the foundations of their subjects for fear that there is some elementary error they have made that will be brought out by the signals that the extraterrestrials are sending, and that, of course, is good.

I think there will be a worry on the part of many belief systems that something different will be favored by the extraterrestrials, but I believe that the overall effect of a confirmed extraterrestrial signal, even before it is translated, is to bind up the planet. The differences that divide us will appear insignificant compared to the differences between *all* humans and the extraterrestrials. Beyond that, a great deal depends on the message content which, needless to say, no one is in a position to predict.

Who are the outstanding people in SETI today?

In the United States it's clearly Frank Drake and Philip Morrison. Barney Oliver has played a major role in the instrumentation development. Paul Horowitz of Harvard is one of the brilliant new people of the next generation. There are a bunch of people like Sebastian Von Hoerner who've made important contibutions to our understanding of it.

There have been observational searches, the most elaborate by part of the people I've already mentioned, and by Zuckerman and Palmer. In the Soviet Union the key names are Shklovskii, Kardashev, Troitskii. That's at least the first crack. But I think Freeman Dyson has made an important contribution. I think every one of the SETI pioneers that is a signatory of the SETI petition is important.

What scientists or other people in the past do you admire?

It's an enormously long list, much longer than the one I just gave you. How could you *not* admire Newton or Darwin or Einstein or Aristarchus, Copernicus, Kepler—or Christian Huygens, who I think is perhaps the first person to speak with a modern voice on the extraterrestrial intelligence issue? There's clearly hundreds, thousands of people. Even Newton said that if he had accomplished anything it was by peering over the shoulders or standing on the shoulders of giants. Think of how many people after him fit that category. So it's a very hard question. To give a serious answer I'm sure I would have to write down more than a thousand names. Science is a collective enterprise.

You were mentioning some of the important people during the past twenty years or so, the real pioneers in SETI. How would you describe the role each has played?

Frank Drake did the first modern SETI search using radio astronomy. He did it essentially by himself. He had absolutely necessary support from Otto Struve, who was then Director of the National Radio Astronomy Observatory. If not for that bit of luck—well, in many observatories of that time Frank would not have been given permission to go ahead. But Frank was the person who put it together, proposed doing it, carried it out, and knew how to do it. And in several other areas, including the so-called Drake Equation and the idea that you could transmit pictures by having the number of bits be the product of two prime numbers, Frank has made essential theoretical contributions to the subject.

Philip Morrison. The Morrison and Cocconi paper, which was roughly contemporaneous with Drake and independent of Drake's actual observation, set the stage for radio observations and the 1420 megahertz hydrogen line as the first frequency to look on, and he has continued to play an absolutely major role in a whole set of studies since then.

I would say that those two, of the people we're talking about, have made clearly the major contributions among American scientists. And then in the Soviet Union, Shklovskii and Kardashev have played somewhat similar and independent roles.

What would you say has been your most memorable moment or event in your life?

I just can't relate to questions like that. There are many memorable moments in my life. It's very hard to pick out just one as the most memorable.

What about your greatest satisfactions in life?

Again, there've been hundreds—and greatest disappointments: hundreds. It's very hard to say, you know, what's the best book you've ever read? What's the tastiest dish you ever had? I think the right approach to those questions is a much deeper biographical foray than we can make in this conversation, so I hope you forgive me.

Certainly. What else should future historians know about you that would not otherwise be preserved?

I hope I'll be making some further contributions for some decades hence and maybe I'll make my most significant contribution ten or twenty years from now. So it's hard to tell. It's so hard to know about future historians: who are they, what prejudices will they have, what's their angle on the subject, and so on.

I tend to think that if the people that we've been talking about had never lived, if there had been no Philip Morrison or Frank Drake or me or Joseph Shklovskii or Nickolai Kardashev, then everything would've gone more or less the same, that some other people would've seen that the technology was now right for this and would've written comparable papers. Maybe there would've been a few more years before we got going, but in areas like this I really don't believe in key people coming along. Einstein's insights clearly, but not *these* insights. These insights are tied to the technology. Lots of people understand the technology.

So while I certainly understand why, in writing histories, it's convenient to write it about the people involved, I think this is a very good example of an area where the subject would have gone on independently of the people who are associated with it. In some alternative timeline where the world was slightly different and this handful of scientists were never born, SETI would still be in roughly the same shape it is in right now.

That's an interesting view. Are there any other things that you think we should get down on the record?

You know, I'm happy to have spent a little more time than we said; you're doing an important job and I certainly want to cooperate with it. But, no I don't have any other thoughts. The subject is just about to take off. Although maybe on the other hand that's just the moment you ought to be writing what you're writing. Maybe that's right.

John D. Kraus

RADIO PHYSICIST, ENGINEER
AND ASTRONOMER

Born June 28, 1910

Ann Arbor, Michigan

John Kraus is the Taine G. McDougal Professor of Electrical Engineering and Astronomy at Ohio State University and the director-founder of the O.S.U. Radio Observatory. Searching for extraterrestrial civilizations since 1973, the observatory's discoveries include twenty thousand radio sources, the farthest known objects in the universe, and a powerful ETI-like signal which remains unidentified. He edited and published *Cosmic Search*, the first journal dedicated entirely to SETI.

His research interests are antennas, electromagnetic theory, and radio astronomy. His textbooks on these topics have been translated into Japanese, Russian, Spanish and other languages.

Majoring in physics at the University of Michigan, he received a B.S. in 1930, M.S. in 1931, and Ph.D. in 1933. He then worked at the university and as an independent consultant on noise reduction. From 1940 to 1943 he worked at the Naval Ordnance Laboratory, Washington, D.C., on a project to protect ships against magnetic mines. He also developed an effective radio telephone. Moving to Harvard Radio Research Laboratory in 1943, he worked under Frederick Terman to develop radar countermeasures. He joined the Ohio State University faculty in 1946.

During the 1930s physicists working on the University of Michigan cyclotron met in his home for weekly short-wave radio conversations with their Berkeley colleagues Ernest Lawrence, Edward McMillan and Luis Alvarez. Other physicists who visited his station included Werner Heisenberg, Enrico and Laura Fermi, Wolfgang Pauli and Nils Bohr. Doctors also used his station to provide medical information to a hospital in the Belgian Congo.

He invented many types of antennas, including the "corner reflector," of

which millions were subsequently sold as UHF television antennas, and the helical antenna which is the workhorse of space communication.

His work became known world-wide through such activities and hundreds of
reports, beginning in 1933 with "Some Characteristics of Ultra-High Frequency
Transmission," *Proceedings*, Institute of Radio Engineers. Later topics included
"The Radio Position of the Galactic Nucleus," *Astrophysical Journal* (1955); and
"Maps of M31 and Surroundings at 600 and 1415 Megacylces per Second,"
Nature (1964).

His awards include the Edison Medal of the Institute of Electrical and Electronic Engineers, and the Sullivan Medal of Ohio State University. He is a member of the National Academy of Engineering, the American Physical Society, and
the American Astronomical Society. He is a fellow of the Institute of Electrical
and Electronic Engineers, and an honorary member of the Society of Wireless
Pioneers.

Interviewed February 1983 at Delaware, Ohio, by telephone

Where did you live during childhood and when you were growing up?

Ann Arbor until I graduated from the University of Michigan. Ann
Arbor at that time was a small town, a small university town.

What were your childhood interests?

Radio, baseball. Baseball was my favorite sport. During my earlier
years I played it summers in vacant lots as often as I could assemble a
gang of boys. My ambition was to become a major league baseball
player.

I was also active from a very young age building radios, receivers
and transmitters. In 1920 radio broadcasting was the new sensation.
As a ten-year-old boy I was intrigued with reports that nightly broadcasts had started from a new station, WWJ in Detroit, only forty
miles from my home in Ann Arbor. There were stories that WWJ
could be picked up in Ann Arbor with a simple crystal receiver. I
wanted to try.

The *Popular Science Monthly* explained that all you needed was a coil
of wire, a galena crystal with a fine wire or "cat whisker," and a pair
of earphones. These were fastened together with a few pieces of wire
and connected to an outside antenna. The coil consisted of several
hundred turns of wire wound on a cylindrical oatmeal box or a mail-
ing tube. The cheapest source of wire was a Model T Ford spark coil,
which cost a dollar. With hatchet and screwdriver the coil, encased in
wax and wood, could be chopped and pried open and the wire
unwound.

Earphones were more expensive so I removed the earpiece from our household upright Bell telephone. The earpiece was easily disconnected, and a hammer then hung on the hook in its place so that "central," the operator, would not be alerted. The telephone was inoperative, of course, but this was only a slight inconvenience for a few hours in the evening.

My father was a minerologist and he gave me some pea-sized scraps of shiny cubical black galena crystals. For a "cat whisker" to make contact with the crystal, we went to a manufacturer of scientific instruments, including one Dad had designed. The shop foreman gave me several inches of fine hair-like phosphor-bronze wire and a small brass clip. The wire would serve as the cat whisker and the clip would hold the galena. Returning home I arranged the parts on the table next to my bedroom window and eagerly hooked them together with short pieces of wire. This was my crystal receiver.

For my antenna I bought a roll of copper wire and several glazed ceramic insulators. I strung the wire with insulators from our house to an electric power pole at the back fence of our lot.

Next I connected the antenna to the receiver. Then I held the telephone to one ear and used my free hand to probe the galena crystal with the "cat whisker." Searching for a sensitive spot was tricky and often took many minutes. When I suddenly heard music or a voice, I stopped searching and held my breath, lest anything disturb the delicate adjustment. Someone moving about in the same room, a door slamming in another part of the house, or even a passing truck, might dislodge the cat whisker from its precarious position and the set would go dead.

However, the thrill of listening to the early broadcasts compensated for these frustrations. It gave me an eerie, out-of-this-world sensation to hear the faint but clear voice of "Ty" Tyson announce: "This is WWJ, *The Detroit News*. The next selection will be . . ." It was coming without wires through apparently empty space from forty miles away!

A year before I had listened over a radio receiver for the first time. It belonged to Bob Swain, who lived a few houses from me. He was a senior in high school and his knowledge of wireless seemed limitless. He had built a multiple-wire antenna stretching from a tall pole above the roof of his house to a huge guyed tower in his backyard. I was fascinated by Bob's antenna, and regarded it as no less an engineering feat than the Brooklyn Bridge.

I once sat all afternoon at Bob's receiver with the headphones pressed tightly against my ears, hearing nothing except an occasional click of static or a sputtering noise as an electric trolley car rounded a bend near Bob's house.

"Just wait until after supper," Bob said. "Then you will hear the fellows with their spark transmitters."

The "fellows" were radio amateurs or "hams." Bob had learned the Morse code and built a spark transmitter. Activity did increase after supper, but just before ten everything halted as all receivers were tuned to NAA, the powerful 2000 meter U.S. Navy station at Arlington, Virginia, which regularly transmitted time signals. Clocks were checked and reset when the long dash finally came through precisely at ten o'clock.

I could hear only a few broadcast stations with my crystal receiver, but soon I built my first vacuum-tube receiver. It had a tube mounted behind a bakelite panel adorned with knobs and binding posts. The tube filament was lighted by a six-volt automobile storage battery, which had to be recharged frequently, so about once a week my father and I loaded it into our four-cylinder Crow-Elkhart touring car and took it downtown to the battery service station where we left it overnight. Since the battery weighed about fifty pounds, the weekly trip was a chore.

A second battery was needed for the tube's plate voltage supply. This was a bulky 45-volt Burgess or Eveready type. A new one was needed about once a month because this battery could not be recharged, and old ones had to be discarded.

When the batteries were low, reception was weak; but on the evening after the storage battery had been charged and a new plate battery installed, the excitement of good listening returned. WWJ was comfortably strong and much more distant stations were also audible.

With the coming of fall and cool, clear, crisp nights, the static disappeared and radio stations came through in profusion. With headphones glued to my head I could hear voices and music at every position on the receiver dial. It was a modern Babel, a kaleidoscope of jumbled sounds. A voice said, "This is WBAP, Fort Worth."

Another mentioned Jefferson City, Missouri, and others gave Schenectady, Chicago, New York, Pittsburgh, or New Orleans as their locations. Station identifications were announced frequently, and in an hour I could identify and log dozens of them. As the evening wore on I could hear stations farther and farther west.

Then one night I listened past midnight, speculating whether I might even hear California. Most of the eastern stations had signed off. I searched the dial, listening intently. I heard music. The station was clear but weak. It was almost three o'clock. I waited and waited. Minutes dragged by slowly.

"Won't they ever announce?" I muttered.

Then a voice: "This is KFI, Earl C. Anthony, Los Angeles." Anthony was a Packard automobile dealer and a well-known pioneer broadcaster. My heart pounded. I had heard a voice all the way from California, over 2000 miles away!

I also recall the first time I communicated with a radio amateur in Australia. That was rather a unique thing at that time. It had only been possible for a few years because shortwaves had not been in use that long.

When was that?

1925. 1926. Right in there somewhere.

Later, when I was working on my Ph.D. degree in physics at the University of Michigan, my dissertation problem involved the study of propagation of radio waves at short wavelengths. These wavelengths are now used widely for television but at that time were used for nothing at all.

You mentioned one time a childhood activity of making a telephone out of two empty soup cans and connecting them with a string so you could talk to the boy across the street. I assume that was before you got involved in actual radio?

Yeah, that was even earlier. I just was interested in communication, I guess. First we built telegraph lines and then we converted them into telephone, then into radio, first receiving, then transmitting. So I have been a communication person from a very early age.

What else do you recall from adolescence?

I read a great deal. I read books about inventors, discoverers of science. I read everything I could find on Edison and I was very interested not only in radio but also in aviation. I aspired to be either a radio engineer or an aeronautical engineer. I was a member of a glider club at one time, as well as a radio club at the University of Michigan.

I loved geography and I found mathematics interesting, and the subject in high school that intrigued me the most was physics. I really became fascinated with it, and that determined my later course in the university. I graduated with a Ph.D. in physics from the University of Michigan in 1933. I was twenty-two years of age at the time.

Would you say that elementary schooling gave you a good start on your scientific career?

Yes, reasonably good. I read a lot of other things. I read books on radio. They didn't call it electronics then but it would be equivalent to that. And aeronautics. My favorite magazine was *Popular Science*.

What about secondary school?

As I said, the course that was most fascinating to me was physics. I had an excellent teacher and a full year course in physics which was very good. I also took two-and-a-half years of Latin which I, you might say, endured. But it gave me an appreciation of language.

When did you first think about the possibility that extraterrestrial intelligence might exist?

I don't know. I read about Percival Lowell, who was highly publicized at that time. He had built the observatory in Flagstaff, Arizona, with the best telescope available at that time, to observe Mars. He saw things that persuaded him that there were intelligent beings who had done things to shape the surface with geometric lines or what might be canals. It was a very popular subject in the press, with much speculation.

I remember very well some pictures and diagrams in *Popular Science* magazine showing a big bank of mirrors set up in the Sahara Desert which would reflect the sun's rays back to Mars. By tilting the mirrors they would be able to signal, with Morse dots and dashes or flashes of light, to the Martians. That seemed to be just the accepted thing; this would be the way it would be done.

When was that?

I would say 1924.

Did you discuss the possible existence of extraterrestrial intelligence with other people?

No, I don't recall. If it was, it was just casual. People may have mentioned it. You mentioned Martians just as casually as you'd mentioned Europeans or South Americans. It was natural to bring them into a conversation. You didn't necessarily think too much about them but you were at least cognizant of the possibility.

When you did happen to mention this, how did the other people respond?

Well, I don't know that I initiated it. That is, it was just give and take. Somebody else may have mentioned it. It wasn't necessarily *my* suggestion or bringing it into the conversation. It was just like talking about catching long flyballs in baseball. You just talked about this, and Martians were talked about in conversation just as casually. I didn't really think seriously about it and nobody else did. People just seemed to take it for granted: of course there were Martians!

What about communicating with, or at least searching for, extraterrestrial intelligence? When did you first think about this possibility?

It was kind of considered as a joke. There was a column called the DX column in the amateur magazine *QST*. It showed an amateur with headphones in front of his receiver tuning dials. He was obviously intently listening, and from one headphone came a little balloon above his head in which the cartoonist had written the word "Mars." That gave kind of a humorous context to this column in the magazine which had to do with distant stations on the Earth that were communicated with and being heard in Europe or Australia or places like that.

Was that also in the 1920s?

Yes.

During your college years, were there any ideas, books, teachers, classes, other students or events which stimulated you to think about extraterrestrial intelligence?

No, none at all that I can remember. Nothing of that kind. We were concerned strictly with matters on the earth and maybe observations of the sun. I took a course in astronomy and enjoyed that very much but the idea of extraterrestrials was never mentioned at all; never even came up. I don't think anybody ever thought about it.

After you completed your schooling and began working as a professional scientist, who or what has influenced your thinking about ETI and SETI?

Nothing at all for a long time. My interest was in antennas. I had become intensely interested in developing antennas while I was at the University of Michigan, and I invented a number of new types of antennas that became very widely used. I was a member of the Institute of Radio Engineers and I published many articles on these antennas.

After World War II when I came to Ohio State I continued the work. I invented still more types of antennas, including the widely used helical antenna, and had been interested in the idea of radio astronomy. I had heard of Karl Jansky and Grote Reber. In 1934 with a colleague at the University of Michigan we had tried to detect radio waves from the sun, unsuccessfully of course, because our receivers were much too insensitive.

But the whole idea was right on target as far as there being some radio activity there, and my colleague Arthur Adel had speculated that these might be regions close to sun spots, which is exactly what was later found to be the case. When there were some big spots on the sun in 1947 (I was at Ohio State then) I remember hurriedly putting up some antennas and trying to see if I could detect them with a radio receiver, but again was unsuccessful. It wasn't until 1951 that I started building our first radio telescope. I picked up the sun and then successively more and more objects and developed a radio astronomy program.

I was very busy for many years at Ohio State developing courses in antennas and wave guides and electromagnetic theory. I wrote a number of textbooks. I just didn't have the time to also construct, almost by myself, the antennas and necessary equipment to carry out these experiments.

What do your colleagues think about your interest in SETI?

I don't know. Some of them seem to be interested, some of them not.

Do they ever discuss it with you or ask you about it?

Well, the ones who are into this work, of course. But I'm interested in many things, and with certain groups I just talk about red shifts and radio spectra, and with people who are interested in SETI searches I talk about multichannel bandwidth systems and things of that type. So I talk different languages to different people, depending on their interest.

What about your family? What do they think about your ETI interests?

I don't know. They just seem to take it, just along with looking for high red shift objects and all of the other explorations of space. I wouldn't say that it creates any different perspective at all. It's just scientific research of one kind or another.

What about your superiors, bosses?

I think that a lot of the administrative officers do not have the vaguest idea what it's all about. And even if you explained it to them, they still wouldn't know. Once in a while you find one who has some interest. But in general university administrators have very little interest in this. They're mainly concerned with getting some money to keep their institution afloat.

What about your friends?

I really don't know. I have so many interests that some of them may not even be aware that I'm involved in it at all. I don't even discuss it with them. I talk to them about the things that they're interested in. I talk to them about computers. I have an Apple II computer. To other people I talk Earth stations—I have an Earth station—and SETI is never mentioned. Then with Bob Dixon, I talk a lot of SETI with him because he is Mr. SETI. He has a lot of interesting ideas on it.

How do you think most scientists today view SETI?

I don't know that I could answer that question. You would have to ask all the different scientists and take your poll. I have not polled them. The only thing I can say is that there is now a Commission No. 51 of the International Astronomical Union which is the SETI commission. The search for extraterrestrial intelligence has been sanctioned by this highest professional astronomical group in the world.

Professor Michael Papagiannis of Boston University was the first president. He was also very instrumental in getting this commission set up so that SETI now has official recognition by the world's astronomers. Many astronomers now have come out very, very clearly and stated emphatically that they feel this is an important phase of astronomical research.

Do you think the public is more enthusiastic than scientists, about the same, or less enthusiastic than scientists are?

I wouldn't know. That's a tough question to answer and I wouldn't want to say without doing my own Gallup poll on it. I think you'd find all shades of attitudes on this.

Before going on to discuss your work in SETI, let's get a little more background information. What was your father's occupation?

He was a mineralogist; professor of mineralogy at the University of Michigan. Later he was dean of the largest college there—Literature, Science and Arts—for many years. He was on the faculty there for something like forty years.

He must've had a Ph.D. then.

Yes, he got his at the University of Munich in Germany in 1897 or something like that. At that time, if a person wanted to get the most advanced education he went to Europe, and in science particularly he went to Germany. So he studied with some of the great scientists at the University of Munich. They were the world leaders at that time.

What about your mother? What formal education did she have?

She had education through high school. She grew up in the rural area near Syracuse, New York. She taught elementary and secondary school later in Syracuse.

My father was born in the city of Syracuse.

How would you describe them? Quiet, outgoing, reserved, serious, gregarious?

All of those things. They had many interests. Always had things doing.

How did they influence your eventual choice of science as a career?

They didn't try to. They let me pick it myself. They gave me encouragement and help and a few dollars to spend on equipment when I needed it desperately.

And what about ETI; how did they influence you in this respect?

I would say not at all. I don't think it was ever mentioned.

Were there other people we haven't mentioned who influenced your ideas about science and ETI?

Of the teachers in high school, my physics teacher was definitely very influential, and I had a number of excellent professors at the University of Michigan who had a great deal of influence on things. This was a very formative time. I found these people were very, very stimulating. The University of Michigan is a great institution in that they've had a tradition of excellence for a long time and their faculty had been and continues to be absolutely tops.

I'll tell you a little story. The name Karl Guthe was a household word in our family. He had been a physics professor at the University of Michigan. I think he had come from Germany. He was absolutely outstanding. He was a legend in Ann Arbor; his name was a household word. You heard Karl Guthe mentioned—well yes, he's the physics professor who was at the University of Michigan around the turn of the century and really started Michigan in physics in a big way.

The next time I heard the name Karl Guthe came in a rather interesting connection and I really didn't make the connection until quite recently. K. G. Jansky of the Bell Laboratories was the father of radio astronomy. In the early 1930s he built a rotating antenna to study the direction of arrival of thunderstorm static; and he picked up some weak signals that he traced to the center of our galaxy, a remarkable piece of research.

Later, meeting members of the Jansky family and particularly his older brother C. M. Jansky, Jr., I did get the connection because the full name of the founder of radio astronomy was Karl *Guthe* Jansky. It turned out that his father studied at the University of Michigan around the turn of the century and so admired Karl Guthe, who was one of his professors, that when Jansky's son was born in 1905, he named him Karl Guthe Jansky.

This is the connection between the University of Michigan and the founder of radio astronomy. I think that this is really interesting. You can see how the University of Michigan figures over and over again in influencing people. It influenced me, it influenced Jansky's father, and Karl Jansky bears the name of a great Michigan physicist.

About yourself and your own family; how many brothers and sisters do you have?

I had a sister who was killed in an accident shortly after she was married, and I was the only other child in the family.

What occupation did she have?

She was a historian. She was at the time teaching history at a secondary school in Massachusetts.

What about your friends when you were growing up; did they share your interests in science and ETI?

Nothing was ever mentioned about ETI. I can't remember anything about that at all. I don't think that there was much influence from my young associates except that some of them who were a little older had gotten their amateur license, had put up enormous antennas between their houses and trees, and learned Morse code. I felt that this was very impressive.

What were you like during childhood and adolescence?

Oh, I would say about average. I had lots of friends. We had lots of activities. I organized the baseball team. I was very active in radio groups. We had a radio club at the high school, with a lot of activity in it.

What religious beliefs did your family have?

We were very strong churchgoers, members of the Methodist church. I was brought up in a very religious atmosphere.

That's interesting because Frank Drake mentioned the possibility that SETI pioneers might have had some serious religious exposure.

Well, there was never any thought of conflict between science and religion in my thinking or in my upbringing. Science and religion were simply both seeking ultimate truth but using different ways of going at it. No conflict at all.

What recreations do you enjoy?

I like to spend a great deal of time outdoors. I enjoy chopping wood; sawing wood with handsaws, not with power saws. I'm not interested in how much I can do; it's more the exercise associated with it.

What has been your greatest satisfaction in life?

I think one has a feeling of satisfaction and accomplishment when one has completed a task like building a radio telescope, inventing and studying a new antenna, completing a new textbook, and things of that type. Many different tasks on completion, if they turn out satisfactorily, are sources of satisfaction.

What has been your greatest sorrow in life?

Well, I'd have to think about that one. What's your next question?

What was the most memorable moment or event in your life?

That's a tough one, too. What's your next one?

What else should future historians know about you that would not otherwise be preserved?

Again. I don't know. I wouldn't have any idea.

What scientists or other people in the past do you admire?

Well, I've read a great deal about many of the scientists of the past. I've read a great deal about Galileo, Newton, Lord Rayleigh. I was fascinated by the work of Heinrich Hertz, Marconi, Edison and, of course, I've read a great deal about and tried to understand what Einstein has done.

Who are the outstanding people in SETI today, and during the past twenty years or so?

I think Cornell has the corner on them: Frank Drake and Carl Sagan. And, of course, Philip Morrison and Giuseppe Cocconi were at Cornell when they proposed the search at 21 cm in *Nature* magazine. So it seems like it centers around Cornell quite a bit, although Cocconi is now at CERN in Switzerland and Morrison is at MIT.

What role has each played in SETI?

Drake did the first experimental attempt, and Sagan is an excellent expositor on the subject. Morrison and Cocconi wrote the first definitive article on it.

Some people are obsessed with the idea, and that's all they think about. Then there are others, more like myself, who think about this along with many other things. They find it a very interesting topic and one that's easy to philosophize about, but it is not one about which they are totally concerned.

All these people are very busy, and they think SETI is a good bet and we should give it a whirl. We're playing with very slim chances of success, but if we don't at least look, it is certain we'll never find anything.

It's a matter of being alert and keeping it in the back of your mind that this *is* a possibility, and planning your work so that if you *do* observe something unusual you won't necessarily rule it out. You will do like Jansky did: run it down and see if there's something more than local interference.

Cosmic Search was the first magazine specializing in SETI. How did you happen to start it?

Well, we realized that NASA support for our work, which we had been getting, would not be at a very high level, and that maybe it could be supplemented if we had some extra income from another direction. If we started a magazine and it was successful, we would have a small, modest supplement to support our work and maybe that of radio astronomers at other locations who were also interested in this thing. So the motive behind it was that this might generate some monetary support for SETI research.

But unfortunately, the magazine just lost money. Even at best, it was never able to break even. It lost continuously, and it also became more and more my own responsibility for keeping it going. I just was not able to devote the necessary time and energy, so after thirteen issues it stopped.

I tried to find someone else to take it over but without success. One person took it for a year but didn't bring out any issues. Then it merged with a space-oriented newsletter, but that folded, too.

That's a shame.

Yes. Well, how are your other interviews going?

Quite well. The biggest challenge remaining is to get some follow-up information from the three Russian scientists.

Well, I hope you're successful. I sometimes get a Christmas card or a New Year's greeting out of Russia from some scientist there, about six months later, by registered mail. They've got things so tied up there, so constipated, I just wonder that anything is accomplished.

As I understand it, Shklovskii has backed off from an enthusiastic position that he once had. Kardashev and Troitskii are more gung-ho and actually may be doing some observing work.

Even if that's true, it would still be historically important to interview Shklovskii.

Yes, because he wrote the first book on the subject, and then in collaboration they brought out an English version in which Sagan added comments here and there. This was really the first textbook or extensive treatise on the subject which went into quite a bit of depth. Sagan and Shklovskii collaborated on that book without ever seeing each other. They did it all by mail.

What effect, if any, has your interest in ETI had upon your career as a scientist?

I'd rather lead up to how I even got involved in it at all. I was *not* interested in searching for extraterrestrials. I wanted to do astronomical research. We embarked on a survey of all of the sky that could be observed from this location, mapping radio sources. We discovered and catalogued about 20,000 radio sources. We found that some of these were very unusual.

In cooperation with other astronomers, radio and optical, it turned out that many of these objects that we had discovered were among the most distant known objects in the universe, the ones with the highest red shift. So our telescope was not only a discovery instrument but it was penetrating to the greatest distances into the universe that were possible.

This work went on for many years, and at its conclusion we had available this very large telescope here, the Ohio State University radio telescope, which we had built. It was available, and it was not being used full-time. It was then that my colleague, Robert Dixon, who had come to Ohio State to study with me and who had gotten his doctorate degree in electrical engineering under me and had been working at the radio observatory, suggested that this would be a great opportunity to embark on a search, using narrowband receivers, to look for extraterrestrial intelligence. It was *his* interest in it that got me involved.

I was director of the radio observatory. I felt that he had very good insight into the problem. He had proposed many excellent strategies for the search, in connection with bandwidths and with searching areas, and particularly with the data reduction of the output to be able to readily interpret any signals that were being received. So I gave him my support in making the telescope available for this purpose. My interest in this kind of work has continued from that time.

When was that?

1973.

What part of your time do you spend on SETI?

SETI is a very small part, very small. And it's largely in connection with helping Robert Dixon that I'm into that. I have written a number of textbooks that are used worldwide. My books are in many different foreign editions. They've been out now for over thirty years.

I am involved in writing new books and revising old ones. Some of these are tremendous jobs in that they not only have a text, but they also have problem sets. The one I'm working on now has something like one thousand five hundred problems, all of which have to be solved, and I'm preparing a solutions manual with the answers. This book will run close to a thousand pages in length. So it's a big task.

I do astronomical research, radio research, design and build antennas, so my main task is to decide on a given day what I will start on, selecting out of ten possible things. It's not a matter of wondering what you're going to do; it's wondering which of the many things you may be able to squeeze in on that day. I'm really very busy, and the SETI activity just comes in around the edges when I talk to Bob Dixon and find out what's going on and how I can help on it.

Of course, this work with SETI has caused me to back off and philosophize a little. I like to do that. My attitude about SETI is very simple and very objective: either we're alone or we're not, and either way is a simply mind-boggling concept. And if we want to find out whether we are or not, we should do some observational work, which we are now in a position to initiate.

Maybe we will find some evidence that would be a positive indication but maybe we won't. Nobody knows how long it might be before anything is found. After a long period of searching and nothing is found, then we may begin to wonder if we really are alone. So we're in the period of beginning this type of thing and I think it's very exciting.

Realizing that it is only a guess, when might contact with ETI be made?

I have no idea whatsoever.

What might ETI be like?

I have no idea whatsoever.

What stages has your thinking about SETI gone through? Have your ideas about it changed much since you first began thinking about it?

Well, they're pretty much as I've just indicated. I have thought a bit about what you might call search strategies: What wavelengths do you use? What bandwidths do you use? Do you search the area of particular stars or do you look systematically over the whole sky, since nobody knows where or at what wavelengths any of these signals might occur? It's kind of a guessing game but that's what makes it fun.

What alternatives or what other possibilities about ETI have you considered?

I mentioned Percival Lowell. He thought he had found evidence of extraterrestrials on Mars. Actually he may have been closer than we are now. That is, even though we have no reason to believe that there may be life on Mars (there might be some very primitive life, but we don't believe there is anything we would call intelligent life on Mars),

his thinking that there might be life on Mars may have been no farther off the mark than our thinking now that we may be able to pick up
radio signals or beacons from extraterrestrials. But in any case, radio
techniques do seem to be the most promising at the present time.

On the other hand, if and when extraterrestrials are found, this
discovery may come serendipitously from completely different techniques. For example, not from radio waves at all but in some way
connected with, to give an example, gravity waves. And the person
who makes the discovery may not be looking for extraterrestrials at
all, any more than Karl Jansky was looking for radio waves from our
galaxy; he was studying thunderstorms. Someone doing some experiment without the least idea of extraterrestrials in his mind at all may
be the one who, through serendipity, will make the discovery. This is
something that one should fully recognize.

What would you do differently if you were starting over again to study ETI?

I don't know. I have no thoughts on that.

Why did you choose to investigate ETI?

I didn't. I've just been helping Bob Dixon.

What has been the most difficult aspect of your SETI work?

To keep a large radio telescope operating properly is difficult. The
instrumentation must be maintained. Its sensitivity must be kept high.
As with any kind of equipment, it takes work to keep it going and
performing at its best level. So we have problems in keeping the antenna adjusted properly, the receiver working right, tracing sources
of interference which cause spurious records, eliminating those, and
things of that type. It's a very difficult job doing observational work.

It's like an astronomer who's taking photographs of stars. If there
are bright lights, this causes a problem—or mist or fog or cloudiness
or things of that type—so you have to work against many obstacles in
doing observational work. You may observe many nights without getting good records because of one problem or another, and then
things click and you get good records for weeks on end.

Tell me more about the telescope.

It's near Delaware, Ohio. We began building it in 1956, with National Science Foundation funds and Ohio State University students
doing the work.

At first, progress was slow but by 1963 the telescope was far enough
along to begin preliminary observations. Two years later we started a
systematic survey of the entire sky accessible to our telescope. It continued for eight years, during which we mapped and catalogued some

20,000 radio sources in the largest, most complete survey anyone had made. Radio maps were rather new at the time. Because radio wavelengths are so much longer than light, the radio sky is quite different from the visual one.

In 1964 when Bob Dixon came to the Ohio State University as one of my graduate students, I started him on a mapping project which carried over into the big eight-year survey. One of Bob's early contributions was the development of a method for using antenna pattern-fitting with a computer to produce cleaner, more reliable maps. This method has become standard for improving the quality of telescope outputs.

In 1972 money for our mapping project stopped abruptly, due to a shift in funding priorities, so we found ourselves in a strange situation. Our telescope was one of the world's largest. It also had one of the most sensitive 21 cm receivers in operation anywhere.

Our survey had resulted in the discovery of many objects belonging to a new class of radio sources having enhanced radiation at centimeter wavelengths. Some of these—OH471 and OQ172—proved to be the most distant and most powerful sources ever found. Thus, we had a telescope of great capability but with no money to do anything on a large scale. Would it be possible to do something at a very modest level?

Several years earlier, as a master's thesis project, William "Bill" Brundage had built an eight-channel receiver for studying the hydrogen in the Andromeda galaxy. Bob Dixon now suggested that we use Bill's receiver to begin a "waterhole" search for narrowband signals of extraterrestrial intelligent origin. He reasoned that if we could discover *what* is out there essentially to the "edge" of the universe, maybe our telescope could be useful in finding *who* might be out there.

I was the observatory director, and Bob didn't have to work very hard to persuade me to go along with the idea. So Bill's receiver was reconnected and by late 1973 we began our search on a routine basis.

Within a couple of years we added a fifty-channel receiver to supplement the old eight-channel one, and a dual-horn feed system, giving us two observations of a source on each transit instead of one, and significantly improving the quality of the records. Bob Dixon had also connected a computer on-line to process the data and print the output of all fifty channels in a simple-to-read format.

Even though the observatory funding was now close to zero we were running the survey around-the-clock, and continually making improvements in the equipment, in the data processing and in its analysis, but only because we had acquired a lot of unpaid help. Many undergraduate engineering students designed and built new components for course credit, obtaining valuable practical experience to supplement the theory of the classroom.

Then Richard Arnold volunteered to work an evening or so a week on equipment. He had a doctorate in electrical engineering and was employed by the Bell Telephone Laboratories of Columbus.

Jerry Ehman also volunteered time on weekends; during the week he taught at Franklin University in Columbus. He had a Ph.D. in radio astronomy and had been involved in our big sky survey some years earlier but had to leave when our funds were cut off. We were glad to have him back as a volunteer. Bob Dixon was a volunteer, too, because he was working almost full-time elsewhere. So with very little money but with a number of talented and dedicated volunteers, a very large, well equipped radio telescope was running essentially full-time on a routine SETI program.

Was that the only full-time, routine SETI search?

Yes, at least as of that time. Jerry gradually assumed major responsibility for the observing program and for analyzing the data output. Bob Dixon and I occasionally assisted by scanning stacks of data and discussing results with Jerry.

Beginning in 1973 we surveyed for thousands of hours without finding any signal suggesting extraterrestrial intelligent origin. We did find a considerable number of narrowband signals, perhaps five kilohertz or less in width, which we concluded were cold, isolated neutral hydrogen clouds.

Then *it* happened. In August of 1977 Jerry showed Dick Arnold, Bob and me a section of a new computer printout with all the characteristics one might expect from an extraterrestrial beacon signal. Jerry's amazement was indicated by the word "Wow!" which he had written on the margin of the printout.

The printout format, which Bob had designed, consisted of fifty columns, one for each channel. A single digit printed every twelve seconds indicated the signal level in that channel in units above the background level. The unit used was one standard deviation or one "sigma." A blank signified that the level was at zero. Any number above 4 or 5 might be considered significant and probably not due to some random fluctuation. In order to accommodate levels above 9 with a single character, Bob arranged that the computer run through the alphabet with A for 10 through Z for 35.

What Jerry had noted was a sequence of characters in channel 2 running: 6, E, Q, U, J, 5. Translated into numbers, they were: 6, 14, 26, 30, 19, 5. When plotted up they produced a pattern which matched exactly (allowing for measurement error) the telescope antenna pattern. This indicated that the source was very probably celestial—fixed with respect to the star background—and that it passed through the telescope beam as the earth rotated.

The signal was very strong (30 sigmas or thirty times the background), and it was narrowband (width of 10 kilohertz or less) because it appeared in only one channel.

But even more significant, it was *intermittent*. A steady signal would have appeared two times on the record, a few minutes apart, as our telescope with its twin beam scanned the sky. The possibility that only one horn was functioning at the time can be ruled out because the two horns are balanced and, if one were out, the system would not have worked. So it was an "on and off" signal. Was it deliberate? Was it intended for us?

Bob, Dick, Jerry and I had urgent discussions about its significance. We referred to it as the "Wow!" signal. We continued to scan the same region of the sky on the chance that the signal might reappear. But it never did. After weeks of patient listening we moved our survey to other parts of the sky.

We checked star catalogs for any sun-like stars in that area, and found none. We consulted ephemerides and made inquiries to see if any space probes could have been in the area, but found none. The region is near the galactic plane and not too far from the center of the galaxy. Although it is not far south of the ecliptic—the plane near which most solar system objects are found and toward which a probe might be traveling—the moon, the major planets and even the four largest minor ones were all far away from that position. The Voyager spacecraft had not yet been launched. So we are completely at a loss to explain it.

It certainly came from a very great distance: it was well out into the Solar System. It might have been a probe from Earth, but it was not in a direction where there were any probes that we know anything about. Or, it could have been a signal from an extraterrestrial civilization which was turning on and off, or sweeping by, and we just happened to catch it for one full antenna pattern. But since it has not returned so we could study it further, this is pure speculation.

Since that time we've incorporated many improvements into our equipment, so that now we could tell a great deal more about any signal. But since we can't turn the clock back, we have to settle for the data we have, which is, nevertheless, considerable and impressive. That signal is by far the most suggestive of extraterrestrial intelligent origin of anything we have detected in years of searching.

One of the problems with extraterrestrial signals is they may sound very much like a man-made signal. You might get something and say, "Ah, that's terrestrial interference," and so you discard it; but if you pursued it, you might think otherwise. With all our space probes out there now, the matter is not as simple as it used to be. You can have transmitters way out in space.

The "Wow!" signal is a beautiful example of exactly what an extraterrestrial radio signal could look like, but we only observed it once, and so we can't say any more about it.

It is also a beautiful example of the fact that our equipment works, and it is very, very efficient.

The other day you mentioned a threat to the observatory.

Yes, "destruction" is the word. After twenty-seven years of interuniversity cooperation, the Ohio Wesleyan University sold the land out from under the radio telescope. It's going to be owned by a golf club that wants the telescope torn down to make room for a fairway. The outlook does not look promising. The only thing that seems to be uncertain now is whether Ohio State University or the new owners, that is the golf club, would have to pay for the demolition of the telescope. It is an enormous structure.

It's going to be very costly for either party to destroy it. But the real tragedy is the destruction of an excellent radio telescope.

This telescope, which detected one of the most likely signs of extraterrestrial intelligence, may be torn down to make room for a golf course?

Yes.

That about says it all.

Yes.*

I feel privileged to have had these opportunities to talk with you.

I've enjoyed them also. I'm glad to have made your acquaintance and I hope we can continue in touch. You see, you provide a different slant on things. I think this is good. You're interested in sociological things. There is a professor of communication science here at the university who is very interested in SETI because he said it is the ultimate in communication. I enjoyed talking with him, too. It has given me a new insight into this activity that I look at largely as an engineering problem: bandwidths, channels, sensitivities, antenna gain. To him it was part of a science or art of communication. And you see attitudes, social concepts, and things of that type—and this is all part of the whole picture.

Man is a pretty complicated machine.

---

*After much negotiation the telescope was spared, in return for payment of an annual fee.

John Billingham

AEROSPACE PHYSICIAN

Born March 18, 1930

Worcester, England

Chief of the Life Science Division at NASA Ames Research Center, John Billingham was the first person in the United States to head a government unit officially concerned with extraterrestrial intelligence. He has worked since the late 1960s to establish SETI as a legitimate NASA function. In addition to his ongoing administrative duties he has chaired many committees and conferences on SETI for NASA, the International Astronautical Federation, and the International Academy of Astronautics. He was co-director, with Bernard Oliver, of Project Cyclops, a 1972 design study of a system for detecting intelligent extraterrestrial life.

At Oxford University he received an M.A. in 1951 and his medical degree in 1954. He subsequently served as medical officer of the Royal Air Force Institute of Aviation at Farnborough, specializing in aviation medicine and physiology.

In 1963 he came to the United States as chief of the Environmental Physiology Branch at Johnson Space Center in Houston, Texas, and three years later moved to Ames Research Center as assistant chief of the Biotechnology Division. In 1976 he became chief of the Extraterrestrial Research Division and acting chief of the Program Office for the Search for Extraterrestrial Intelligence. He was appointed to his present post in 1984.

His leadership positions, in addition to his SETI work, include the MOONLAB project, a design for a permanent manned lunar laboratory. He has also been director or cochairman of other American and international projects, sessions, and committees, on man in space, and managing earth's limited resources. He is Lecturer in Stanford University's Medical School and Department of Aeronautics and Astronautics.

He holds two patents related to temperature-controlled pilot garments and is author of seventy articles on topics ranging from "The Demography of Extra-terrestrial Civilizations," in *Space: Mankind's Fourth Environment* (1982) and "A Review of the Theory of Interstellar Communication," *Acta Astronautica* (1979), to "The Effect of Altered G-Levels on the Deposition of Particulates in the Human Respiratory Tract," *Journal of Applied Physiology*, (1975).

His honors and awards include the Guy's Hospital Physical Society Prize in Cardiology, an honorary doctoral degree from Hawaii Loa College, and several awards from NASA, including the Ames Research Center Outstanding Leadership Award, 1983.

He is a fellow of the Aerospace Medical Association, member of the American Institute of Aeronautics and Astronautics, and academician of the International Academy of Astronautics. Other professional memberships include the Society of NASA Flight Surgeons, the Astronomical Society of the Pacific, and the International Astronomical Union.

Interviewed June 1981 at Mountain View, California

Where did you live during childhood and when you were growing up?

Worcester, England. It's a small city. It's the county town or the capital; the town of Worcestershire.

What were your interests in childhood?

I liked both intellectual pursuits and physical pursuits. Everything from cricket, rugby football, and athletics to all sorts of different intellectual pursuits. I was very fond of chess. It really was a wide range; I don't think there was any one thing which stood out.

Tell me about your earliest activities in science.

The most vivid concerned starting science at school. We used to start science early at the Royal Grammar School, where I was at that time. The earliest memories are all of doing quite detailed, biological work with some very good teachers at my school. That's what really got me interested in the biological sciences.

As one went through school, one did more and more sophisticated biology, physics, chemistry and math until by the time I was finished with school I had a pretty thorough grounding in science; it was really very high-class teaching.

What about during adolescence? What activities do you recall from those years?

I do recall that I arranged to give my colleagues in one of the senior physics classes a lecture on the planets in the solar system. That was perhaps the earliest feeling that I developed for extraterrestrial events. That would've been equivalent to my junior year in school.

How was your elementary schooling?

Exciting and stimulating, all the time.

Did it give you helpful preparation for a scientific career?

Yes, in all ways. The teaching standards were very high, people were very interested, and I found myself rapidly becoming interested in all aspects of science.

What about secondary school?

Very stimulating.

How would you assess its helpfulness to your scientific career?

I would say it laid the foundation for the things that I have been able to do. And it was very solid training, very solid background. It's the sort of stuff which I constantly find myself drawing on even today.

Do you recall any thoughts about interplanetary travel or extraterrestrial life or other space topics in those early years?

Not as a child, except every so often I would come across a science fiction story. But nothing very specific; I don't really remember anything very specific about space until I was a medical student.

What do you recall at that particular time?

I recall seeing announcements of meetings of the British Interplanetary Society. I went to occasional meetings and listened to what they had to say. It was a place—"hotbed" would be the wrong word—a place where people who did think about these things would gather to talk about their ideas and concepts, and from which a lot of the ideas about space flight ultimately came. At that time my interest was *not* in other civilizations and SETI. It was much more in things like manned space flight: flying to the moon and journeys to the stars—which were not considered to be "proper" subjects at that time.

This was 1951 through '54. It was the time I first became interested in things extraterrestrial. It was our species going into the extraterrestrial environment.

Then I got my medical degree and I did a year's internship where I was far too busy to think about anything extraterrestrial. Then I had to go into the armed services. We had conscription.

I had been involved in the air cadet thing as a youngster. I read a lot about flying and nearly became a pilot when I was at Oxford. I was just greatly intrigued by flight. I knew that with my background in physiology and so on I could probably make some contributions but I had to get to a place that did good research work. It's no good being an ordinary medical officer at some air force base. You won't learn much about flight that way; you only deal with medical complaints.

So that's why I elected to go to the Royal Air Force Institute of Aviation Medicine at Farnborough. That was *the* place, equivalent to the School of Aerospace Medicine at Brooks Air Force Base. I was strictly interested in flight. I wanted to work on respiratory physiology, the physiology of respiration in high-altitude, high-speed flight, but that was full when I got there. However, they needed couple of people in climatic physiology and thermal physiology. So I went there and became an "expert" in thermal physiology. I dealt with heat exchange between people, particularly aircrews, and their environment. So it was primarily flight rather than space that motivated me. It wasn't until I was at Farnborough that I really became interested in space.

That was 1956 and 1957. At that time, talk of space travel was very much on the increase. Werner Von Braun was here in the U.S. building Redstone rockets. The various societies were beginning to discuss, tentatively, the possibility of manned space flight, which was still looked upon by most people as a fantasy—not by the British Interplanetary Society, of course, and not by Von Braun.

Anyway, as I worked on problems of keeping people at a comfortable temperature in terrible environments (extreme heat, extreme cold) for the purposes of aviation physiology and medicine, I would see occasionally a notice of a meeting about space.

A meeting within the air force?

No, we didn't talk about space much during our everyday activities at the Institute. No, it was the British Interplanetary Society. I remember a meeting in London, which was one of the first meetings of the British Interplanetary Society to be heavily attended by Americans. The reason was clear: there were enough people around London in societies and the government who could see what was going to happen, as I think I could see what was going to happen.

I presented two papers at these meetings. They were accepted and I duly went and gave the papers. They were about heat exchange between man and his environment on the surface of the moon, and about the design of space suits for use on the lunar surface. They were well received and were published, but I don't think it was considered quite a proper activity for an officer and gentleman in Her Majesty's service.

A few remarks were made about, "Maybe you should spend a little more time doing your work," "It seems to be a little far out"; sort of gentle remarks. Nobody clapped me in irons or hauled me off somewhere.

Who made these remarks?

They were made by senior people in the air force, to whom all this was really strange.

I began to think more deeply about what would happen to people when they went into space. I wrote a review for *Nature* of the British Interplanetary Society meetings.

Was this still before Sputnik?

No. '58. It was just after Sputnik. My interests were not SETI oriented; they were manned space flight oriented.

Then, as part of my job with the Royal Air Force, I visited Washington and various Air Force and Navy physiological research centers here in the States. As I did that, I met many of the people who were beginning to talk about space business for real.

Immediately after Sputnik, of course, everything changed. Everywhere I went in the States in '58, '59, and '60, everybody was talking about manned space flight. The Air Force labs, the Navy labs were heavily involved in the physiological and medical aspects. One couldn't help but be infected by the tremendous drive and excitement. When in 1960, I think it was, they formed the Space Task Group at Langley Research Center in Virginia, I made a point of going to visit them on one of my trips to Washington. I found out all about the plans for Project Mercury. They were beginning to build simulators.

Then I got an invitation to join NASA. This was in '61 and I had two offers. One was from Houston, at the Manned Spacecraft Center. The other one was from the Ames Research Center, here in the San Francisco Bay Area. I chose to go to Houston at that time. Both places were doing life sciences in space, but Houston was the place where all the action was, because they were flying people, astronauts. I had a good background for that stuff.

So I went to Houston, became Chief of the Environmental Physiology branch there. I did about three years at Houston. It was very exciting. Space medicine and space physiology. Responsibility for defining the physiological requirements for food, water, acceleration, vibration, radiation protection, noise tolerance, thermal control, space suits—it went on and on.

Then in '65, I was offered a job at Ames, a division job. It was a very good job so I came. I continued to work on space medicine and physiology until about 1972 or 1973.

So I had a lot of years in aviation and space medicine and physiology. It was a dream come true, in the sense that I'd been thinking about these things back in England. I actually got to introduce new heat exchange systems into the Apollo lunar landing space suits. They were called water-cooled suits, which I had invented back in England with a colleague called Derek Burton.

When I came out to Ames I did similar types of things: finding out human limitations and reactions to the stresses of the aircraft and space environment.

All this time I had a research division down on the first floor, which was called the Biotechnology Division. It had to do with aviation and space medicine and bioengineering: the effects of space on humans—it was physiological research. I also knew the people on the third floor who dealt with exobiology, the study of life outside the earth.

They were a different set of people and they had different disciplinary backgrounds, but I ran into them frequently and I became very intrigued with fundamental questions about the existence of other life—extraterrestrial life, not intelligent life—because they weren't considering it. They were only concerned with earlier forms of life: the getting together of chemical compounds you have to have for life; then the origin of life; then the early steps in the evolution of life—research which goes on to this day in this division. I was always greatly intrigued with exobiology.

At that time the big idea was to go to Mars. Everybody was planning to go to Mars and search for life there. I was very interested. My boss here, Harold Klein, really began the exobiology program. I learned a lot from him, and not just about exobiology! Then, about 1968, someone, I don't remember who, said I must read Sagan and Shklovskii's book, *Intelligent Life in the Universe.*

I was very excited by that book. It was very stimulating. There was a reference in it to Cameron's 1963 book on interstellar communication. So I read that, too; and from then on I began to see the possibilities. These books were very wide ranging and raised all sorts of possibilities. They had some substance, in terms of the origin and evolution of life and things like that, but they were speculative.

It gradually began to dawn on me that here I was in the life sciences in NASA, which was charged with the exploration of space. And here in this very building there was a division devoted to exobiology and the search for life in space, albeit microbial life. And here was a book published under the auspices of the National Academy of Sciences dealing with interstellar communication; and here were Sagan and Shklovskii writing about intelligent life in the universe.

I realized that if ever one were to have a real chance of actually carrying out a search and detecting another civilization, one would have to do a very thorough systems study, choose an approach, go and sell it to everyone, and then, within the framework of NASA, which had the mandate and the capability, go and do it. These thoughts just came to me every so often at that time.

Then in the summer of 1969 we got a new Center director. His name was Hans Mark. He is a very interesting person. He's one of these people who take a personal interest in having contacts with research people. He's not a director who sits up in his oak-paneled office and doesn't communicate. He gets down from there, talks to people, and he's receptive to ideas. He's got lots of suggestions of his own.

One of the things I did every summer was to run engineering system design studies with a colleague at Stanford, Jim Adams, in which we would tackle some interesting big design project, like building a base on the Moon. We would have twenty faculty come from all over the country and we would give them this job. We showed them how to work as a team to design something.

Every September Jim and I would prepare a little proposal to go to NASA for next summer's study. I had got the idea for the 1970 summer study—on a conceptual engineering systems design for interstellar communication—as a result of talking to another colleague, Max Anliker, in the Stanford mechanical engineering department. The first thing we needed to do was to try and find out how, if you wanted to conduct a search, how would you do it? And Max said "Look, you're the one that runs these Stanford summer studies. Why don't you make it the topic of the Stanford summer study?"

I thought it was a great idea. So one afternoon in September 1969 I met with Hans Mark to discuss the possibility that we might do a study, in the summer of 1970 on, as we called it then, "interstellar communication." I remember the talk very vividly because he listened to the ideas and concepts. He said it sounded like something we should pursue, but he had one word of caution for me. It turned out to be terribly wise. He said do it extremely slowly. Don't try and come out next year with some fully completed product. He knew perfectly well that it would take years and years and years for SETI to become reputable, organized and accepted.

So, rather than go whole hog and make a full-blown summer study out of it, we would do a mini study. So in the summer of 1970 we did a mini study of interstellar communication. We got in faculty from different places. We had one exobiologist. We had one radio astronomer. We had one electrical engineer. We had one astronomer. We had faculty advisors from here, four of us: Cyril Ponnamperuma, David

Black, Dale Lumb and myself. It was a study of some of the essential elements of interstellar communication, to see whether it was a feasible thing. The results were, yes, it looks to be a reasonable sort of idea.

So I went back to Hans Mark and told him that it looked reasonable. He said, "All right, let's do it."

The next step was to carry out a more comprehensive engineering systems design study of a SETI system. I was no expert in the technology involved. We needed a director who understood the whole business. There were two names at the top of our list. One was Frank Drake, the other was Barney Oliver. I remember discussing it with Hans. He said he thought Oliver would be better because Oliver was both scientist and engineer, whereas Drake was primarily a scientist; and Oliver is local here.

So in September of 1970, just after we finished our summer study, I went to see Barney Oliver to ask him if he would take leave for three months from Hewlett Packard, where he was vice president of research and development, and come and direct a summer study on the design of a system for detecting extraterrestrial intelligent life.

I remember that meeting very well. We met in his club here in Palo Alto over lunch. He was greatly taken by the whole notion. He was made my co-director at Stanford, and we worked together through the year to establish contacts with all sorts of colleges and universities, get the message out that this is what we're doing, and assemble the faculty.

In the summer of 1971, we began that study, which was Project Cyclops. Project Cyclops was really the start. That was the thing that really launched SETI. At that time we didn't call it SETI; that came later. We called it interstellar communication. Cyclops was the name Barney Oliver chose for the summer project.

And we completed it. I remember the minute we finished, Barney went off to a meeting in Vienna of the International Astronautical Federation and gave a paper on Cyclops there. Then he worked with some of the faculty guys, but mostly by himself, through the following year, on honing, refining, and polishing the Cyclops report. Then it was published. It is NASA Contractor Report No. 114445, "The Design of a System for the Detection of Extraterrestrial Intelligent Life."

That was the benchmark of SETI. It was very widely distributed. A lot of people wrote for it. I would say fifteen or twenty thousand copies have gone out in the last ten years, many to technical people. It's a very highly technical report. It is the foundation on which everything happened since. It is the ultimate system, the result of a no-holds-barred study. Our faculty fellows were charged with designing a system that they hoped would stand a very good chance of doing the job. That's what they did.

To summarize, when did you first think about the possibility that extraterrestrial intelligence might exist?

It was in the late sixties, when I began to realize that there were books like the interstellar communication book, edited by Al Cameron, and in particular, the Sagan and Shklovskii book. At the same time I was very much in contact with my colleagues here at Ames, who were in the exobiology business, and thinking about life outside the earth, although at that time none of them dealt with the subject of *intelligent* life. They were concerned with much more primitive life, or microbial life in the solar system. But "in the sixties" is the answer.

It was really the combination of those two things. I was already interested in extraterrestrial life, but not intelligent life, and then I read Sagan and Shklovskii's book and realized that there was an even more important question than the existence of primitive life: whether other civilizations exist.

Sagan and Shklovskii's book did the trick?

Yes, that was the real catalyst, but there was some fertile ground, in the sense that I had become intrigued because I was running into colleagues every day, in the same building, who dealt with exobiology.

I began discussing it with other people after I read Sagan and Shklovskii's book. I talked first of all with Max Anliker, and then I had that crucial discussion with Hans Mark in late summer of 1969, which led to Cyclops.

Where we stand today is that we are doing something called SETI, which doesn't even begin to approach the simplest version of Cyclops. In a sense you could look at Cyclops, although we didn't at that time, as being something which might happen in 1995. We are doing things now on a very small scale, in which we use existing radio telescopes to do the search, whereas Cyclops was a new entity; you would build your own radio telescopes.

What we are doing first is to design search systems for use with existing radio telescopes because we might find the signal with existing radio telescopes. It's the logical first step.

Since Cyclops we've gone through many years of trying slowly to get SETI established as a respectable ongoing research program, and to do such a program with existing radio telescopes, which you fit out with a very sophisticated data-processing system. So that as you search the cosmic haystack, your chances of success are much greater than with any other sort of way you might do it, including accidental discoveries by very tiny individual searches like Ozma. Since 1976–77, when Bruce Murray of the Jet Propulsion Laboratory became interested in SETI, we've worked jointly with a SETI team of scientists and engineers from JPL.

In a nutshell that's the story, and the SETI paper in the *Life in the Universe* book will give you a summary of what we've done in the last ten years. It's the technical outcome of ten years of thinking and ten years of study. Those ten years were packed with numerous technical challenges, problems, blind alleys, arguments, disagreements, emotions, and interactions of different sorts with different people and different organizations.

People back in Washington could give you a feel for how they viewed it from their end: how to deal with this strange, wonderful, new breed in California. Noel Hinners, for example. We had many running battles with Noel but also he did quite a lot for us. He was in charge of the Office of Space Sciences during a crucial period for SETI, which was about 1976 through 1979. Three or four years. Andy Stofan has done much for us recently; and also Jerry Soffen, Director of Life Sciences, and particularly Don DeVincenzi, Chief of the Exobiology and SETI Programs at NASA Headquarters.

Hans Mark left here about three years ago to become Secretary of the Air Force. He is now the Deputy Administrator of NASA. He knows SETI very well because, as I said, he was in it on day one as far as NASA was concerned, which was the day he and I met here in this building in 1969 and agreed, just the two of us, that we were going to do it. For many years he was the Center Director here, until '78. He was the person who had to say "yes" to all my strange doings in SETI. He helped us immeasurably.

I would have liked even more help but a lot of people looked at us very skeptically. It was new and smacked of science fiction and was so different from anything else NASA was doing.

That's an important point. NASA is very conservative. What NASA likes to do is sit down and rigorously work out a mission which will go into Earth orbit or to Jupiter and on to Saturn and through the rings of Saturn and on to Uranus and on to Pluto. NASA wants to be able to say, to the second, when it will arrive at each place, many years in advance, and also to say with uncanny precision what sorts of data will come back:

"We're going to do experiments on plasma fields, magnetic fields, surface temperatures and organic compounds, and so forth. We don't know exactly what those results will be but we know we're going to measure certain things. We know we're going to get data back. The experiment will be concluded on the 4th of September 1997 at six o'clock in the morning." NASA would detail the exact sequence of events leading up to preparation of a report, then do it all as planned, then go on to something new.

But the difference between SETI and typical NASA programs goes deeper than this. SETI is an idea, whereas planets and spacecraft are objects. You can see Jupiter from telescopes. You can see Saturn from telescopes. We know they exist; we know a lot about them. We know about the dynamics of space craft. We know how to send messages backwards and forwards, within the solar system. It's all planned and beautifully worked out and orchestrated.

But with SETI there is nothing you can see. We don't even know if there are any other planets out there. We *think* there are; we think there are billions of planets, but we don't *know*. There is no direct evidence of even the simplest forms of extraterrestrial life. The discovery of a signal would be the *first* evidence of extraterrestrial life of any kind. If we had found microorganisms on Mars in Project Viking we'd be in a very strong position and we'd have a much bigger SETI program—but we didn't.

We know nothing about the other people out there. We don't know where to look. We don't know how long to look. We don't know what frequency to look on. We don't know whether to come back to the same source later on to look again. We don't know what polarization to look in. We don't know what sort of pattern of signal we're looking for. We don't know whether it has any modulation on it or not.

When you get through with this list, it's no wonder people begin to balk. The contrast between this and sending a spacecraft that you know will land on Mars within three seconds of four o'clock in the afternoon on a certain day in August 1994 couldn't be more extreme.

And so, over the years we have developed a concept that SETI is exploration more than science. It's based on science, dwells heavily on science, has an underlying philosophy which is scientific in nature, namely that there is extraterrestrial intelligent life, and SETI has lots of scientific benefits. But it is not science. It's exploration; much more like a Columbus voyage than like flying a mission to Mars. This is just one of the difficulties with SETI: the fact that it doesn't fit nicely into NASA's by now classical way of doing things. SETI is strange, strange. You don't know that you're going to get any results at all. It's a high risk, high return endeavor.

I think that every country that has some spark of adventure and inclination to look into the future must do some exploration. That's what successful countries have all done in the past. That's what Spain and Portugal did back in the fifteenth and sixteenth centuries and what England did in the sixteenth, seventeenth, and eighteenth centuries.

You have to be prepared to risk. You're not risking 50 percent of the gross national product or 50 percent of the annual federal budget. A tiny amount of money should go into things like SETI. I think it should be much larger, but I'm built that way. You should spend 5 percent or 10 percent of your funds in research and development on far out things. They are far out because they're risky and they're far out because if they do come off they promise huge returns. So there is also that side of it, in addition to being so different from a conventional mission.

Really, the bottom line is that we don't know what it takes to detect extraterrestrial intelligence. First, we don't *know* whether there is a signal there at all, though we *think* there are many signals. Secondly, we don't know what its characteristics are, so our detection systems have to cater to a large number of possibilities. The search could take us five years. It could take us fifty. It could take us five hundred. We have to face those blunt, hard truths.

In the last analysis we are most unlikely ever to detect signals from other civilizations unless we make a concerted effort to carry out a search program, using our best detective abilities and our best technology. Then we *do* stand a chance.

And SETI is different also because it tugs at many people's basic ideas or preconceived notions or rather fundamental questions that they've always asked themselves and never have been able to answer. People get very emotional about it. They get polarized. They hold all sorts of beliefs and very strong feelings and concepts about SETI, so you have to deal with that, struggle with that, all the time.

It's politically quite sensitive. One always runs the risk of being classified with the far-out, fringe communities. Establishment figures tend to look upon anything new in that way, and SETI in particular, so we have to deal with that. There's the perennial business of wasting government money on far-out ludicrous projects. Indeed, we got a Golden Fleece award back in 1979.

Last but not least, we have to deal with the UFO. "How are you guys related to UFOs?" "Are you looking for UFOs?" To which our answer is very clear. "We are not related to it. We don't have anything to do with UFOs at all."

So there is a list of things about SETI which puts it in a totally separate arena from the vast majority of things that NASA does and any R and D enterprise does. That's part of the excitement, of course. It's one of the things that makes it so fascinating. It's so different.

Another appealing aspect of it is to successfully get across the notion and the concept and the idea and the approach and the planning to other people. That's what I grapple with, that's my top priority. When it comes to action, how do you get things done? What do you need to do? The top priority is to convince critical numbers of people, in critical walks of life, that this is a reasonable enterprise for an advanced technological society.

What effect, if any, has your interest in ETI had upon your career as a scientist?

Well, it produced a fairly drastic change. Up until about 1972, my job was chief of the biotechnology division here, which was the division which undertook research and development activities in aerospace medicine, aerospace physiology, bioengineering and aeronautical human factors. Those were the things I was working on until I began to develop the interest in extraterrestrial intelligence.

Then in '72–'73 I took a year off from that job to specifically undertake the study of how we might get a SETI program put together and underway in the agency, and to study some of the science and technology of SETI. During that time it began to emerge that it was a not unreasonable thing to try to do. And at the end of that time it dawned on us that it would make a lot of sense to meld any future work that we might do in SETI into all the existing work that had been solidly established at the Ames Research Center, and in the university community on the outside, on exobiology. Exobiology is the study of life outside the earth. And the exobiology division had been working steadily for fifteen years building up the underpinnings of other good solid scientific programs.

At that time, of course, the great excitement was in the Viking mission to Mars, which was in preparation. I was invited in '74–75 to head up the exobiology division here and incorporate SETI studies into its structure. So, that was a very interesting change: from aerospace medicine and aviation medicine to extraterrestrial life.

What was the origin of the acronym SETI?

Early in 1975 Robert Machol, on leave from Northwestern University, with SETI at Ames, launched a campaign to decide on an acronym for the project. After much discussion during that year, we brought the matter up at the SETI Science Workshop at Arecibo in December 1975. Out of a number of submissions, the one which received most support was proposed by John Wolfe: SETI with an S, standing for the search for extraterrestrial intelligence. It finally got the blessing of everybody at the meeting, and we adopted SETI as the NASA acronym.

At that time the acronym in use internationally was CETI with a C, for communication with extraterrestrial intelligence. We left that alone, since it was already in popular use. Then, over the succeeding years, what happened was that in the international sphere it was also discussed, at international meetings and the like, and so that is changing, too.

What do your colleagues think about your interest in SETI?

Well, as is the case with many people who have thought about SETI, opinions vary enormously. I think some people still feel that it's a rather far-out topic and really not quite that respectable, and indeed that it doesn't fit terribly well into the properly established exobiological establishment. Others feel the reverse. That is, that it offers the only real way, at this time, to carry out a search; in other words, to achieve a detection. There are now no space programs on the books to go back to Mars to look for further evidence of life. So it's very mixed, opinions that people have about SETI.

What about your family?

They're quite intrigued by the whole business. It's more, I think, from the fact that I occasionally appear in magazines or on the evening television shows. But they have come to some of the lectures I've given.

What about your father?

He's greatly intrigued by the stuff we're doing. Of course, the technical side of it is a mystery to him. He's now retired, a senior English teacher. In fact, I recently sent him a copy of the *Life in the Universe* book, and he was just fascinated by it.

What about your mother?

She died a long while ago, 1943.

What about your superiors, bosses, and so forth?

There are different opinions within the structure of Ames and NASA, as always happens. The majority, who are in the line organization, are in favor of us undertaking a modest SETI program. They've been convinced over a period of many years that it is a valid scientific enterprise and it's a very appropriate thing for NASA to do. The result is that, in spite of the Senate's denial of funds for SETI last summer (which means that at the moment we're in the termination mode of SETI) the agency has gone back to the Congress and asked that SETI be included in the budget in the next fiscal year, '83.

It takes a lot of very strong support within one's agency for them to do that, when Congress specifically denied money the previous year. That probably gives you the best idea of the strength of the support that exists in the agency, although people's opinions vary all over the map. Yes, some are very enthusiastic; some are just mildly enthusiastic; one or two others are less than enthusiastic. But they're at the extreme end.

Openly hostile?

No, no.

What about your friends? What do they think about your ETI interests?

They find it greatly intriguing. It's a very unusual thing to be involved in. Remarks vary all the way from people who are so fascinated that they sit there and stare in astonishment and say, "You really get *paid* for doing that?"—in other words, it's something that's so exciting you should really be paying to do it—to people who still sort of think it's out on the fringe of respectability, to people who can't understand how one can actually spend time studying an area where there is no subject matter, to people who think that it's the greatest thing since sliced bread. Again, it's a complete mixture.

Before going on to discuss your ideas about SETI, let's get a little more background information. What was your father's occupation?

He was a schoolteacher. He taught in a grammar school which I attended. It was a very high-quality English grammar school. He was the senior English master.

How much education did he have?

He went to Oxford just as I did, and obtained his master's degree there in English.

What about your mother?

She did have some college courses. But I don't remember just exactly how far she got.

Did she have occupational skills or work experience other than that of the typical wife and mother of the time?

No.

How would you describe them?

Fairly much on the quiet side. Not really outgoing and gregarious people. Although, I think, very normal people.

How did they influence your eventual choice of science as a career?

They didn't. They just encouraged anything that I sort of latched onto. They did encourage me, of course, to do well in school, and I had all the usual support and inducements to do well. I recall that as a child, age five, six, seven, I would spend a tremendous amount of time reading encyclopedias. That was very stimulating because in the encyclopedias was a wealth of fascinating and challenging knowledge about all walks of life.

My decision to go into medicine was due to something quite different. I had an uncle who was a gynecologist and obstetrician consultant in the north in England. We used to go there for vacations and he was one of my boyhood heroes. I decided at the age of twelve, thereabout, that I would go into medicine, to try and emulate some of the things that he could do.

Were there other people we haven't mentioned who influenced your ideas about science and ETI?

We could divide it into two categories, one being science and one being the ETI stuff. In regard to science, there was my biology teacher at school, who was just one of the best teachers that I've certainly ever known. He was Walter Wheeler, and he edited the standard textbook in England on intermediate biology, and happened to be a biology master at school. He's one of these people who inspires his pupils; and he was probably the greatest influence. That was intermediate and high school.

My tutor at Oxford was Dan Cunningham. He helped me to understand the research world, and that was very valuable. And then there was this long period where I was a specialist in aerospace medicine, both the aviation business and the space business, where the ETI thing didn't come in at all. No, I honestly think it was Sagan's and Shklovskii's book that did the trick.

What religious beliefs did your parents have?

Church of England, but really not terribly strongly religious. There really wasn't very much in the way of religion.

How many brothers and sisters did you have?

I have one brother. He's younger than I am. He's in the British Royal Army Educational Corps. He looks after education of all types within the army.

What about your friends when you were growing up; did they share your interest in science and ETI?

No, I didn't have any interests in ETI at that stage. But interest in science, yes, very much so. I tended to associate with friends and colleagues who, although not exclusively, were fascinated by biology in particular, but also by chemistry and physics and science in general, both at school and in college.

What were you like during childhood and adolescence?

I would say somewhat on the outgoing side. Reasonably gregarious.

What other professional interests do you have, apart from SETI?

My professional interests are in the whole subject of life in the universe. SETI is one facet of that, only one facet. In fact, the division here that I look after has this large number of other approaches to the whole question of the origin and evolution, the nature and distribution of life in the universe. So I have many professional interests in those other things; namely, chemical evolution, the origin of life, the evolution of simple and complex life, and so on.

In addition to that, of course, I've professional interest in the things which I used to do before doing exobiology, and that is aviation medicine and space medicine and things like life support systems and human factors—even sociology.

What hobbies or recreations do you enjoy?

Hiking, cycling, skiing, photography. Those are the main ones.

What has been your greatest satisfaction in life?

Well, on the professional side, the greatest satisfaction has been to be able to contribute a few things to space science, as a rather broad thing, and to have had the opportunity to do it right at the time when I was ready and the space program was ready.

As for other kinds of satisfactions, there are the usual things like having a stable marriage, raising a family and seeing them grow up and graduate and get jobs. There's a tremendous amount of satisfaction from that.

What has been your greatest sorrow in life?

There again, there are these different aspects. On the professional side, my greatest sorrow is that I did not study more of the hard sciences, by which I mean mathematics, physics, electrical engineering—those sorts of things. It's simply that I find myself not as well-equipped as I would like to be to deal with some of the broad questions that we're asking in the space sciences, and in exobiology and in SETI in particular.

What was the most memorable moment or event in your life?

There are really far too many to single one out; it is very difficult. Once again it depends on whether we're talking about the professional or the personal side of life. Getting married and the children being born and the children graduating, and these are all the usual milestone type things.

On the professional side, though once again there's a series of things, I guess the one which stands out from early days is winning an open science scholarship to Oxford University against enormous competition. And receiving the letter: that was really quite an achievement. Getting one's degree at Oxford, and going through all the formal ceremonies, which are really quite something there. Winning an exhibition to Guy's Hospital; an open exhibition, which is like a scholarship. It's one of the great London Hospitals, where I went to do my clinical work.

Then, of course, getting my medical degree. You'll find that all M.D.s have very vivid memories of their first year dealing with patients, so there were a lot of memorable occasions there. Being accepted to go to the Institute of Aviation Medicine at Farnborough. And then many memories of exciting research activities there.

Then the great transition to the States; in a sense, that's one of the most memorable things. When you leave your country of origin and you decide to emigrate, it produces quite a lot of memories, obviously, and it's a time of some stress. And then being involved at Houston at the onset of manned space flight, and being responsible for a lot of things that went on there.

Turning now to SETI, about what fraction or part of your working time do you spend on it?

Fifty percent.

What stages has your thinking about ETI and SETI gone through?

Other than what?

The present plan to go with multichannel spectrum analyzers in the waterhole.

Oh, O.K. We call that "microwave observing program," or MOP. We have taken a look at other possible ways of detecting ETI's, including things like observing in the ultraviolet, observing the visible part of the spectrum, looking for streams of particles, and last but not least, looking in the infrared region. Of those, the most promising is the infrared, as far as we can see. We've given quite a bit of thought to that, but at the moment there's a lack of really suitable instruments or detectors for the sorts of infrared signals that we would be looking for.

And in any case, we still think, by and large, that it compares in its probability of success with working in the microwave region. So we find ourselves always coming back to the microwave program as being the top priority. But we are very careful not to discount other possibilities and not to completely throw out quite different approaches in the future if they should emerge as good candidates.

But we have been studying this now for ten years, and we don't really find anything else that looks very promising.

What would you do differently if you were starting over again to study ETI?

I really can't think of anything, other than a few fine details, that we would do that was significantly different. I think we've done the right sorts of things. We started with a very thoroughgoing, broad study, which was Cyclops. It showed that the job could be done in a very thorough fashion if you really wanted to go all out.

And then we began to face the reality of translating the SETI idea into a real program, funded by a real agency, and with real people working on it, and leading towards a real search. You always go through a similar sequence of steps when you try and do something like that. And the first thing you do is establish a nucleus of people within an agency, technical people. The second thing you do is try and establish some support from the outside, that is, from the scientific community. And the two groups then work together.

First you have a series of science workshops to explore, exhaustively, every angle, which we did in '75 and '76, and that's the SETI Blue Book. And then after that, what you try and do is put a program plan together in the context of the agency structure and the agency language, which we have done. A long time ago, we got our support within the agency, that is, from NASA. But as you know, we have been having problems with the Congress over a number of years. But I hope that that won't last too long, and we are going to be trying again. So, I can't see anything drastically different that we might have done.

Why is ETI important to you?

It's the most exciting of all human endeavors, and it's very challenging because we have to do everything indirectly because we have no direct evidence. So it's that magic combination of exploration and science, which is the type of thing that Darwin did in his voyage. It seems to me that such a discovery would be one of the achievements of our own civilization. And to be a part of that and to contribute to it in some way is the thing which appeals to me.

What has been the most difficult aspect of your ETI work?

That's an easy one to answer. The greatest problem has been to persuade everybody in the system, whose support you need to get the program funded and backed with the proper resources, that this, in fact, is a valid and exciting enterprise. I don't just mean a percentile within the agency, I mean within the community in general, that is, the scientific community; and, of course, in Washington.

It's an extremely tough problem because a lot of the people one is dealing with do not have the technical knowledge and do not have the scientific knowledge, and have to be persuaded through the various channels of the pluralistic system in which we live. Nobody can arbitrarily make the decision, "We are going to do SETI, and it shall be done." In this country you have to get a base of support which is fairly wide.

And since this is still such a new idea, and since the technology is still brand new, it just takes a long time to sink in. And it's a complex situation; it's not an easy one to explain to people. The result is it takes years and years of careful, patient explanation and argument to convince people that we should be doing this.

The technical side, by comparison, is enjoyable and easy and straightforward, even though it's breaking new ground all the time. Our biggest problem is people.

How do you think most scientists today view SETI?

I think most scientists will agree that SETI is legitimate and important provided it's done at a reasonable funding level. What they do not want to see, and this has become abundantly clear in the last several years, is a very large-scale SETI program, on the scale of Apollo, or some very large national enterprise. *Most*, not all, but most are now reasonably content to see SETI supported and flourishing at this modest level.

In fact, that is exactly what the Field Committee has just said in its report. The Field Committee is the Astronomy Survey Committee for the 1980s. Each ten years the astronomers of the country get together and have a two-year planning marathon on what should be done in astronomy over the next ten years. And they, without being requested by NASA, undertook to study SETI as part of the variety of different possible programs that they were considering for the next ten years. They've come out with a very positive report, which is why I'm able to say the things that I do now about where the scientists stand.

What about public attitudes toward SETI?

It's been our experience that the public is a good deal more enthusiastic than scientists are. Scientists, by and large, are very conservative people and want to be very sure before they embark on something. The public, on the other hand, doesn't have too much patience with that sort of hard scientific attitude and is a great deal more interested in things like exploration.

SETI is only part science. It is literally part exploration. Therefore, I think it has quite an appeal to the public. We've always found in talking to the public that it is of great interest, and is now widely recognized as being something which has been contemplated by different countries as being an international program.

It's very hard to quantify, but the various Gallup polls, which they have at about three-year intervals, confirm that there is an increasing number of people in the United States who believe in the existence of extraterrestrial intelligent life. It's now above the 50 percent mark, and climbing.

The interest in the scientific community is considerable but you have to distinguish between interest and support. In the case of the general public, there is both interest and support. In the case of the scientific community, there is great interest. Let me give you an example. The recent COSPAR meeting, which is the big international space meeting, had a paper on SETI by Frank Drake. Their meetings went on for two weeks. They had hundreds of papers given, and all sorts of sessions on them: space science and the planets and astrophysics and everything you can think of. And the best attended was the SETI paper. That's been our experience again and again and again.

That's why I say "interest." But if you turn around and ask different groups of scientists to *support* SETI, then it can be a different story, because then they suddenly get conservative, or can get conservative, some of them.

There's always this element when you're trying to introduce a new program into the National Science Foundation program. The disciplines closest to the new one that you are trying to introduce fear competition. And those are the ones to which most people turn for an assessment of whether your new program should be introduced or not. It's a psychological response to the introduction of new programs which, in a democracy, is one of those factors which holds back the introduction of new programs.

You see this in academia all the time. We see it all the time, too. When you're out on the cutting edge of science and technology and trying to introduce new ideas as opposed to bringing in rather standard science, which is well within the established norm, you continually run into this problem. "Don't let's vote for anything which would stand the slightest chance of taking even $10 away from the funding of some well-established, ongoing program."

It's difficult to know exactly how important that's been as compared with the other problem, which is simply that people have to be educated about SETI and haven't really understood it properly.

There's one situation we face year after year: If you try to explain SETI in a half an hour to any group of scientists who have not previously been acquainted with SETI and its ideas, their reaction is one of mind-boggled astonishment at the whole thing. It's simply that the concepts are so deep and broad and new, that it's hard to get hold of.

I've seen it so many times now that we are very reluctant ever to go and talk to a really important scientific audience for the first time, and spend only a half hour on SETI. So we try, if we possibly can, to make it half a day. Or we try and prepare people beforehand by sending them some of the publications and the reviews and reports. But it is an interesting phenomenon.

You're referring to a scientific audience?

Yes. We don't have that trouble with the lay public. In fact, the public seems to grasp the broad picture much more easily. I think it's because the scientists are always trying to find out what your evidence is for the concepts that you're presenting. But the public is less critical, by and large, and will accept the concepts which you're describing, without asking, "What control did he use?" "Has he proved that?" "Did he do a double blind study?" This is the classical scientific approach to things; it is hard science, which is fine.

But SETI is different. SETI is more exploration than science. But the scientists tend to think of it as being science, because they're being briefed by a scientist and they are scientists, and that's the way they think all the time. It's hard for them to see things in terms of exploration as well as science. But not always. There's always some who understand this right away.

We deal with three constituencies, really: the general public, the scientific community, and the Washington political community. The general public is much the easiest; the other two are very difficult. The political community want to know first of all whether the scientific community endorses the thing. That's usually the first question, which I think is a good way to operate. Now, at long last, after ten or twelve years, we're in a situation where we can say, "Yes, there *is* support for this as a legitimate area of scientific endeavor." Now we have to convince the politicians that this is not a waste of public money.

This may be a good time to clear up the recurring issue of UFOs. You mentioned it briefly. What do you think UFOs are?

I believe UFOs are exactly what the words say, and that is something which is flying in the atmosphere which is not identified. That's literally what the words mean, and that's my own personal interpretation of UFO.

The feeling that I think all of us have is that if you look into it in some detail, the vast majority of sightings of UFOs can be explained as being ultimately IFOs, which are *identified* flying objects. In other words, they turn out to be balloons, aircraft, spacecraft, strange lighting conditions in the sky, rockets which are blown up as they are drifting off course, and on and on.

You would say that they are basically misinterpretations of known phenomena?

Yes, and there are other things to add to the list. For example, the eye plays tricks on one. We know that very well. Occasionally, we have circumstances where it's something in the eyeball. The brain also plays tricks. There are a number of different categories there, all the way from something which people think they see but which is not really there; something which is fabricated or manufactured by the brain— this is a common enough thing in psychology and psychiatry, and it's called illusion—all the way to the extreme end of the spectrum, which is a frank hoax.

Going beyond that, let us suppose that clear evidence of extraterrestrial signals is received; the kind of evidence you've been seeking all these years. What should a scientist do with that evidence?

We have been doing a little work on that because the question has been asked of us many times in recent years, and the frequency with which we face that question is going up. Our answer is fairly straightforward.

If you have a signal you believe is likely to be of extraterrestrial origin, the very first step is to bring all your scientific, technological knowledge—the knowledge of many people—to bear on confirming that it really is of extraterrestrial origin, because the vast majority of signals that we see are going to be something which is masquerading as an ETI signal: all the way from weather balloons to a new spacecraft which has just gone up, to reflections from the moon on radio signals, to new astrophysical objects not yet discovered, to glitches in one's electronics, to finally, at the extreme, a hoax.

We think the worst possible thing that anybody could do would be to announce that they have discovered another civilization only to find out, as Barney Oliver says, that the Cal Tech students have chalked up another success.

We are very serious about all of those, particularly the hoax thing, and as we gradually complete the design of the signal detection system, we begin to make quite sure that it includes as many techniques as we can possibly incorporate to exclude the possibility of a wrong diagnosis.

If we've gone through all these steps, then the next major step is to call another observatory 5000 miles away and say: "We have a signal possibly of intelligent origin; the coordinates in the sky are so-and-so, the frequency is so-and-so. We know that your telescope could detect this; you have the right receiver system, the right signal analyzer to detect this. Could you please confirm?"

It is possible that we may make such arrangements in advance with other observatories. It would make a lot of sense.

Then, if *they* confirm, it gets much more likely. Ultimately, if we are really convinced, the ground rule that we've adopted for ourselves, although it is not yet formulated in any elaborate document, is that this information is distributed as quickly as possible to everybody. That includes the scientific community. It also includes the public, it includes leaders in all nations. One of the ground rules is that the story must be told specifically and clearly and accurately, because people can misinterpret things, and everybody should get to know about it. So that's our position.

What do you guess would be the public's response to this?

It will vary considerably between different individuals and different groups. A very large fraction of the population, certainly in first world countries, is well enough educated to have had some exposure to the possibility that we are not alone, and to those people it will come as no great surprise. It will, of course, evoke great interest.

To certain individuals in certain groups it may come as a surprise, and then be accepted as an interesting new development.

Some people may have difficulties integrating it with their own philosophy of life or their own religious beliefs.

One of the difficult things is that the signal may come in a number of different possible forms. It may come simply as a bleep, bleep, bleep as from a radar beacon, or it may come with a complex, lengthy message attached to it. Both would be fascinating new knowledge, but one would be much more than the other.

We did discuss this when we had our original science workshops on SETI, and there are some remarks in the SETI bluebook about this. If it were a message, it most likely would take a long time to decode, and even to receive; it might come slowly.

Yes, there would initially be enormous interest. It would clearly be a dramatic new discovery, and it would affect people's lives and people's thinking. However, it's a very remote thing. It is not something which is going to change our lives here overnight or within a week or a month. After the initial excitement has died down, people will by and large return to ordinary terrestrial problems, of which we have more than enough, that they have to deal with in their daily lives.

It won't be until years or decades or even centuries later that the full impact will be felt and the full influence of the new knowledge, whatever it might be, is absorbed and assimilated and appreciated and has soaked into either our own existence or the body politic of the people of earth. In that sense it's a little like the Copernican revolution, which did not really affect the lives of ordinary people very much until many decades, or even centuries, had gone by.

Looking a century or two ahead, do you have any thoughts about the long-term effects of discovering ETI?

Yes, I do. Again it's a personal opinion. The first aspect of this is that any civilization we detect must be older than ourselves. It is likely to be very much older.

Supposing they are ten million years older. It is likely that they will be in a condition vastly different from the condition in which we find ourselves. They have achieved a modus operandi or modus vivendi which has allowed them to develop a society which is long-lived and clearly stable, because they have existed as a civilization for ten million years.

That knowledge in itself would be tremendously useful to the way in which we think here, about our own future. The reason is that it tells us it is possible to survive as an intelligent species over very long periods of time. That would change the opinions of many people who feel very strongly that we only have another twenty-five years to go here, or that we are subject to a long period of conflict and disease and war and pestilence and the like, because we can't somehow manage to live together in peace and harmony in some sort of organized society which people can agree on. So that's an important thing.

But I think we can go one stage further. If we indeed can understand what their history was over their last ten million years, it's in a sense a way of looking into some possibilities for our own future. We could perhaps learn from people who are more advanced than we are what sorts of societal structures hold out the best hope for stability and longevity, and conversely, what sorts of societal structures are likely to result in the opposite: the disappearance of a society.

I say that because it is possible that the civilization we detect may know of other cases, may be in contact with other civilizations, and may have learned something of the history of different types of societies in different places in the galaxy, some of which were destined to be very successful and some of which, as Carl Sagan says, "suddenly went off the air" and met an untimely end.

Some people are a little afraid of new knowledge, and they say, "We have enough problems here; why should we go messing about with advanced civilizations that we know nothing of? Let's just stay here and mind our own business."

That's a little bit like closing our eyes and ears and acting dumb. We never hesitate to go to experts in a certain field to learn more about ourselves, because we want that knowledge and they have it. A student doesn't hesitate, by and large, to go to the professor and say, "I want to learn your opinions about so-and-so." I look at it in the same way.

Another aspect of it is very interesting: the very knowledge that there is another civilization, another type of species, or many of them, might do something to bring our own species together and encourage a more cooperative approach than the very disparate types of behavior we see today. I'm not saying there is any magic answer, but some people are thinking that this other civilization might constitute a threat, and that this threat could be a unifying factor for us. While the thought of somebody seventy or a hundred light years away being a threat is not very realistic, it might have some unifying effect on us.

Those are some thoughts on the long-term implications. There is one other and that is the question of communication. What is meant by *communication*? We tend not to use the word "communication" because it is so often misinterpreted. It is used in very different ways.

Our purpose right now is simply to *detect* signals from the other species. If we are successful in that, it would be one-way communication, them to us. And if we detect a signal, after we had a chance to absorb it, questions are going to arise as to whether we should reply. And if so, what do we say? Who decides whether we should reply and what we should say?

We've been thinking a little bit about these questions lately and had some discussion about them in the one international forum where SETI has been discussed for twenty years, and that is the International Academy of Astronautics. We have raised with the people in that group, and the affiliated group on international space law, the question of whether it is worthwhile exploring any type of international discussion, perhaps later to be followed by some sort of international agreement, on the question you asked earlier: how to deal with the information once it has been obtained? Is there a mechanism one wants to get everybody's agreement on, for wide dissemination of that information?

And the second question is one I just raised, which is whether it is worthwhile having international debate on how to deal with the question of whether we should reply and, if we do, what to say, and who decides what to say?

We're having some preliminary discussions. This is not a NASA thing; it's an international society where we're beginning to bring up such questions. We will have a session devoted to these questions at next year's meeting of the International Astronautical Congress at Innsbruck. It's still not something we devote much time to, because we are so deeply involved with the scientific and technical aspects of detection.

Realizing that it is only a guess, when do you think contact with ETI will be made?

That depends entirely on the resources that are put behind the program and the skill with which the SETI program develops over the years, so it's hard to say. If nothing's invested in SETI, then contact will not be made, or the probability will be vanishingly small. If there's a reasonable program established, a few million a year, as an ongoing activity, then I think the chances are good for contact before the end of the century. The chances are high enough to make everyone involved in SETI enthusiastic about something happening within their lifetime. I'm not saying it is a big chance, but it's big enough to make people enthusiastic about maybe being able to see the results of their endeavors.

What might ETI be like?

First of all, as chemists and biochemists, at the moment we have no models for life structures which are based on anything other than DNA and protein and the sorts of things we know and understand because that's what we have here. So our supposition is that if you have another Earth, going around another Sun, somewhere else, the origin of life will take place in something like the same way it happened here. And then you might finish up again with similar types of molecules, based on carbon.

We take this view for three reasons. One is that we're not clever enough to imagine anything else. The second is we know it happened here. The third reason is that we know that the chemicals out of which we are made, and all living systems on Earth are made, are also abundant elsewhere in the galaxy.

You can put those arguments together and say that life based on carbon is self-replicating with some DNA type of molecule, and protein types of molecules are very likely to have developed somewhere else. Once that's happened, then Darwinian evolution takes over, and you gradually get an increase in complexity, and you get the formation of some sort of nervous system, increasing control of the internal environment and the external environment. And ultimately, of course, the gradual association of individual organisms to make up a culture and cultural evolution, which is the standard story.

Something like that has to happen elsewhere. It doesn't have to be exactly the same, of course. But, if those things happen, and if the other Earth is something like our Earth, then the intelligent creatures are going to have some of the characteristics that we have, at our stage; they must have.

They must have, first of all, mechanisms for sensing the environment, and they have to be fairly sophisticated, with touch, some type of hearing, some type of vision, smell, whatever. It doesn't have to be all those things, but some combination of those things.

Then the creatures must have some way of manipulating the environment, just as we do. So they must have some sort of end effectors, or equivalents to hands. We don't know how many hands, we don't know how many limbs, but you have to be able to operate on the environment, if you're going to build things and do things and collect food and so on.

And then you have to have a brain of some size to do all the things which brains do. So, those qualities must be present in the other creatures.

External shape and color and all those sorts of things, one can't even guess at. We know that biology is capable of producing organisms of incredibly wide diversity; all you have to do is look around here. And so we are always very careful not to try and speculate on external appearances. We tend to talk about fundamental qualities, or fundamental subsystems or structures certain organisms must have. We try strenuously not to be talked into drawing them or coloring them or describing a whole individual, because it can't be done.

There's one other wrinkle in this whole situation, and that concerns artificial intelligence. There is a lot of discussion at the moment whether one can ever manufacture a device which is not a human being, but which has the same sorts of characteristics and behavior patterns, or even more advanced characteristics and behavior patterns, in terms of intellectual capacity and ability to do things, and so on.

Some people say it's impossible, it can never happen. We could never, ever reproduce the sublime complexity of the human being in pieces of machinery. The idea is ridiculous.

Others think quite the reverse. They think that biological systems are pretty lousy in many ways, and very slow at processing information, and have lots of drawbacks, and that properly designed machines could do the job much better.

I personally don't see any reason why an advanced civilization couldn't put together very sophisticated machines which are intelligent. These would be nonbiological systems, although they are built by biological systems. And indeed, I basically don't see why such machines shouldn't ultimately be capable of propagating themselves, as Von Neuman described in his theoretical work, and as people like Tipler and Hart imagined such machines to perhaps colonize the galaxy in various ways—although *that* I'm very skeptical of. But I don't see why you couldn't put together such machines. I don't see any fundamental laws which deny that capability. It may initially be a very large machine. I'm reminded of the evolution of the computer.

So I can see the possibility of a far-out scenario. The Earth will be destroyed in about seven billion years, by the sun becoming a red giant. Life as we know it here will also be destroyed. Everything will be fried and all water will vanish from the Earth. Conceivably, in such an environment, a machine designed to withstand such wildly different environmental circumstances might continue to exist as some sort of intelligent structure.

It's conceivable that if we detect a signal from somewhere else, it could be a signal being transmitted by an intelligent machine. The machine might have been left there in situ, with the job of transmitting and sending out its message when its biological predecessors have left for some reason, or even become extinct. Or, it's more likely that such machines would be operating in parallel with their biological counterparts.

I don't see any inherent reason why you can't make machines which would be just as capable as people are. If you belong to one particular school of biologists, you look at human bodies as machines anyway. So it can clearly be done; it's just that it took four billion years to do it nature's way. And the big question now is whether it can be done in a rather different way, artificially, by those same beings.

Speaking of unusual beings, what scientists or other people in the past do you admire?

I think Darwin has to come first. He was responsible for the biggest change in human understanding of ourselves and our environment and evolution. It's a bigger understanding than any other single thing. Of course, there's such a long list of scientists, but one has to think about Copernicus, Galileo, Newton.

As for non-scientists I have to include Shakespeare, who had probably a broader grasp of psychology, sociology, and the understanding of human behavior than almost anyone else. So those are a few names for the past.

Are there people today that you particularly admire?

Phil Morrison comes very high on my list; a very unique individual in the twentieth century. He's a man for all seasons, a Renaissance man, which is extremely rare. To the point where, as an astrophysicist, we invited him to be the chairman of a science workshop, which we held recently, on the origins of life. He's that broad.

There are a few other people like him. I would personally put Barney Oliver in the same category. I'm terribly fortunate to work with people like that.

Who else do you consider to be among the outstanding people in SETI today?

The people at Stanford who are building this first machine for us. They're very important, absolutely central to the whole thing because that's where the real action is, in the sense of the hardware that you have to have to do the job. They are a little bit like the guys who built the Pinta and the Santa Maria.

I think we're all hesitant to view ourselves as being part of Columbus's crew brought up to today; the analogy isn't quite the same. Nevertheless, I think deep down we probably do view ourselves this way. I think Barney Oliver views himself a little like Columbus, and it's not a bad analogy. Morrison may be the person who develops the same sorts of ideas that Columbus really drew on, way back, in constructing his own concepts of his voyage.

I view Morrison as a sort of a current-day analogue to the earliest scientists, who worked in the centuries before Columbus, who began to get that first glimmering of the fact that the earth was round, and who worked with measurements of the stars and sun and moon to deduce the fact that the earth was round, and make those first early suggestions.

Drake I view more as being like John Sebastian Cabot and Leif Eriksen, the people who did the very first voyages of exploration.

I view Barney Oliver as more like Columbus himself: the leader who is going to get something done and is going to do a good job at it, and who is not going to stop at explorations but who is going to establish it, to see the thing all the way through to settlement.

The analogies are not exact, by any means. Oliver and Morrison are both Renaissance men, and there are very few of those today. Morrison is the thinker and Oliver is the doer, but he's also a thinker.

I see quite a bit of Morrison at various gatherings and meetings and so on, but I've never really had a chance to sit down with him and ask him these same questions you are asking. They interest me, too. We did one time invite a historian of science from Harvard to come to Ames to talk with us, and he was fascinating. We just had a pleasant meeting and he went away. We didn't communicate again. He writes about things like the development of clocks.

Yes, historians of science often focus on things long past, such as the emergence of Greek science and the devolopment of clocks.

I do know the picture is changing because I happened to have lunch at the Smithsonian the other day with a key person back there called Kerry Joels. He made a request which to us was a little bit astonishing; not so much to me but to my colleagues at NASA headquarters in Washington. The request was to have an exhibit of SETI at the Smithsonian. My colleagues were very startled. They said museums exhibit things which happened in the past, but Joels said, "No, we are also looking to the future." They're looking to see what things will, in the future, be viewed as important, and hence to put in a bid in advance for relics. "Relics" is the wrong word. Artifacts is the right word. He said, "For example, your first multichannel spectrum analyzer, which is the heart of the SETI. I want it."

It's not even built, but they were already doing a SETI exhibit.

How would you characterize your role in SETI?

It's a mixture of things. It's more as a midwife to the concept or idea. I guess the bottom line, or the heart of the matter, is to change the concept or notion of SETI from being merely an interesting academic study, with a few isolated attempts to actually carry out scattered searches, which it was in the sixties, to something which is much more coherent, much better recognized, much better supported; an entity which has a critical mass, which has enough structure and maturity to enable it to be born as a widely recognized entity and on the road to being a program.

What has been happening is that we've realized, all of us in SETI, that there's no way you can do SETI with any reasonable chance of success if you do it as a spare-time activity, a sort of hobby, with individuals doing it on their own, a one-man show, a fragmented endeavor with a few hours here and there, and a few radio telescopes here and there, and with equipment that's primitive compared to what you need in order to have some reasonable chance of success.

One of my roles is to change SETI from its former situation. It's not a big program at all, but now, at least the way we're trying to put it together, it offers an opportunity for a much more cohesive, integrated approach. Correspondingly, that offers the opportunity for chances of success better by many orders of magnitude, compared with the original project Ozma, carried out by one of the original SETI people, Frank Drake, in 1960. Frank is very much involved with us as we do all this. I'm sure he would agree with what I've said.

We have tried to quantitate the problem. It's very difficult but we have tried, with a diagram that Jill Tarter has drawn, based on Frank's own original thinking, called the Cosmic Haystack.

The Cosmic Haystack is a way of trying to show how primitive those early attempts were and how we think, by doing what we are currently doing, that we're going to make a much bigger dent in the search of space, in terms of probability of success. The person who worked out all those details was one of our SETI people here, a Berkeley person, Jill Tarter. She is a key person and she's been with us for some while now.

Drake mentioned the difficulty of attempting to reach into the haystack and pull out the needle.

What we're trying to do is to make the needle more clearly identifiable, so that our chances of actually finding it are much better than they were before. We're trying also to figure out which parts of the haystack are more likely to contain the needle than the others.

We have written a key paper called SETI, put together by our group at Ames and our colleagues at JPL. It's coming out in the *Life in the Universe* book based on the conference we had here a couple of years ago. The paper is the essence of SETI. It is the translation of all our long-term plans into one very cryptic paper which is designed for general distribution.

For laymen?

Not really. It's *Scientific American* style. I'm the editor, but Charles Seeger and Vera Buescher did a lot of the work assembling this book. The key SETI paper is the next to the last one. It is, I think, the landmark paper because it represents the results of years of extensive thinking about SETI.

In the sixties people were thinking about some of these things but they never sat down and asked one crucial question, which is what I asked in 1969: "Supposing you wanted to carry out a thoroughgoing program to search for extraterrestrial intelligence. How would you do it?" And behind that was a second question: "Once you established what it was that you wanted to do, and that it seemed to make sense to do it, then why not put it into effect?"

I think that's really the role I played. It's primarily not a technical role. Although I understand the technical side, I'm not an electrical engineer or radio astronomer. I understand enough about it to follow it pretty closely but I think my role is a catalyst, or midwife. Just to bring things from that rather vague stage of the sixties—fragmented, intriguing, very academic—to something much more coherent and well recognized, as an entity of its own.

Finally, what else should future historians know about you that would not otherwise be preserved?

After twelve years, it's pleasing to be able to report that SETI is gradually being accepted by people in all walks of life as an exciting and acceptable venture in space exploration. Significantly, we have the endorsement of many in the scientific community, often in formal planning documents. Now we even have a SETI Science Working Group, chaired by John Wolfe of Ames and Sam Gulkis of JPL and composed of distinguished scientists who work closely with us on SETI.

Philip Morrison, chairman of our SETI Science Workshops back in 1975 and 1976, said that people did things in the past when they had the capability to do them. We have reached a moment in time where there is a confluence of ideas and capabilities in the science and technology of SETI. On the science side we have a logical story of the origin and evolution of life in the universe, and on the technical side the ability to be able to communicate over the vast distances between the stars.

SETI is an idea whose time has come.

Charles L. Seeger

RADIO ASTRONOMER

Born October 10, 1912

Berkeley, California

Now working in the SETI Program Office at NASA Ames Research Center, Charles Seeger is the only pioneer devoting full-time to SETI. He contributes to it in many ways, from furnishing electronic expertise in design, construction, and refinement of SETI equipment, to presenting colloquia, public lectures, and TV and radio reports.

He received a bachelor of electrical engineering degree from Cornell University in 1946, and pursued graduate studies at Leiden University in the Netherlands under Jan Oort from 1951 to 1960.

Active in radio since the 1920s, when World War II began he set up pilot production of radar pulse tubes at Western Electric Company, and designed and put into operation a high-power pulse test. Moving to Cornell, for the rest of the war he taught courses in radar, networks and antennas, and participated in various military research projects. From 1946 to 1950 he established the Cornell Radio Astronomy Project and served as its chief scientist. He is credited with a number of instrumental and observational firsts.

During the 1950s he designed and built systems for UHF continuum observations, observed solar eclipses and lunar occultations of radio sources, and produced the first relatively high resolution survey of the UHF sky, showing multiple spurs from the galactic plane.

He was co-developer of the first ultra-stable, high-gain, low-noise parametric amplifiers for radio astronomy and in 1960 obtained the first clear detections of the plane polarized component of the galactic background synchrotron radiation. He led the initial studies for the first very large distributed radio telescopes. He has been a consultant for Hughes, Zenith, N. V. Philips, NEC and other corporations, and his ideas have appeared in many research and engineering reports.

He has held university radio astronomy appointments at Stanford, the University of California, Berkeley, and at Texas, New Mexico, Leiden in the Netherlands, and Chalmers in Sweden. Since 1974 he has been adjunct Professor at San Francisco State University.

He served on the National Academy of Sciences Committee on Radio Frequencies for Scientific Research, subcommittee for Radio Astronomy, 1961–1970, and on NASA's astronomy subcommittee, Space Science Steering Committeee, 1964–65.

He is a member and often delegate of the International Scientific Radio Union and the International Astronomical Union. He was the chief representative of several such organizations at the 1959 United Nations conference which gave radio astronomy international radio frequency allocations.

Other professional memberships include the American Astronomical Society, Institute of Electrical and Electronics Engineers, Nederlandsche Astronomen Club and Nederlandsch Radiogenootschaap, and the American Association for the Advancement of Science.

Interviewed June 1981 in Mountain View, California

Where did you live during childhood and when you were growing up?

After six years, which left me permanently attached to California, we went east and lived in New York City, suburban New York, and Connecticut, where I went to boarding school. I lived mostly in cities except when away in school or in the summertime.

What were your childhood interests?

Beyond the usual ones, I got interested in radio when I was living in New York in about 1922. My allowance was spent entirely on Hugo Gernsback's *Radio News*, and I could often be found rollerskating to the nearest radio store window to look at all the knobs and dials and vacuum tubes and batteries and all the other magic there. It developed into a hobby: building radios, visiting transmitting stations— even the old Navy station at the Brooklyn Navy Yard, where they had spark transmitters until an ungodly late age.

When I had polio in '28 I took a year off from boarding school to recover. My being idle for nine months did not appeal to my mother. She found me a job as an office boy at RCA Communications on Broad Street in the City, with the help of a distant cousin who had at one time been a vice-president for law at RCA. I had a chance to work around the group of early radio engineers that built RCA Communications: Taylor, Beverage, and so forth. They were way ahead of the rest. I learned about office politics, complacent secretaries, and snotty office boys, and the pleasures of Eskimo Pies that hit the streets that year.

At the same time I joined a radio club at the Midtown YMCA in New York. There I met three boys my age who were already earning their living in the radio business. Coming from poorer backgrounds than I did, they had to go to work after the eighth grade. They were already very good radio sevicemen, and working in the Cortlandt Street area.

Cortlandt Street was absolute, pure heaven. It was *the* place in the world to get radio hardware, new technical ideas, and bargains galore. It was also the wildest sales place for radios you can imagine. It was peopled by geniuses, entrepreneurs making fortunes or going bust, con men, auctioneers and rip-off artists.

These friends taught me a lot, and the connection got me semi-skilled work when I badly needed it. I graduated from high school in a wheelchair, having had double pneumonia, parents divorced, father out of work and broke—the whole Depression bit. I had to support myself. After recovering my health, one of these friends got me a job at twelve to fifteen dollars a week, installing radios in expensive cars in a small shop on the Grand Concourse, in the Bronx. I drove a lot in the process; drove some of the most exotic cars, since we always had to test the installation on the road in those days. That was back when cars were made of wood as well as steel, and generated a lot of radio noise.

Radio was my chief hobby and later a livelihood, but my brothers and I did many things by hand. We played musical instruments, built model sailboats, made a hay barn into summer living quarters, and so forth. I worked with metal as well as wood. Father was superb at woodworking, and good with mechanical devices.

In summer 1934 I worked as a guard at the Chicago World's Fair. Connections had something to do with it: I had an uncle who was a press agent for the English village there. When that was over, I joined my radio club friends in Washington, D.C., an island of prosperity as a result of the governmental expansion under President Roosevelt and the New Deal. The four of us worked for the same chain of radio stores and I learned the trade of radio serviceman—so well that when I left in 1937 for New York City, I found a job on my first application, right in the middle of the Depression.

I worked throughout the Depression on my ability to repair radios. I was a natural-born troubleshooter. I had inherited a critical, fault-finding proclivity, and it served me well. I was curious and read all the manuals and such other technical literature as I could make a stab at understanding. Many of the people I worked with were not very well educated and had trouble understanding English and reading circuit diagrams. I soon found myself in charge of the expensive radio installations; in yachts and the like—the cream of the crop. This supported me until a major layoff after Christmas 1938. They laid me off because I was the last hired in the shop and had also been assistant shop chairman in a successful strike for seniority and higher wages.

At that point I was recently married. My mother-in-law asked if I had ever thought of going to college. I had been captain of the fencing team in high school and had received the promise of an athletic scholarship at Yale, but had to forego it because of the pneumonia. When I had recovered, Yale was in financial straits and the financial aid was lost.

So in the summer of 1939 I went to Cornell summer school to see if I could still study. (In those days it was common opinion that one largely lost the ability to learn as one passed into manhood.) I creamed two courses to perfection and in the fall of 1939 enrolled in the School of Electrical Engineering. My wife found a full-time job for five dollars a week, and I found a part-time radio repair job for a radio store just off the campus.

I soon found out that the best job on the campus was operating the transmitter of the university-owned AM broadcast station. During the Christmas holiday I passed the federal exam and received a first-class ticket.

From June 1940 on, I received a dollar an hour to study for twenty hours a week, take care of emergencies, turn the transmitter on in the morning and off at night, and make occasional local-remote switch overs. And once to shut off an announcer who lapsed into inexcusable, blue language.

The station management received its due. I kept my scheduled hours on time, dealt with emergencies correctly, and learned preventative maintenance. A dollar an hour during the Depression; that was almost incredible pay! Then, too, I got to know well some of the electrical engineering faculty who became my teachers, and whose courses I also taught when they were away during the war; even a course I had yet to take for credit. That produced an interesting situation.

In the fall of 1940 I received the only available scholarship in electrical engineering: two hundred dollars, or half a year's tuition. For the summer of 1940 Western Electric/Bell Labs took me on as a junior engineer, in the increased war preparatory effort. I was put to work on a one-man production line to see if we could introduce a newly designed high-power radar pulse tube. I had a very busy supervisor whose main contribution was to steer me through the local politics of the Hudson Street Vacuum Tube Shop.

I was left almost alone for four-and-a-half months before Pearl Harbor, to have a wonderful experience working sixteen hours a day inside one of the world's great bureaucracies, AT&T. I learned to work with glass; with the temperamental oxide-coated cathode; with mercury and with high voltage; to understand something of vacuum tube technology and cryogenics; and to build test equipment as needed.

On my return to school late in the fall of 1941 I was asked to instruct in the Army pre-radar training program. Before the war was over, I was an instructor in the regular undergraduate curriculum. All this delayed my actual graduation a year. But I managed to take more physics than was usual at that time for EEs. In fact, the war provided me with opportunities and experiences that would have been totally beyond the pale in peacetime. Above all, it got me into research.

Because of a tendency to read widely, I became interested in what is now called radio astronomy. While holidaying with members of the "Y" radio club I happened to hear Karl Jansky make his New York announcement of the reception of signals from the direction of the galactic center. Walking down Broadway someone said there was an IRE meeting that evening; shall we go? And we did. And there was Jansky, announcing this breakthrough in astronomy. (The IRE, or Institute of Radio Engineers, was the liveliest of the two predecessors of the Institute of Electronic and Electrical Engineers, the IEEE of today.)

In 1940 a charming young Canadian couple had become our neighbors. Don MacRae had just received his Ph.D. in astrophysics from Harvard, and knew of and was interested in Jansky's and Reber's pioneering work. We agreed that, as soon as the war was over, we'd try to get a project going to explore this new phenomenon. Then a second young astrophysicist appeared: Ralph Williamson from Yerkes Observatory of the University of Chicago. He was also interested in the new radio observations. When the war was over, one was at an atomic energy establishment, one was at a university in Illinois, and I was still at Cornell. But through long-distance mail we managed to get a project going.

In 1946 the U.S. was far behind England and Australia in radio astronomy, but with the help of E. R. Piore and the Office of Naval Research we started from scratch. First we had to build a laboratory out in a field by the local airport. Grote Reber of the Bureau of Standards and I were the first to take up radio astronomy after V. J. Day. Neither group had the wide interest and cooperation and access to facilities that helped the Brits and the Aussies. Nor did we have so many scientists and students participating in our work.

In England and Australia, Appleton and Bowen turned the mostly young radio physicists loose on the new phenomenon. We lacked such leadership in the U.S. Our astronomical establishment was the world leader in optical astronomy, but it generally looked askance at this new and inexplicable radio radiation. Then too, in the main, only the youngest Ph.D.s knew enough electromagnetic theory and physical optics, or anything about the burgeoning new electronic technology to trust and be interested in the new field. As a result, radio astronomy was not taken seriously by the U.S. astronomical fraternity until the 1950s. Hence England, Holland and Australia led the way.

I graduated at the end of the war, was appointed assistant professor of electrical engineering, and shortly had the only two federal research grants in the EE school. Such eminence, and I was so naive. This was not good, as it turned out, for the school had a new director, appointed in spite of a strong faculty vote of disapproval. Within a year I was crosswise with him. It never occurred to me that anyone would want anybody else's research work for their own papers, or anything else along that line. But this cuss came from Bell Labs with his private agenda.

It became clear that I either had to give up my research to him or leave Cornell. The professors I had worked with all through the war and afterwards came to me and gave me that message perfectly bluntly. If I would give up my research, they would guarantee my promotion and tenure. Otherwise I was out. I left at the end of my appointment, in 1950.

The National Bureau of Standards made me an offer early in 1950 but I felt I couldn't leave until the end of the school year. That meant that I could not get emergency clearance. NBS security pointed out that it would take at least six months to a year to clear me because of my family connections. The war was over but a Red Scare was underway.

So I couldn't wait for that, and accepted a fellowship from Professor Olaf Rydbeck and Chalmers University in Gothenburg, Sweden to do radio astronomy. From Gothenburg I lectured at a number of places in Europe, including Leiden, the Netherlands. Professor Oort invited me to return to Leiden, which I did in the fall of '51, and remained there for ten years.

Leiden was a wonderful place to be. Leiden and Cal Tech were the two hubs of astronomy in the fifties, and I was happy to be in Leiden among a remarkable group of astronomers, visited by astronomers from all over the world. Holland was a poor country when I arrived but radio astronomy flowered there. It is still a leader in the field.

From Leiden, in 1961 I went to Stanford University for two years with Ron Bracewell, and then for two years with Harold Weaver at the Berkeley campus of the University of California. With the Hughes Research Laboratory I developed a very stable, high gain, low noise maser for an 8 GHz observing program. Due to lack of funds this had to be shelved in favor of helping to develop the receiving system used to discover the interstellar OH lines.

By the time the OH line was detected I was at the University of Texas, in Austin, helping Harlan Smith and Charley Jones design the 105-inch Fort Davis optical telescope. I was in charge of the overall system design. It was an exciting time. I looked on the dome and the telescope as a single, integrated instrument, not as a separate building housing a telescope. When we finished we could dismount the mirror in the morning, clean and silver it, and be observing again that night. I believe this was a first for large telescopes.

On the completion of my share of the telescope development, I received an NSF grant for lunar radio occultation studies and joined Jim Douglas's University of Texas radio observatory group outside of Marfa.

This work was terminated abruptly, and Jim's set back several years, by the unexpected launching of the LES-6 military satellite. It put signals right smack in the middle of the band for which Douglas and I had been building all our equipment, and for which it was tuned. They wiped us out.

Since Douglas held the only radio astronomy position at Texas, I set out to help New Mexico State University develop a new graduate department of astrophysics. This was successful, and many of the faculty I hired at that time are still there. It is a pleasure to follow the careers of some of our first grad students.

Five years at NMSU and I decided that I'd rather be a small frog in a large pond. I shocked a number of colleagues by resigning a tenured position without having secured another one first. At this juncture my marriage crashed quite unexpectedly, and I found myself the sole acting parent of two lovely little girls and with some responsibility for their mother, who was in effect having a nervous breakdown and had left for the Haight-Ashbury on very short notice.

So I moved out to Palo Alto where I had friends and would be able to keep myself and the children in touch with their mother. I felt this would be better for the girls than just letting their mother disappear into a void. It seems to have been the right decision.

While recovering from the shock of the breakup and a sudden and seriously elevated blood pressure, I was asked to sit in with John Billingham's Interstellar Communication Study Group (ICSG). If I had been anywhere else than Palo Alto at this time, heaven knows what I would be doing now.

To backtrack for a moment, I had met Barney Oliver when I was at Stanford. We were both members of the Maxwell Society, an informal discussion group chiefly made up of physicists with unusual interests. In 1971 Barney invited me to take part in the ten-week Stanford Ames ASEE summer study that produced the Project Cyclops.

It was a great summer on Project Cyclops. That's where I got really interested in the possibility of searching for extraterrestrial intelligence. At that time this was not a popular topic in astrophysics departments, though one might occasionally mention it over the years, as one does in coffee conversation. Now here was a real chance to demonstrate that we were capable of doing something about the question. It was obvious to me that it was more a matter of proving to the rest of the world that we could carry out a meaningful search, rather than doing anything technically very new.

The events of that summer were curious and interesting. In sum, we laid the groundwork from which Barney constructed the remarkable Project Cyclops report (NASA CR 114445). Hewlett-Packard gave Barney a summer's leave for this effort. Considering his H-P responsibilities, this was quite a gift from Bill Hewlett and Dave Packard.

So in the Fall of '74, when asked if I wanted to sit in with the informal NASA group, plus Barney, trying to get a project actually started, I was delighted. Shortly after this the ICSG was funded to run a two-year workshop on the question, and Billingham asked me if I cared to connect myself with some university and become a NASA grantee in order to help get the program going. Does a fish like water?

NASA, through the consortium office at Ames, had cooperative arrangements with a number of universities whereby faculty could work with Ames on mutually interesting research projects. Through the courtesy of Dean Hensill and later Dean Kelley, I became the equivalent of an adjunct professor at San Francisco State University, taught occasional courses and gave some seminars, but was able to work full-time at Ames on the program soon to be called SETI.

We did not invent the name SETI until 1975, in the Morrison workshops, when we realized we needed an acronym. We chose SETI instead of the older CETI (Communication with ETI) to emphasize that we did not intend to transmit to ETI, but just to detect any radio signals they might transmit.

Going back to earlier days, how was your elementary schooling?

Highly irregular. I didn't go to school until I entered something that might be called the sixth grade. I had learned to read and do elementary arithmetic at home. I could count, add, multiply, divide and do fractions. My parents moved and travelled a lot, and there were no school attendance laws as there are now.

We were living in New York City and I went to Miss Parkhurst's Children's University School on 72nd Street. This was the origin of the Dalton School in New York that, since then, has become famous as the school for the sons and daughters of progressive people—"liberals" they would be called today.

It was exciting for me because I had French, Chinese history, and the like as well as arithmetic, English, and geography. I completed about a year and a half before I caught the measles and a severe mastoid infection. With an operation and recovery, I was out of school for about half a year.

Then came two public school years, seventh and eighth grades in Nyack, New York, where we had good old-fashioned teachers who damn well ran their classes. There was no question about discipline in the classrooms. They didn't care whether you were dirty or clean; you behaved properly in class. They taught the old-fashioned subjects and I'm very grateful to them. It helped me get to boarding school in spite of having missed so much instruction.

I went to Loomis Institute in Windsor, Connecticut, where I received a rather fine education except for one thing; they had me tracked for engineering because of my radio hobby. That was an educational disaster as I look back on it: less English and less languages in general. I remember being jealous in my last year because with only two years of Latin, I couldn't do much with it. Those who had four years were able to converse and joke in Latin. They even did it at graduation.

My parents never paid much attention to my schooling, once a school was picked, and it never occurred to me to talk to them about such matters. I did what came my way as best I could.

I was, in a way, an outcast in the school. Almost all the students came from very well-to-do families that led normal lives. My short life had been highly irregular, being the child of unsettled musicians. I was a shy and silent adolescent at Loomis. I learned to like team sports but was too lean and tall for anything but tennis and fencing. I graduated from boarding school essentially without a friend. I have seen, and by accident, only two or three of my classmates since. I got along with them, by and large; only a couple of fights in the first two years. I was captain of the fencing team, and sixth man on a five-man tennis squad.

Loomis was a school where the teachers showed considerable favoritism to athletes, particularly in football, soccer, baseball, and tennis (the schoolmaster's pet.) It was somewhat of a lonely life for me and some others who were not denizens of the favored classes.

I worked hard at tennis but received no coaching. Trying to make the team, I probably overdid it. Got double pneumonia and graduated in a wheelchair, excused from final exams. Embarrassing. Rather dramatic as well. Mother gave a lovely concert in the chapel by way of thanking the school for caring for me so well.

I built demonstration radio equipment for the physics course, knowing just a bit more than the teacher about this new technology. I had one of the better secret radios. You weren't allowed to have one, but I had one that had been given to me by my friends back in the radio club. They built it into a cash box. I'd have it now except that it went to the bottom of the Farmington River while out canoeing.

While in New York City after the polio, I learned to shoot. I practiced in the basement of the Bronx Armory, where policemen and gangsters as well as amateurs maintained their skills. We were quite used to seeing bootleggers there, keeping their hands in tune, as it were. Fred and I were good enough marksmen at twenty-five or so paces to hit matches held at arm's length between thumb and forefinger. (I have thoroughly proved that it takes exceptional luck to grow up.)

Back at school I bought a Savage .22 repeating rifle and often went plinking with others along the Connecticut and Farmington rivers. I soon sold the rifle after a sad accident. I aimed and fired casually at a distant bird flying high in the sunset sky. A beautiful blue heron dropped like a stone into the fast-flowing river near to me. Only Annie Oakley might have done it on purpose. I was just getting rid of a last shell in the magazine.

I read anything I could understand on radio, and a lot I could not understand: the technical radio columns that appeared regularly every Saturday in the *New York Sun* and the superb, original *Electronics* magazine of the thirties. These didn't require as much mathematics as the *Proceedings* of the Institute of Radio Engineers. I built Loftin-White amplifiers after designs in the *Sun*. These were powerful, truly high-fidelity "feedback" amplifier designs that appeared before the theoretical works of Black and Nyquist were well known.

I sold some of the equipment I built. One item, built for a Chinese ethnographer and anthropologist, was a battery-powered portable recorder using aluminum discs. (Coated disks were not yet available.) It was required to play forward and backward. Dr. Chow felt this helped him to distinguish some of the sounds in the dialects he was studying.

I heard, many years later, that the recorder worked well until after the war, when tape recorders became available.

I built an electronic transcriber for Edison phonograph wax cylinders. This was used at the New School for Social Research to transcribe priceless African music recordings from before the turn of the century.

And so it went until I reached the Cornell Electrical Engineering School. Its reputation was greater than the facts justified. MIT would have been better for me. But the war helped. It stiffened the curriculum. I was coopted by some young professors who had set out to modernize the curriculum. It was a real academic revolution, and an exciting and educational experience.

I was classified 4-F by the local draft board because I was teaching radar and other essential courses to Army draftees and Navy volunteers in the V-12 program. If drafted I would have gone willingly since I felt that if there ever was a justifiable war for the U.S., this was it.

In sum, I had highly irregular education which encouraged my appreciation of independence. At the same time, I learned that I liked to work cooperatively with people, rather than be a loner.

When did you first think about the possibility that extraterrestrial intelligence might exist?

I don't know when it was. My father brought up the subject, possibly before I was six years old. Father was chairman of the U.C. Berkeley music department and 'one of his good friends was Tolman, the cosmologist. Father used to take me out to look at the stars and talk. This was back when there was argument over whether there was a single galaxy, of which we were a part, or whether there were many island universes. Such philosophical matters appealed to Father. He and his friends were on the side of a plurality of intelligent species.

I was his first child and he treated me as an adult from about the time I was ambulatory. Though I had never complained, in later years he apologized for not having given me a decent childhood. He was an erratic parent. But I had as good or better time than my siblings. Father lived an affectionate, intellectual life and he always included me in.

I discovered Edgar Rice Burroughs about as soon as I could read with some facility. Science fiction became my favorite childhood literature. From early on I tacitly assumed we were not alone in the universe, and was quite unaware that the idea had been a matter of hot discussion for twenty-five-hundred years, or more.

Doing anything concrete about it was another matter. I was not familiar with some of the relevant discussions of the early thirties, such as Hiram Percy Maxim's book *Life's Place in the Universe*, which I recently found in the Congressional Library, fifty years out of print. I have made copies for Philip Morrison and Frank Drake, in case they don't already have them.

Maxim was one of the idols of my youth. He was the chief founder of the American Radio Relay League (ARRL). He founded the amateur radio monthly *QST*, and often wrote for it under the pseudonym "The Old Man." He had a delightful, crusty way of upbraiding ill manners on the air and not abiding by the rules and regulations. He also wrote amusing short stories. Some were pure science fiction. I liked his sense of humor. I wasn't reading *Scientific American* at the time or I'd have known of his other interests, and I might have caught on to the explicit idea of SETI earlier than I did.

Life's Place in the Universe is a small book, about 175 pages. The whole last chapter was on the question of life out there and the radio communication possibility it brought up.

The nice thing about Maxim was he was trained at the tail end of the Age of Reason. There was none of this mystical certainty about unknowns that you find in the works of Carl Sagan, Frank Tipler, Frank Drake, and so many others, where emotion and intuitive belief are so prominent. Maxim was objective and down-to-earth about the matter. Nor did he suffer personal guilts about the imperfections of human society.

It is a greatly changed social universe since WWII and I find it fun to read Maxim, since I grew up in the lingering memory of the Age of Reason. I was twenty-seven when I went to college and I suppose my psychological mindset was pretty well determined by then.

Radio equipment was so very obviously not adequate for interstellar communication before WWII that I believe those who might have thought of it could see it only as something for the very distant future. The formative years of radio astronomy began in the forties. It took the explosive developments of electromagnetic and electronic technologies, and radio astronomy, in the two or three decades following the war, to make the observational idea attractive.

The postwar years were incredibly interesting. Up to 1963 I think I knew or had met every single graduate student or worker in radio astronomy, but after that it was a hopeless operation. The numbers working in it had grown explosively. The initial period was over, the main phenomena had been explored and laid out, so it became a question of more sophisticated instrumentation coming in. Transistors were coming in, and we'd been through the period of the invention of the parametric amplifier and the maser and the absolute calibration of antennas, and all the rest of it.

By 1960 it was clear, as Jan Oort remarked, that the invention of the radio telescope was as important as Galileo's invention and use of the optical telescope. And now the process is being repeated in the infrared, ultraviolet, X-ray, and gamma-ray regimes. All are new ways to observe the universe, and truly most fortunate are those in a newly opening field.

So your father got you thinking about ETI when you were very young?

Just the general idea was there. The decisive event occurred when Barney Oliver and John Billingham managed to establish the Cyclops summer study in 1971. The advertisement for applicants was all that was necessary. Here was a chance to prove the obvious to the scientific community and the world at large, and get in at the beginning of a new field of research. The obvious element was that technology had developed to a point where we could design a powerful means to look for signs of hypothetical cousins in our galactic neighborhood.

The Cyclops study was memorable for many reasons. One that comes to mind now is the difference between the electrical engineers, physicists, and astronomers on the team, and the members from more conservative disciplines: civil and mechanical engineering, and economics. The latter were leery of large numbers like ten to the twentieth or ten to the thirtieth, of designs several orders of magnitude beyond current practice, and of possible ultimate costs of several billion dollars. In fact, some of them thought the entire proceedings daft. One of them contributed little to the study, spending most of his time on private interests. Some of the others did not want their names attached to the final report in any way that would imply their approval.

Thinking in large numbers doesn't mean you're going to *do* it, but it is a useful intellectual exercise for calculating the limits of a technology, which is what we did. So it was a very exciting project for me.

The report Barney produced is a recognized classic in this field and esteemed today, even though progress in solid-state technology has superseded some of the technical solutions necessarily chosen in 1971. Over ten thousand copies of the report have been requested, and requests are still coming in.

But back at New Mexico State in the fall, it was made quite clear that it was too freakish, and in the absence of an obvious source of outside support, I should have to forego this "dubious" research direction and tend to more immediate matters. Anyway, I was thoroughly occupied with teaching and other university activities. Thus little was done to further SETI until I left NMSU in 1974. I did, however, talk about SETI before many public groups, and found out how attractive the exploration was to the general public.

When I joined Billingham and Oliver in 1974 I did not foresee how long it would take to get SETI launched. It seemed like such a logical thing to do, and it was a natural for one of my temperament. Since there wasn't an obvious good fit for me in radio astronomy in the United States, since I had more old-fashioned views that were much more sympathetic with what went on in Leyden than what went on over here in radio astronomy, I just moved over into this new field when the opportunity opened.

Complex life, particularly intelligent life, is such a dramatically different form of physical matter that it is worthy of being classed as a fifth form of matter—the others being solid, liquid, gaseous and ionized matter.

I have always read far afield from my main activity, bridging many fields: engineering, physics, astronomy, and to some extent the life sciences. Thus I understood that SETI was but one part of an overarching problem of great interest to many scientific disciplines: an exploratory arm in the quest to understand the origin and prevalence of life in the universe. Billingham had also recognized this, and well before I came on the scene. SETI is a technique for perhaps getting an answer sooner rather than later. It is certainly our hope, our expectation at any rate. Our terrestrial data can tell us nothing about whether we are the only ones, one of a few in the galaxy, or one of many—one example of a common phenomenon in the universe.

I have no certainty about what the answer is, but because of the data now on hand, I know where I would place my bets. I'm convinced it's a natural phenomenon; that life on earth is the result of natural processes operating in a suitable environment. In all cases tested so far, basic physical processes found on earth are also operating elsewhere.

Descended from three generations of agnostics generally conforming with the ethics of the surrounding Christian societies, I have never felt a need to believe in a willful, divine origin for our universe. I'm happily content just to exist, but I would very much like to know if our species has galactic cousins nearby.

What about communicating with, or at least searching for, ETI?

If we discover an ETI signal, trying to communicate with them is the next and obvious challenge. I think we certainly will be able to, if they and we care to, because we will share so much common knowledge about the physical universe. Hans Freudenthal (mathematician at the University of Utrecht) has shown it is possible, even using a very limited linear medium, to develop broad communication between two intelligent, quite separated beings who knew some mathematics but were otherwise totally ignorant of each other's language.

All that is required is that ET have at least our understanding of the physical universe and our ability to manipulate it. Their production of a coherent radio signal that we could receive over interstellar distances should guarantee that they had a mathematical discipline and physical knowledge that paralleled much of ours. This one-to-one correspondence should provide an obvious opportunity for the development of a mutual understanding by interested parties.

I should mention one caveat, however. I am, as all humans are, intelligent by self-definition. We know nothing about any possible intelligent species elsewhere in the cosmos. I am assuming that the fifth state of matter is recognizable and understandable throughout the universe.

Because of the time it takes to communicate over interstellar distance, I have implicitly assumed a long future for the human species. I am basically optimistic. We have evolved over billions of years, and survived in our present form for perhaps fifty thousand years or so. I bet on winners. I am optimistic even though we have been our own worst enemies for ten thousand years or more, and particularly so now.

Human beings do not do everything of which they are capable, so I do not share the paranoid fear of the imminent end of our species. I just do not believe it is a reasonable reading of the situation. It is a tough one, certainly. Dangerous? Of course. It's always been dangerous. However, if a sufficient number of people deeply believe Armageddon is just around the corner, then we may not survive. A case of self-fulfilling prophecy: you get what you ask for, by and large. I do not share that vision of doom, and I raised children who do not share it.

Have we covered the main influences that gave you these ideas of the possibility that ETI might exist?

It was just an idea presented to me when I was very young, and it fitted in very well with my increasing understanding of the cosmos as I grew older. The idea of *doing* something about it, though, came at the time of the Cyclops summer study. In the sixties I was aware of Frank Drake's experiment and the 1959 Cocconi and Morrison paper, but I was then too busy in the forefront of several other technical areas, such as building optical telescopes.

Did you discuss the possible existence of ETI with other people?

Seldom. Even for some time after Cyclops, professionals split into two extreme positions: those who said, "Of course," and those who said, "No way!" I rarely found someone who said, "I don't know, but I would very much like to know"—which is my position.

What effect, if any, has your interest in ETI had upon your career as a scientist?

Enormous. I am not practicing inanimate radio astronomy, though I have a continuing interest in the field and keep a steady eye on developments. Yet as I see it, I am still in radio astronomy. SETI merely adds another parameter, coherent intelligent signals, to a field defined by the technology used. And I believe the Ames/JPL team is designing some excellent radio astronomical receiving systems. Of course, these receivers do possess unique on-line capabilities.

What do your colleagues think about your interest in SETI?

My older colleagues in radio astronomy but outside SETI? Many are equally interested. In some cases I think I detect an odd respect for my good luck. Others are less interested and kid me as a lost soul.

What about your family? What do they think about your ETI interests?

Family? Good heavens! Well, since I make a pass at running the immediate family, of course they are interested in it and supportive. As for the wider family, we keep in touch. Only one expressed any great concern in connection with SETI. For a long time my brother Pete, the musician and folk singer, thought I was wasting my time, that I wasn't doing anything very useful for humanity. But he wrote not long ago that he had thought it over and changed his point of view. He now thinks what I'm doing *is* worthwhile. That pleased me. I suspect my other siblings are just glad that I am enjoying life and taking care of myself and mine.

What about your friends?

Some pro, some contra. After all, it is not a common topic of conversation: this dogged effort to get a SETI project going at a really significant, exploratory level. And there are so many other interesting things to talk about. Friends do ask me occasionally how my work is going and I tell them at least something. Generally, we try to avoid shop talk. If the friends are also SETI colleagues, then SETI may dominate the conversation.

What about your superiors, bosses, and so forth?

They're all sunk in it as deeply as I am; even more so and for a longer time.

Before going on to discuss your ideas about SETI, let's get a little more background information.

Father and mother were both dedicated musicians. Father was born in Mexico City, where his father was in business at that time. Father was tutored at home until he went to boarding school back in the States and then on to Harvard. There he graduated with B.A. cum laude in music, much to his father's disappointment; business was the only proper occupation for a gentleman.

Father wanted to be an orchestra conductor. He went to Germany for postgraduate studies and found out, while doing a stint as back-stage conductor at the Cologne Opera, that he was too hard of hearing. He couldn't hear the cues. This was a bitter disappointment.

He turned to musicology and became a major force in music in the U.S. for the rest of his life. Active to the end, he died quickly in his ninety-second year.

Mother was born in Denver, Colorado. She had almost no formal education. She wrote well and was well trained in all the social graces and spoke French like a native. She almost never mentioned her childhood, and then briefly. It was not a happy one. She was spotted as a child violin prodigy quite early. In her teens she was granted the first scholarship ever given to a female violinist at the Sorbonne. Later she studied with Leopold Auer and Fritz Kneizel.

She was a beautiful and graceful waif, a superb violinist and violin teacher, and born three generations too soon to have a public career. She could perform in wealthy private homes but almost never on the stage and still remain a lady. What she learned, she learned as she lived. A teacher at the Juilliard for many, many years, she would never help train a prodigy. She felt it too destructive and quite unnecessary. Father and Mother were both born in 1887 and were married in 1912.

How would you describe them?

Oh, quite gregarious. Quiet, by and large; definitely reserved; definitely serious. Father and Mother were different types. I think they were married, after a whirlwind courtship, because they had music in common and, of course, she was beautiful and he was handsome (all his life) and a thoroughly spoiled, naive, and talented brat. Their marriage was rough: in and out of love with each other a number of times, but seldom in phase. They had nervous breakdowns of the twenties variety, separated on and off several times, and when finally divorced, it merely put an official end to the matter.

Father married again, shortly after, and for the last time. All of this went on generally very quietly and beyond my awareness. I heard raised voices only once.

Father was far the greater influence in my early life, once I could talk. And I was the eldest. That I fell into science was in large measure his doing. He had a logical, questioning mind and was a good observer, greatly interested in society and the cosmos. He saw below the surface. Well educated in ideas on the humanist side, he spoke or read well ancient and modern languages, but claimed he was only passable in English. His basic outlook was scientific as well as artistic. His greatest handicap came as a result of early innocence and lack of guidance. He never studied any natural sciences, and math stopped with Algebra I, which he promptly dismissed.

How many brothers and sisters did you have?

Seven. Father first sired three boys, then married another fine musician and composer, and then there came a boy and three girls. We were all naturally musical, but some were indelibly coded amd showed clear talent for music. Out of the seven, three are professional composers and musicians. But music is essential to all our lives and shows up in the next generation as well.

What about your friends when you were growing up? Did they share your interests in science and ETI?

I had few friends. I make a distinction between friends and acquaintances. A long-lasting friendship began at the "Y" radio club. We meet occasionally and compare notes. We are mutually fascinated with the other's ideas and doings. He is largely ignorant of my specialties, except that we share appreciation of the beauties of the universe. We both have read science fiction when we had the time, which is seldom nowadays. He is Greek and married to a fine Pennsylvania Dutch woman.

What were you like during childhood and adolescence?

Noisy and bashful. I was outgoing and timid, if you can imagine the combination. I was gregarious if I received any kind of an invitation. Otherwise I tended to be quiet.

We were a reading family. Father read, and read aloud to us. We spent much of our spare time reading. We did not have distractions like TV; no radio until I built one; no phonograph until Mother came home with a wind-up portable and Van Camp's recorded setting-up exercises. They were awful music to our ears.

Father immediately brought home records of exciting music by Debussy, Ravel and others. Father played the piano more often. He was a gifted pianist who always exhibited a remarkable musicianship. Even as his piano technique declined, for lack of practice, he could fake a difficult piece so that one hardly realized it at the time. With the radio we listened to concerts with the scores in front of us. Mother practiced the violin daily and often gave lessons at home to special students.

When one couldn't go outside or it was dark or you were living in the country and you didn't want to work in the garden, if you weren't building something by hand you were reading. And so we read voraciously. My paternal grandfather's house, where we used to spend the summers, about sixty miles from New York, had a gorgeous library with leather-bound volumes of *Harper's* and *McClure's* magazines going back to the Civil War years. There were other well-illustrated collections as well, and I learned a lot of history. At first I scanned them for dramatic pictures. Then I read hither, thither, and yon in them; whatever caught my eye at the moment.

During adolescence, being the eldest, I was allowed to stay quietly in the background while Freud was being discussed by Father and his friends, so I never looked at Freud as though he had a halo. I must have heard every side of Freudian analysis at one time or another, and seen it practiced by American psychiatrists on friends of the family, with dubious results. I think Freud was remarkable and broke open a field in psychology, but my God, what came along after!

What other professional interests do you have, apart from SETI?

Biology, geology, anthropology, everything affecting man. If I had had the knowledge I have now, I might have chosen cultural anthropology rather than astronomy for a career, simply because in the study of mankind one faces the most difficult problem of all. Trying to understand the nature and evolution of the universe is only the second most difficult task in sight. I've always been fascinated by the reasons my colleagues in physics and astronomy and engineering do what they do, the way they do it. I look on cultural anthropology as the encompassing topic. All the social sciences, historical and current, are but much narrower branches of it.

So I spend a lot of spare time on such things. I have taught a lot and have developed some firm views as a consequence. If I were to stop working full-time on SETI right now, I think I would try to publish some pretty strong and challenging statements on the subject of teaching, our educational system, politics—you name it.

What recreations do you enjoy?

Outside of the above, and having turned one hobby into a profession, my recent recreation, pleasure, and duty has been raising my later children. Too many people in this country don't understand what raising competent children entails; some do it well instinctively, and some well with forethought. You're *growing* children, and it's a long-term matter. If the parents don't do it, it is left to the random action of extrafamily contacts and events.

Growing children is also a good way to learn about one's self. People learn about themselves by interacting with others and thinking about it, not by going off and communing with their belly buttons in isolation.

What has been your greatest satisfaction in life?

Just living, I guess. It's great to be alive and doing.

What has been your greatest sorrow in life?

Greatest sorrow? I mark that the failure of my first marriage after twenty-two years; tragic because it so hurt the three boys also involved, and my ex-wife, who spent years getting out from under hate. I've got absolutely no use for wasting a moment's time hating either myself or somebody else.

Turning now to SETI, about what fraction or part of your working time do you spend on it?

When not paying attention to private matters I spend all my time on SETI. Let's put it that way. It is a full-time job plus; a job and an avocation.

What would you do differently if you were starting over again to study ETI?

I don't "study" ETI; no one can. I don't think there's anything I can say to your question, because it is pointless to go far back in time and ask, "What if?" Circumstances, ideas, and motivations bring about action, and if you change one bit, you change the whole thing.

I did say that if I'd known what many of my currrent interests are now, I might have become a cultural anthropologist. That statement was more to indicate the degree of my interest in that area; in no way did it indicate a wish that I had chosen otherwise in the past.

Why is ETI important to you?

Why I am interested in SETI is simple enough. I would like to understand better the origin and prevalence of life. We have one powerful datum, the history of life on Earth as we understand it today. It strongly suggests the hypothesis that earthly life originated and evolved in a natural, physical way. To be accepted as truth our hypothesis requires more than one supporting datum. If earthly life happened to be unique in the cosmos, I strongly suspect that there would be no sure way, using purely earthbound data, to reach an ironclad decision between this *physical* hypothesis and any of the other, often longheld *miraculous* understandings. But uniqueness seems to be the most improbable of all possibilities.

There is a saying in physics, "If it can happen, it will happen." In this astonishing, huge universe, every time we have been able to make the proper test, a concatenation of basic physical processes producing an observable phenomemon in the solar system will also have produced a plurality of similar results elsewhere in the universe. Thus we expect life—and intelligent life—to exist elsewhere, beyond the solar system. And our expectations are greatly enhanced by our understanding that the solar system is a relatively young stellar system centered on one G2V star among billions of similar stars in our galaxy alone.

There are other reasons as well that attracted me to SETI. I like to work in the founding of a new research area that promises something of lasting value to humanity. Of course, it has to be related to my capabilities, if I am to be of some use. I like controversial problems. They are often the important ones needing solution.

Then, too, I am by nature curious. When I go out hiking, I'll go any place for the fun of it. When I drive off my by myself for a day or two (which I often do), I'll follow dirt roads just to see where they go. What's over that hill? Around that curve? I'm quite safe; I'm not an idiot. But what's over the horizon is quite interesting. What I recognize as unknown to me, and perhaps to others, catches my interest almost without fail, and makes life interesting.

Finally, one can imagine the fascinating possibilities following on the discovery of an extraterrestrial species capable and willing to communicate back and forth with humankind. On the basis of what we know now, that is a distinct possibility.

What has been the most difficult aspect of your ETI work?

Putting up with incompetence. Good lord, that's the most frustrating. Our project has literally been set back by people who haven't learned how to think and do their job properly. Incompetence is hard to tolerate when it impinges on you. I know there's a lot in the world and I'm certainly incompetent in a lot of things, but when it interacts in my areas, I resent it like mad. But I'm very patient about it, for a very simple reason: there's not an awful lot you can do about it. You gotta take it. Evasive action, yes, on occasion. And sometimes programmed aggressive action.

When we found out that the astronomical community was worried that any money spent on SETI would come out of the astronomical research budget, as small as it was, several of us set out on a campaign to disabuse that point of view; in a friendly way, to educate them that the money would never have come to astronomy if we didn't get it; that SETI was, first of all, astronomy, an extension of it. We opened the gates to whoever wanted to partake in it, and generally got them familiar with it.

It took three or four years to persuade the astronomical community (actually longer than that), that we were a serious, legitimate exercise in the hard sciences. Some of the people working were from our group and some people were outside. It was a combination of Philip Morrison, Frank Drake, George Field, Carl Sagan, Barney Oliver, and myself.

We came out with that final statement that essentially said we were right down the center line of exploratory science. That was done by persistent effort on the part of a dozen people. It wasn't highly organized, but their instincts were right in all cases. We had to do the same thing with NASA headquarters, and do some hard design engineering and whatnot on the side, as well.

That's been the most frustrating because it's taken so long. When I joined the group here in 1975 I felt it would take maybe five years. It's partly bad luck. We had NASA headquarters more or less convinced, somewhat against its will, when a key person in the structure above us died by accident on Mount Everest, and that set us back.

But we recovered and we received the help of Proxmire. And I mean 'help'. Getting the Golden Fleece was just what we could use. It brought us to the attention of a large number of people who resent Proxmire, and hence became our friends, literally. He's a good politician and a powerful one, but in this area he didn't have good advisors. He put a motion on the floor which robbed us of support for a year, so we theoretically had to close down the project.

That was the doing of a reporter, Robert Schaefer. He wrote an article for an ultra-conservative publication in Santa Barbara, and sent a reprint to Proxmire. The article accused NASA of being very underhanded and hiding things from Congress. NASA hadn't been that at all; it was pure imagination. But in the throes of the elections and Reaganism, the editor made Schaefer's article worse than Schaefer did. Proxmire waved it around and passed a resolution that no money could be spent on SETI, not even on studying how to do it. That set us back seriously, a year or more.

But it got NASA's dander up, because NASA wasn't about to be told by Congress what they could or could not study in an R and D project. It had never been done before; Congress had never meddled at that level. It can't. It can't control the imagination of research scientists. It *can*, and justly, say what projects shall be carried out at a significant level of cost, yes, but *not* what you may study on paper. So NASA was bound and determined to have that not repeated.

Proxmire consented, because he became impressed by the fact that the scientific community thought he was dead wrong. Then he realized he'd made a mistake. We did not attack him at all, because he had to be given a face-saving way out of the embarrassment he'd gotten himself into. That's the only way to handle a political situation like that, because he could've been unpleasant if he cared to be. It was a matter of being rational with him and giving him a way out at the same time, and we did.

How do you think most scientists today view SETI?

One can divide the scientific community into three parts. There are those who feel that we are not alone—I suppose because of the way they were enculturated from birth. Others, for the same reason, are positively sure that we *are* alone in the galaxy; and some members of this group think it's our bounden fate, even duty, to colonize the galaxy within the next million years or so. There's a third bunch that just doesn't give a damn. If you pin them down, you'll get one of the other views, but there are plenty who just don't care.

The outspoken opponents of SETI seem few in number. However, their attacks on SETI brought the matter to the attention of the scientific community at large. This seems to me to have been largely responsible for the Sagan petition. Part of the reason that so many prominent scientists who had no connection with SETI signed that petition was because they recognized that this was a clear-cut case of a scientific exploration looking for new knowledge, harmless or not, important or not, and that the political process had done science a stab in the back. So they signed it.

It is only fair to note that a number of the signers do not expect striking success except, perhaps, in the long haul, if ever. But they support SETI because this currently seems to be the best way to tackle the basic and unusually interesting question of the prevalence of life in the universe.

What about public attitudes toward SETI?

Talking often to the general public, as I have over the past fifteen years, has led me to believe that when the topic is new to the audience, there is an instinctive and favorable response. They ask a lot of questions, and many people call us and want stuff sent to them; they want to know where they can find out more about it.

This should be expected. For millenia we have written evidence of human concern about the origin of the perceived universe and humankind, and curiosity about our "place" in it. Religion and philosophy have always aimed to answer these universal questions. Now physical science has entered the arena more powerfully than ever before.

Surveys and popular lectures, TV, radio, and the movies all indicate a widespread public interest, even hope, that intelligent species exist.

In this context, a very odd peer review of a SETI proposal is worth recall. One of the responses of this questionable peer was this: "This proposal [a request for public monies] should not be supported because it is popular with the general public."

It is common to find a belief that NASA SETI is underway. There is surprise, and an understanding acceptance of reality, when one gives an account of what is necessary to get SETI launched.

In my experience there is little fear of extraterrestrials on the part of the public. That may surprise some professionals in both humanities and sciences. I tend to agree with this public instinct. When and if we discover an intelligent signal will be time for humanity to study the matter and make up its mind about the situation. In the meantime, I suppose, we will continue to advertise ourselves in ways that any species as ept as we are could discover us.

What is your explanation of UFOs?

Unidentified phenomena. In some cases they remain unidentified. In many cases we know, with high probability, what they were.

This is a "not understood phenomenon." It's extraordinarily amorphous when you try to get any solid detail. I've had sufficient contact with people actually working in the area to know that a fair amount of it is psychological. It's in the minds of the people who saw it.

I've talked to Jacques Vallee, a communications scientist. He has actually visited a thousand or more sites, and in some cases found unexpected things on the ground, things that would make you think immediately that lightning has hit, or something like that.

In some cases we understand what they were. They were clearly aircraft, meteors, lightning, inversions, mirages, weather balloons. I remember one case. We were in Leiden and all of a sudden Oort was flying down the street. We joined him because up in the sky was this bright object. Oort thought it was a supernova, so bright it was visible in daylight. It did not appear to be moving. We got on a telescope and there it was—a balloon. The air was very still so it was hardly moving at all. There was a basket of instruments under it. There was an *enormous* flap for fifty miles around Leiden. Telephones were jammed. UFO. "We are being visited!"

I don't go as far as Philip Klass in calling it all nonsense because I think it is a fascinating aspect of humanity. But there is no reasonably respectable evidence that we are being visited by extraterrestrials— that is, no evidence that we can get any agreement on. Most of it is not reproducible; most of it is hearsay.

But we haven't searched the earth for signs. People who say they have *never* been here have about as much proof as those who say they *are* here. Let me put it this way. If you saw the movie *2001* there could be thousands of those megaliths buried six feet below the surface on earth and we'd never know it. How much of the earth have we examined? Would we recognize an extraterrestrial if we saw one? We don't know what we are looking for. So it's an open question.

Dick Haines here at Ames is a psychologist, working in the area of human-machine interaction. His hobby is recording strange events seen by aircraft pilots. These are an unusual group of people. There are certain common themes running through their sightings. Some sightings had multiple witnesses. I don't understand what they are; I don't think anybody does. Most of the pilots say they don't understand it either.

Hysteria, some of it. We get letters from people who are in contact with extraterrestrials. Oliver got a letter not long ago: if he wanted to know the origin of the universe, get in touch with this number. Others are earthly representatives of people from the Pleiades and various other stars. Astronomers have always gotten letters like this—a routine thing—and SETI has gotten a lot of them.

If and when we do receive verified signals of extraterrestrial intelligence, what should a scientist do?

The first thing is to be reasonably convinced that your observatory has received extraterrestrial signals. You've reached that point, having run through a series of rather straightforward tests, to the limit of your ability with a single radio telescope, to sort out the direction of origin of the signal.

The next thing to do is get others to observe it; other observatories with totally separate equipment, and separated so far that it is unlikely that a fraud could be committed on both places simultaneously, by a Cal Tech or Stanford student. I've known some jokers who would do anything to commit a fraud like that.

We are arranging our equipment so that it would be difficult to do, unless he or she is one of us. But even that would be difficult, although some genius might be able to. That's one of the reasons why we would get other observatories to look at the signal. Because of the unlimited imagination of human beings, there's no ultimate defense against fraud.

That's the first thing to do. Then, the news that an observatory has a signal not made by human beings will be almost impossible to keep secret, so the obvious thing to do will be to say we've gotten a curious signal. We're studying it. We don't know whether it's extraterrestrial or not, so we're getting other observatories to look for it and verify it.

Verification can be done in a way that practically guarantees it is fraud-free. It would take such horrendous effort to fake it, say here *and* Bonn, Germany, simultaneously. Almost anything we could get at the limit of detectability at Arecibo could be picked up in Bonn with the knowledge of the frequency, the bandwidth, just where to look for it.

And if *two* observatories suddenly get paralyzed and keep looking at the same point in the sky, that would be known to anybody around who knows anything about the functioning of an observatory. It's not a thing you can keep quiet; it's not possible. If you tried to, you'd have to clear everybody out of the observatory except some of your trusted people, and you wouldn't let anybody within twenty miles in an airplane look at the observatory. That's impossible, at least in this country and Europe.

As soon as we get verification from other observatories, it should be announced as an absolute certainty, and the whole world told about it.

A lot of people don't realize that the Catholic church solved the problem long ago of whether there were likely to be extraterrestrials. This question had run through all major civilizations. At the end of the Renaissance it became a hot argument among the Catholic clergy, and they finally resolved it by stating that it would diminish God's obvious omnipotence to imply that we were the only people He had created.

When the signal has been obtained and verified and given to the public, what will the public response be?

Mild excitement in most cases. Because there's been a well publicized search for many years, it won't come as a surprise. Half of the western population, if you can believe the surveys, expects to find other intelligent beings elsewhere. The idea is so much in the current domain that it won't be a shock for most people, but it undoubtedly will call forth fits of paranoia from paranoid people. I remember the now-deceased Royal Astronomer of Scotland, who, when he read Cocconi and Morrison's paper, said, "If you get a signal, for God's sake don't answer!" He meant it. He was a thoroughly convinced Catholic.

So most people won't be upset by it?

No. Though some will be very upset, many won't. I think most people realize that we can't hide here. We already have the technical power to seek planets around the nearest stars. We know how to build a telescope large enough in space. The detailed engineering we'd have to work out, but that's run-of-the-mill stuff. The theory is well understood. It would take billions of dollars but we could actually see, with our eyes and photographic plates, planets around another star and find out if any are blue like the earth, and things of that sort.

We could put antennas out as far as Jupiter and two or three in orbit, and let them do some three-dimensional triangulation of another galaxy, to establish a beautiful scale of brightness and distance within one or two percent instead of fifty percent. Even putting a telescope on the moon would give us a terrific leg up.

Our technology has reached a point where more things are feasible than we can do as a practical matter, because of limited resources. Exploring the solar system will be a lot cheaper once we have a moon base. Exploring seems natural for the human race.

What do you think will be the long term effects of evidence that we are not alone?

It will eliminate certain arguments that have been going on for thousands of years, and some recent arguments as well. It will tend to broaden the perspective of human beings. The idea that a civilization can survive for a length of time will do a lot to counteract the pessimism about our own civilization wiping itself out. If we can establish contact with them, it will be possible to start communication. It would be a long time between questions and answers but a wealth of information will be exchangeable.

Realizing that it is only a guess, when do you think contact with ETI will be made?

I haven't a clue suggesting when we might discover signs of intelligent life. So-called "contact" could only be many years after their existence is discovered. Interstellar distances are measured in years of signal time. Human indecision may add more years of delay. The estimate that was made in Barney's Cyclops report is about as reasonable a guess as you can make. I see no reason to change my own mind that we may have to look at a million stars like the sun before we find a signal; it may take quite an effort. And this is derived from very rough estimates of the number of sunlike stars there are, and how much energy it takes to transmit over interstellar distances unless you know exactly where your receiver is going to be.

Science and technology are developed to the point where we could surely hasten discovery, but it would require funds far, far greater than anyone would have the nerve to propose seriously today. But it is a small effort to outline much more powerful instruments to look for evidence of life elsewhere.

What might ETI be like?

I do *not* expect them to mimic humans, to look like us. I wouldn't be surprised to find them significantly symmetric or with a carbon/water chemistry reminiscent of ours. On earth we find strong symmetry right down to the level of elementary bacteria, and we have no idea if there is any other chemistry that is capable of generating evolutionary life.

I would expect them to make tools and have a reasonable manipulative ability. Beyond these anthropomorphic hunches I have no preconceptions about how ETI species might look. The whole subject seems rather fruitless and boring for now. We have nothing to go on beyond the above hunches.

Drawings, paintings, and three-D models of hypothetical ET species that are presently a fad evoke little interest on my part. The bulk of this work appears to fall in a narrow range more indicative of the artists' psyches or the psyches of those who buy the pictures. In their monotonous way they seem quite parallel to much commercial art. Seldom is there much truly appealing in them. Humor, affection, expectation, disappointment and other complex emotions innate with our intelligence rarely show up.

Must the pictures tend so often to the gross or still worse aspects of our society? Is it not possible that ET might even appear beautiful to us? Or at least attractive?

What scientists or other people in the past do you admire?

I have a growing pantheon, selected as I come across them. It includes some of the usual heroes, real and imaginary. It runs from Prometheus to Darwin, Bertrand Russell (when past writing books on logic), Peter Oosterhoff [professor of astronomy at Leyden] and Bronowski. The pantheon continues without break through today with the still living: Jan Oort, Stephen Jay Gould, Hanbury-Brown [British pioneer in radio astronomy].

When it comes to not necessarily scientists, but human beings, Billingham is exceptional. He is a remarkable, subtle, persistent, imaginative person. He's the only reason I've been here ten years. I've never stayed or worked for people I didn't particularly care for.

SETI's been brought down the pike largely by Billingham. I'd put him number one, because he spent the better part of his time from 1970 to date doing everything in his power to get it going from a base of operations where there was a hope for cash.

And Barney Oliver is a remarkable person. He has worked very hard on his own, from where he is, as a hobby to help Billingham, to help Hans Mark who was also interested in it. The reason there is a SETI project now stems basically from those two: Billingham number one, Barney number two.

Frank Drake does experiments on it, talks about it a lot. He's an astrophysicist with many interests.

With respect to influencing outsiders, particularly the general public, Carl Sagan and Frank Drake. You bet! Philip Morrison certainly has had as much effect as Drake and Sagan, except on the general public. Morrison, his lectures and whatnot, are attractive to the educated public, while Sagan speaks to the entire spectrum of the public, particularly the uneducated.

Jill Tarter has joined us in recent years. She's been very active in the scientific community and in getting people to examine old records and things of that sort.

Though Ron Bracewell and I got along fine, we were obviously divergent personalitites. He's a fine scholar, and I respect him very much. But his whole view of the universe is so totally different from mine, and his interests so different from mine, though they impinge in various places, that I worked for him for two years, then left.

So they are the outstanding people in SETI today, and during the past twenty years or so?

I think I've listed all the important ones: Morrison, Drake, Sagan, Oliver, Billingham—those are the stellar performers. There are other people who have lent a hand in various ways: Otto Struve, Michael Klein, Sam Gulkis, Donald DeVincenzi, John Kraus, Robert Dixon, Michael Davis, Paul Horowitz, and I. S. Shklovskii. Some of them are fairly called inspired. By their writing, speeches and hard work, they have greatly fostered SETI.

On the whole, it's strictly an American operation. The Russians have been interested in the subject, but they don't have much input.

I, and the people who are in 'origins of life' studies in general, are pretty well convinced of how it will come out, because it seems to fit a natural physical, chemical pattern. If we are unique it will be the first time a major process in the universe has been found to be unique. That's one of the biggest arguments for our not being unique. Of course, there has to be a first occurrence in the galaxy. There was a first star. Things have to start somewhere. We *could* be it.

There might be better ways in the future to look for life elsewhere. Right now we're forced to look for signs of a very restricted group: the intelligent, radio-using fraction of the life out there. That's going pretty far out on the limb, but for now it's the best we can do. And in the true spirit of explorers we know about here on the earth, you don't wait for a jet plane to discover America. You take what is available.

Have you ever been on Columbus' Santa Maria? I've been on a model of it in Barcelona. I wouldn't go *anywhere* in that boat! It's so damned cockleshell crowded; even the short people of that time couldn't stand up in it. Good Lord.

Maybe our technology is that primitive, but I think it is pretty good. If there is anything out there, we are likely to find it. There may even be a strong signal, a beacon. I don't indulge in speculating about motivations on the other side; I merely recognize the possibility. It would be rational. If they were interested in what existed anywhere else, they might construct a beacon and try it. It is a possibility. We know *we* could do it.

We're looking for *all* signals, but we would consider ourselves astonishingly lucky if we ran into a beacon or its equivalent: an unintended beacon, not set up to discover anything, but as a beacon for their own use.

How would you characterize your own role in SETI?

One aspect of my role is a facilitator. I like to work with people, and I do my best to see that the conditions are right for them to work together. It's the business of getting the most out of people you're working with, and getting the most out of yourself at the same time.

Billingham's very good at that. He keeps pushing people to the limits of what they can do. He doesn't expect miracles, but he does expect hemi-demi miracles. We used to talk about the "miracle-machine" here, because deadlines would come, sometimes two hours away, and only once or twice did the magic fail to work. But sometimes those were excruciating hours. We have both worked for a number of days with minimal sleep; the kind of thing I did when I was twenty, not sixty.

We have much the same approach to working with people. And we tend to draw that kind of person in. Jill Tarter is the same way. Ivan Linscott, who is masterminding the final stage of equipment, is that way also.

There've been people on the project who've been pessimists; people who have been in and out of it, who've put in a year's effort or two years' effort and were disappointed in the results. They've felt that it was hopeless. They usually gave up.

SETI is a good idea, and one should carry it on until *absolutely* sure that there's nothing useful in it. Self-censorship is one of the most dangerous things on earth. More good ideas are not carried out because people are afraid that somebody else will not approve. Good heavens, we wouldn't have our society today if that attitude was dominant.

Any additional thoughts about your role in SETI?

Among the people in SETI, I am the oldest. Barney's the next. I spent my time doing what I thought needed to be done. And it's been everything from editing volumes to doing nonsense; doing some engineering, some science, trying to haul people down to earth when their flights of fancy get a little too much.

People within the SETI program, or outside of it?

Both. Up until recently, when the load was taken over by other people, I was giving three or four public talks a month. Billingham and I were the two main speakers for SETI. You get tired of it after a while. I was talking about it from the time Cyclops was over, whenever I had a chance. In part I wanted to find out how the public reacted.

In the last several years it's become a rather popular topic, and that was the result of our activity here, and the activity of people like Drake and Sagan. At one time, around 1970, there might've been one textbook that mentioned the idea, an introductory text on astronomy. Now it's in most of them, and each author expresses his own views.

Our group, inside and outside of NASA, has talked so much, so positively, that naturally we have gotten people to rise in opposition, and they write their books and keep the whole subject alive. Nothing wrong with that. All I try to do is keep everybody remembering that observation is the ultimate test, so let's not waste too much time theorizing.

What else should future historians know about you *that would not otherwise be preserved?*

To be in on the founding of a brand new field is simply one of the world's greatest things! I've had so much fun with the breaking of radio astronomy. First of all, there are a small number of people. I like to work with individuals and not with journals. It's much more enjoyable to develop things rapidly with people in the beginning. And here was SETI, as near a parallel as you could want to the breaking of radio astronomy.

Anyway, it was a natural, to my way of thinking, and I think I've been useful to the development here. It's exciting now to be getting down to some observing; we can look forward to it soon, beginning seriously at a level that is worth doing. I don't believe in doing *anything* at all; I like to have whatever I do *last*, at least a few years. I don't like to just make work for the sake of a paper or something.

In my life so far I have had a wonderful time, an absolute ball much of the time. Through accidents of abilities, training, and propinquity, I have sometimes been in the right place at the right time: sometimes way out of synch. I have, at times, been able to carry my own weight with people I thought were the best.

There have been less productive, even dark times, too, of course, but they never lasted unduly long. There is no point to indulging in self-pity because one has contributed to an almighty foul-up, or just been in the wrong place.

On the whole, life has been really a lot of fun.

Life in general or SETI in particular?

Life in general, and my hand in it. You can't separate them because one flows into the other. SETI is just a small part, and in my later years. While I was a power in a university faculty senate I found those activities interesting and rewarding. Distracting from research, yes, but I did something needed, and learned a lot in the process, and I enjoyed and used what I learned.

Nothing ventured, nothing gained. An excellent motto. We need to encourage people not always to worry about security, but to take a chance and try something constructive.

Before I am completely dotty I want to write about some of these things. I would like to try my hand at reminding people that the human race has an ancient wisdom that is as relevant today as it was long ago.

Freeman Dyson

PHYSICIST

Born December 15, 1923

Crowthorne, Berkshire, England

Professor of physics at the Institute for Advanced Study, Princeton, Freeman Dyson formulated a new theory of quantum electrodynamics and contributed to nuclear reactor technology, atom-powered rocket design, and astrophysics.

In his approach to SETI he differs from some of the other pioneers. He recommends searching for passive indicators not deliberately sent, such as heat given off by the activities of extraterrestrials. For example, he suggested that advanced civilizations might conserve the energy from their central star by building a gigantic sphere around it. We might detect the heat from this "Dyson Sphere" by searching in the infrared as well as the radio region of the spectrum; and we also could watch for "skid marks" left as alien spacecraft traveling around the galaxy at high speed apply their brakes, releasing energy in order to slow down.

Beginning his higher education at Winchester College, England, he attended Cambridge University from 1941 to 1943. For the next two years he performed operations research for the RAF Bomber Command. He received a B.A. in mathematics from Cambridge in 1945, and from 1946 to 1949 was a fellow at Cambridge's Trinity College. Coming to Cornell University in 1947 on a Commonwealth Fellowship, he joined the faculty as professor of physics in 1951. Two years later he moved to his present position at Princeton.

An advocate of colonizing space, he is also very concerned with terrestrial matters of war and peace. Previously opposed to a nuclear test-ban treaty, he changed his mind after being a consultant to the Arms Control and Disarmament Agency and the Defense Department. Today he campaigns actively for nuclear disarmament, the subject of his 1984 book, *Weapons and Hope*.

His other books include *Disturbing the Universe* (1979); *Origins of Life* (1986); and *Infinite in All Directions* (1988). In addition to technical reports he has written popular scientific articles for magazines, particularly *Scientific American* and the *New Yorker*, and has appeared on PBS television.

He was awarded the Hughes Medal of the Royal Society, the Max Planck Medal of the German Physical Society, the Lorentz Medal of the Royal Netherlands Academy, the Robert Oppenheimer Memorial Prize of the Center for Theoretical Studies, and the Harvey Prize by the Technion, Haifa, Israel.

Honorary degrees have been conferred upon him by Princeton, Yeshiva, Glasgow, York, the New School of Social Research and Rensselaer Polytechnic Institute.

He was Chairman of the Federation of American Scientists, and is a fellow of the Royal Society. His other memberships include the U.S. National Academy of Sciences, the Bavarian Academy of Sciences, and the American Physical Society.

Interviewed January 1984 at Princeton, N.J., by telephone

Where did you live during childhood and when you were growing up?

Most of the time in Winchester, which is a small town in England. It is an old cathedral town and that's where we lived from the time I was one until I was thirteen.

What were your childhood interests?

I was always interested in mathematics. I loved to calculate from a very early age. I don't think I was particularly scientific; I just loved numbers. But I *did* like to go star gazing. One of the things which I remember from those times is being taken up on to the roof of a building and looking through a telescope at some planets. So clearly, I was interested in astronomy, but not with any serious passion. I've never made telescopes or anything of that kind for myself.

Roughly how old were you during those years?

I was about seven or eight years old when I started drawing pictures of the solar system and the orbits of planets and that kind of stuff. And I certainly was aware of the astronomical surroundings. I think that came mainly from reading books, rather than from doing anything myself.

What incidents involving science do you remember from those early years?

The thing which might be most relevant is a piece of writing which I discovered stashed away in my mother's papers after she died a few years ago. She had preserved a piece of my writing which I had totally forgotten about. It was a story in a style copied from Jules Verne, about an expedition to the moon. The purpose of the expedition was to observe a collision between the moon and the asteroid Eros. Looking back on the dates, Eros came very close to the earth in the year 1930, and I must have read about that in the newspapers. It was probably being talked about a lot.

So I had this story in which the astronomers calculated that Eros and the moon were actually going to collide, so they sent an expedition to observe the event. Unfortunately, the story breaks off at that point. This was very much in my mind at the age of eight.

I was an avid reader of Jules Verne and H. G. Wells. I think my first introduction to ETI was Wells' *First Men in the Moon*, which I still think is a magnificent story. That started me thinking about the subject.

This was when I was about ten, because I went off to boarding school, as English children do—at least middle-class English children. At the boarding school I found a more or less complete shelf of Jules Verne and Wells. And I read the *First Men in the Moon*. It made an extremely vivid impression; I can still remember it almost word for word, with its description of lunar landscape and lunar vegetation growing on the monthly cycle, and things of this sort. It's a marvelously written story and the ending is very dramatic. I recommend it strongly. I don't think anything better has been written since.

Another book I had as a youngster was a glorious book called *The Splendour of the Heavens*. I haven't been able to find it since. It was a huge, fat thing, about a thousand pages. It was a compendium. It was edited, I think, by Crommelin. It had chapters written by everybody you can think of, mostly on a high level, informative. It had chapters about the whole astronomical universe, starting with the earth and going all the way out, as far as it was known.

It was a splendid book, and I'm sorry it has disappeared from sight. I don't even have my own copy. I wish I had kept it for my kids, but I didn't. It was probably published around 1930. It was given to me by my favorite godfather; I treasured it for many years. It may have had things about extraterrestrial life in it also, but I'm not sure. It was extraordinarily well written for a teenager level, but also pretty sophisticated. I don't think anything as good as that exists today.

How was your elementary schooling?

I would say that schooling meant very little. The great advantage of our school was that they left us a lot of time. They didn't keep us sitting in class all day long, so we had time for reading and extraneous activities.

Did it give you helpful preparation for a scientific career?

Not in any active sense. By itself, school wasn't particularly exciting; the school just left us alone.

What about during adolescence? What activities do you recall from those years?

Nothing much. The childhood phase of my life ended around the age of twelve, and during adolescence I was much more concerned with mundane questions.

What about secondary school?

I went to a very good high school and got a good education, but not particularly scientific. I wasn't spending a great deal of time on science. There again, we had an excellent library. I was certainly doing mathematics seriously at that point.

Was mathematics part of the curriculum or was that on your own?

Both.

When did you first think about the possibility that extraterrestrial intelligence might exist?

I think the answer is Wells' *First Men in the Moon*. Subsequently, I read a great deal of other science fiction; I've always been a bit of an addict. I read Olaf Stapleton later on. It had a very big impact: *Last and First Men, The Starmaker*.

Did you discuss this possibility with other people?

I don't remember, but no doubt we did. I'm sure I wasn't the only one who read this stuff; this was part of the general discourse at that time.

What about communicating with, or at least searching for, ETI? When did you first think about this possibility?

I make a sharp distinction between searching and communicating. The idea of communicating I never had considered until I read Cocconi and Morrison. Cocconi and Morrison raised that question. So the idea of communicating, as far as I'm concerned, dates from 1959.

The idea of *searching* I had thought about earlier. I don't remember just when or how. The basic notion that infrared is the thing to look for if you are dealing with high technology had been in my mind for a long time, but I don't remember just when.

During your college years, did anybody or anything stimulate you to think about ETI and the possibility of searching or communicating with it?

I don't believe so. I was reading Olaf Stapleton at that time, but I wasn't thinking about the subject actively.

Since you completed your schooling and began working as a professional scientist, who or what has influenced your thinking about ETI and SETI?

The answer to that is that I only started thinking about it in 1959, more or less the same time as everybody else. The person I talked to at that time was Arthur Kantrowitz. I don't know whether he was at Cornell then or had recently left. Anyway, he was closely in touch with the Cornell group, and I found him stimulating to talk to. I think my ideas on the subject were worked out through conversation with Kantrowitz.

What effect, if any, has your interest in ETI had upon your career as a scientist?

I would say zero, as far as my professional career is concerned. It's had quite a lot of effect on my career as a guru when I go around talking to students. In my career as a teacher, it has some effect. Last month I was talking with a group of Cub Scouts here in Princeton. They only wanted to talk about E.T. I've found it very helpful as a subject to talk about to all kinds of people. It has broadened my horizons by getting me in contact with people who normally I wouldn't speak to.

What do your colleagues think about your interest in SETI?

My colleagues are very polite. They never said anything nasty about it. They may consider that it's a bit childish; especially some of these younger colleagues who are very serious about their work. I think they feel all this is a bit foolish. My older colleagues, the people I know well, are quite sympathetic.

Do they ever discuss it with you or ask you about it?

Oh, yes. I'm not doing much myself in this area. To me it's only a hobby. But we frequently chat about this over the lunch table.

What about your family?

It's a plus, as far as they're concerned. It's one of the few parts of my professional life which they can comprehend. I think it helps from that point of view.

What about your superiors, bosses? What do they think about your ETI interests?

I have only one boss, who is the director of our institute. I don't think I've ever questioned him on the subject; he's never raised any objections.

What about your friends?

There again, I have a lot of literary friends (I am a rather literary sort of scientist myself), and for them it's a point of contact. It's a part of science which literary people can sympathize with and understand. So there it's quite helpful.

Before going on to discuss your ideas about SETI, let's get a little more background information. What was your father's occupation?

He was a professional musician who made his living as a school-teacher through my childhood. Afterwards he became director of a conservatory. He also was a composer. He was also very much interested in science. Lots of the books I read were taken off his shelf. He read Eddington and Jeans and Whitehead. On a popular level he was very well informed.

It also happened we had a family friend, whose name was also Dyson, Sir Frank Dyson, who had been Astronomer Royal. My father knew him, although we weren't directly related. That had also some influence. Astronomy was in the family. Sir Frank was held up to me as a role model.

How much education did your father have?

Not very much. He was a music student at the London Conservatory. Apart from technical training, he had only a rather sketchy high-school education.

What about your mother; what formal education did she have?

She had much more. She was an LL.B.; she had a law degree from the University of London.

How would you describe them?

They were very social in the musical world. My father loved conducting and running a choral society and things of that sort. He had a lot of public activities, but they also had a very private life at home. They enjoyed it both ways.

How did they influence your eventual choice of science as a career?

They supported this strongly. It was obvious from the start that I wasn't going to be a musician, and it was obvious that I had some mathematical talent. So it was a question that answered itself.

And what about ETI? How did they influence you in this respect?

I don't remember any particular influence. I think my ideas came out of books more than from them. I don't remember ever having any serious arguments about that.

Were there other people we haven't mentioned who influenced your ideas about science and ETI?

No, I don't think so. I suppose the reason why ETI appealed to me so strongly was because it is a bridge between the two cultures. I've always been halfway between science and literature; and this is exactly where ETI stands.

What religious beliefs did your parents have?

They were not very well defined. They belonged nominally to the Church of England. But this was a background without any dogmatism.

How many brothers and sisters did you have?

I have just one sister, who's still going strong. She's three years older than I am. She used to be a medical social worker. Now retired.

Did she influence your thinking about science and ETI?

No, on the contrary, she's always been more interested in literature, and so with her I talk about literature. I don't think the subject ever came up with her.

What about your friends when you were growing up; did they share your interests in science and ETI?

I had friends, but I don't think we ever talked about it. That was not a time when ETI was very much in my mind, either. This came later, after I was in America.

What were you like during childhood and adolescence?

I suppose average. Everybody's different at different times. I certainly had times of shutting myself up, but mostly I was reasonably gregarious.

What other professional interests do you have, apart from SETI?

SETI is hardly a professional interest. It's more a hobby for me. My main professional interest is astrophysics, particularly the dynamics of star clusters and galaxies, at this moment. But I'm interested in all kinds of applied mathematics. Any problem in science where one can find something useful to do with mathematics, I find interesting. My personal interests are mostly in astronomy. But in addition, I do a good bit of writing.

What subjects have you been writing about in the last several years?

I've written about arms control, which is the main thing I've been doing recently. And I like to write popular science. What I've done in ETI has been writing popular articles about it rather than working.

What recreations do you enjoy?

Most of my recreation is bringing up a large family; there's not time for much else.

What has been your greatest satisfaction in life?

I would say raising a family has been the largest satisfaction.

What has been your greatest sorrow in life?

Greatest sorrow, it's hard to tell. I've had on the whole such a fortunate life. I would say the greatest sorrow was the time I spent at Bomber Command in WWII. Seeing that bombing campaign going wrong and not being able to do anything about it. That was my introduction to tragedy.

What was the most memorable moment or event in your life?

Figuring out how to do quantum electrodynamics on the bus coming home from Berkeley. That was the most dramatic event; solving the quantum electrodynamics problem.

I had been working very hard on this problem of trying to understand the interaction between light and electromagnetic fields and atoms. It was at that time one of the central problems of physics, and I'd been working very hard on it for months. Then I went for a vacation to California. I didn't think about it for several weeks. Then I got on a Greyhound bus to come back east. And while I was sitting there on the bus the whole thing suddenly became, somehow, transcendentally clear. It was one of these moments of illumination which every scientist hopes for. It only happens once in a lifetime, but anyway, it did happen.

Turning now to SETI, about what fraction or part of your working time do you spend on it?

Less than 5 percent. It's not "work" anyway. If I'm doing anything with it, it's writing, and it's certainly less than 5 percent.

What stages has your thinking about ETI and SETI gone through?

My ideas stayed pretty much the same. I've always felt that one ought to be looking at the uncooperative society rather than the cooperative one. The idea of searching for radio signals was a fine idea, but it only works if you have some cooperation at the other end. So I was always thinking about what to do if you were looking just for evidence of intelligent activities *without* anything in the nature of a message.

And so from the beginning I had thought of infrared as being the preferred channel, since it's a consequence of the second law of thermodynamics: that any high technology must radiate away waste heat, and the place to look for that is in the infrared. So that was my main contribution to this field: to suggest a second channel. After you've done the radio search, or at the same time as you're doing it, carefully search the sky for infrared sources.

I still think this is a good idea, although infrared astronomy has developed more slowly than radio astronomy. Finally, now, we have the IRAS data, which is going to give us a more or less complete sky survey in the infrared, so we'll be able to identify sources and make a catalogue, and see which ones might be of interest. That gives you a catalogue of places to look in other channels. This is still a sensible way to go, although it's taken a little longer than I expected. I had thought we would have an infrared sky survey much sooner than we actually have.

What alternatives, what other possibilities about ETI have you considered?

My thinking has been to try to make the minimum of assumptions. To try to devise a strategy which will work more or less independently of details. That's why I like the infrared; it doesn't require assuming anything special about the nature of ETI, except that it should be technological and carrying on activities on a large scale. The reason for emphasizing the large scale is, of course, that the larger the thing is, the more chance you have of seeing it. Automatically, the largest-scale activities will be selected as the ones you observe.

I've looked at a few other things. I've been interested in looking for passive signals. That is, signals which are not deliberately aimed at us. One possibility which I was thinking about was interstellar brakes. Suppose that an extraterrestrial society has gotten to the point of interstellar travel and is cruising around the galaxy at high speed. Then it will need to have some system of brakes to slow down. There would be a large dissipation of energy in the interstellar plasma; no matter how you do it, you've got to dissipate large amounts of energy. So the idea is to look for skid marks on the road. That should be done in the radio channels.

I don't take this very seriously, but it's something you might look for if you think that there are any high-speed interstellar vehicles around. You'd expect that you would occasionally see straight-line radio emission, that is, straight tracks in the sky where these vehicles have put on the brakes and are pumping out energy into the interstellar gas at a high rate. You'd get then a large concentration of radio emissions, synchrotron emissions in the interstellar magnetic field, concentrated along a straight track.

That would be something interesting to look for. I don't know if anybody's ever taken the time to look for it. I've never published it formally. I've talked about it occasionally to radio astronomers. But radio telescopes nowadays practically never look in the sky in an unprogrammed fashion. They are always looking at some particular object. All kinds of interesting things could be going on in the sky and nobody would notice.

What does this suggest in terms of programs or activities? How can such a situation be modified, or should it be modified?

I don't know; I've never thought about this. But it would be nice to have a general scan of the sky in the radio band, corresponding to the sky survey plates in the visible. Of course it's a huge project—so much sky and with such tremendous range of frequencies—you'd have to think very hard to find a cost-effective way of doing it. Just displaying the data is an expensive business, but you might turn up a lot of interesting things.

I was enormously impressed with these few pictures that we've got from IRAS of what the sky looks like at 100 microns. Just seeing a big patch of sky at 100 microns gives you a completely new picture of things. All these cirrus clouds or whatever they call them, floating around; it gives you a different feeling of how the distribution of interstellar matter really looks.

It'd be nice to have pictures like that in the radio, which so far as I know don't exist, using lower resolution than is customary, but covering a wider field. It would be interesting to look for short-lived objects, but that is always difficult.

What would you do differently if you were starting over again to study ETI?

I wouldn't do anything differently myself. I think the program has actually gone ahead rather sensibly. Only one bad mistake was made, and that was the Cyclops Project, which I think set back the subject by about ten years. It gave people a false impression that in order to study ETI you had to do something on a grand scale, which just wasn't true. That's the thing to avoid at all costs: to give people the impression that this is going to cost a billion dollars. But apart from that, I think we've done pretty well.

Why is ETI important to you?

This question almost answers itself. It's just one of the fundamentally exciting questions, and it always has been. The only thing about it is: How can you possibly plan to answer it? As soon as Cocconi and Morrison came along it was obvious that this was something that one would have to do.

What has been the most difficult aspect of your ETI work?

I don't think I've had any difficulties at all. What I have done was a very simple and elementary calculation. And I had no difficulty selling it to the public.

How do you think most scientists today view SETI?

I think most scientists consider it legitimate. But there are very few who would be willing to sacrifice their own telescope time to do it, for very understandable reasons. If you've got a limited amount of telescope time, you want to do things that will really pay off. Most of the astronomers that I have spoken to about this would say they are very happy that other people do it, but they don't want to do it themselves.

Do you think the public is more enthusiastic about SETI than scientists are, about the same, or less enthusiastic than scientists are?

I've not taken any polls on the subject. I would say it's not a very meaningful question. Both in the scientific community and in the public, it's a rather small minority that is serious about SETI, and they're about equally serious. The majority is not concerned one way or the other. Maybe a somewhat larger fraction of the public is concerned, but I'm not sure.

Realizing that it is only a guess, when do you think contact with ETI will be made?

I dislike the word "contact." I think that it's the wrong question. I think it's quite likely—naturally it's purely a guess—that we will discover evidence of ETI, good enough evidence to be fairly conclusive, within one hundred years.

I don't think that this would mean contact in the sense of communication, because it's likely that we will find evidence of their existence long before we're capable of communicating. So I would not say "contact." I wouldn't use that word; it's ambiguous. I would say we have a good chance of *discovering* ETI in one hundred years, but to establish real communication may be a thousand years; probably much longer.

It all depends on how far away the creatures are. The universe is so big, any place that is uniquely interesting is almost certain to be far away. If ETI are scattered around in the universe doing things on different scales, there is a good chance that the first one we find will be big and far away rather than small and nearby.

The scale of it can be more or less arbitrary. If we find things that are very far away, then it may take millions of years to make contact, but still, it's an important discovery nonetheless.

What might ETI be like?

I had an interesting discussion on this point with the Cub Scouts a few weeks ago. I was talking about E.T. in the movie. What I was discussing with the Cub Scouts is how to distinguish sense from non-sense, what is reasonable and what is not reasonable in that movie.

The conclusion I came to was that the whole thing was rather reasonable, except for a few details. If there were an extraterrestrial, there is no reason at all it shouldn't look like E.T. The only thing it would not do would be to come down in a glorified juke box. But in all other respects, the story is plausible. There's no scientific reason why that particular story is incredible.

There's no limit to strangeness. The most likely form for E.T. is something we never imagined.

I should also mention the description by Bernal of the ultimate form that human life might take. Bernal was looking into the future when he wrote a little book called *The World, the Flesh and the Devil*. He published it in 1929, stimulating a lot of people's thinking on this question. He describes the final stages of life as more or less totally dematerialized, as having got loose from matter and inhabiting pure radiation. You see echoes of that in Arthur Clarke. His stories are always tending toward dematerialized creatures. That may very well be right. There is no limit to how strange it might be.

Any other thoughts on this?

There again, my main concern is the narrowness of the points of view that have generally been followed in this area. The paper of Cocconi and Morrison was beautiful and a great contribution, but somehow everybody's got stuck with that. It is assumed that the possibilities that Cocconi and Morrison were talking about are the only ones to consider; that, I think, is far too narrow. For people to think of extraterrestrials as looking, technologically, essentially like us and communicating with radio signals in the way that we do—that seems to me to be very dubious. That's why I like to think of other possibilities, and particularly to think of the consequences of high technology which follow only from the laws of thermodynamics and not from any detailed assumptions.

Are there some particular things here that we should be thinking about that we haven't mentioned before?

There's one nice thing which I don't know whether you are aware of. The Air Force took some satellite pictures of the Earth at night in the infrared band. This was passive infrared, in the two-micron band. Everything you see is man-made. It's spectacular.

It is quite extraordinary that as you look at the Earth, absolutely nothing you see is natural, except for tiny little traces of the Aurora Borealis around the North Pole. Everything you see is human technology. And in particular, what you see most of is oil fields. The brightest spot on earth, it turns out, is Kuwait. All about the Persian Gulf is brilliantly lit.

I find that very significant. It shows that this is the right band to look at, if you're looking for artificial objects. It follows from thermodynamics that that must be so, independent of details. The two-micron band is a band in which it is almost inevitable that there would be a lot of interesting signals if there's anything artificial going on. But the thing could be done just as well anywhere out to ten or twenty microns; the preferred wavelength depends on the temperature of the source.

What scientists or other people in the past do you admire?

I come back to H. G. Wells, who I think was extraordinarily successful—not in predicting technology, which on the whole he did badly—but in imagining in a plausible fashion what an alien society might be like. I don't think that anybody's ever done it better. I've read a good deal of science fiction. Wells was, in a way, the grandfather of all this.

Who are the outstanding people in SETI today?

They're the people on your list. The people I've been talking to who may not be on your list are Paul Horowitz and Arthur Kantrowitz.

What role has each played in SETI?

Arthur Kantrowitz played a role at the beginning, getting people excited, and talking with people in the early days. Paul Horowitz is the youngest of the recruits. He's now building data-processing systems for doing ETI in a very narrow band. I think he's very good. He's one of the people who is doing the most professional work now.

I would be curious to have you go over some of the more familiar names. Starting with Morrison; how would you characterize his role?

Very creative. Cocconi and Morrison really did supply the initial kicks to this enterprise. Morrison is one of the most imaginative people I know; I would put him number one in this whole thing. I don't pretend to divide the credit between Morrison and Cocconi; I think that should be shared.

I have a very good relation with Kardashev in Russia. I saw him last about five years ago. But I'm happy to see he got to Samarkand. He loves central Asia. He was the one who, besides myself, has always emphasized looking for the uncooperative targets, and his ideas about Type One, Two, and Three civilizations are very reasonable. It gives one a language in which to talk about things on a large scale.

How would you characterize your own role in SETI?

My role's mostly been a cheerleader. I haven't really done anything, but I've encouraged the people who do. And I'm always hoping that somebody will take seriously this idea of an infrared search. But first of all we need a sky survey to pick out interesting sources, which we don't yet have. That's going to take a long time. Otherwise, my role has been mainly popularizing, going around talking about this to people and telling them what's going on.

What else should future historians know about you that would not otherwise be preserved?

There's one aspect of things which I find amusing: the flow back and forth between science and science fiction, which has been an important part of my life. I started out reading science fiction and then became a scientist, and that set the slant on my scientific work. I like to make connections between life and cosmology and astronomy. Science fiction raises all these interesting possibilities and has had some influence on science in the last twenty-five years—not only in the area of SETI, but also in other ways.

And now I have the experience of seeing the thing go the other way, because of my modest little contribution to SETI, the only thing I've published on a professional level: a one-page letter in *Science* magazine about the search for infrared sources. That was picked up by the science-fiction community, which gave it the name of the Dyson Sphere. So I'm known to the science-fiction fans as the originator of the Dyson Sphere, which to me was only a joke.

But still it's an interesting fact of life that the traffic goes in both directions. It may well be that, after everything I've done in legitimate science has been forgotten, this is the one thing I'll be remembered for.

Could you say something about the Dyson Sphere?

What is the largest feasible technology? This is essentially the question I asked. Since I was interested in looking for infrared sources, clearly the sources you're likely to discover are those that are as large as possible. So the question is, "What is the largest feasible size that a technology could reach?"

If you think of a technology based on a single star, your answer is fairly clear. You have exploited the whole energy output of the star, converted it all into waste heat. Which means then, you have to imagine a biosphere which has been constructed all around the star, with habitable living space in which there are creatures living and machinery and all the rest of it, exploiting fully the starlight and radiating away the waste heat on the outside in the form of infrared.

I suggested that this is what one should look for if one is looking for large-scale technology. So this notion of the artificial biosphere, completely surrounding a star, is a thing that the science-fiction people picked up. They called it the Dyson Sphere; I didn't.

What are UFOs?

Mythical animals. The contemporary reincarnation of phoenixes and unicorns.

What should a scientist do upon receiving convincing evidence of extraterrestrial intelligence?

Call up the International Bureau of Astronomical Telegrams.

What would be the public's response to this information (about the existence of ETI)?

Skepticism, probably justified.

What would be the long-term effect upon humanity?

Wait and see. We will get used to it, just as we got used to angels and devils.

What do you think of this fact that several of you SETI pioneers have been at Cornell? Was it pure coincidence?

Well, clearly it's not a coincidence. Cornell always was a place where imaginative people felt at home. I don't know to what extent Tommy Gold was involved in this, but certainly a lot of us talked to him. I did anyway, and he acted as a catalyst at Cornell during this time. Kantrowitz also. I don't remember whether Struve was at Cornell, how it happened that he got involved with this.

The fact that Cornell was the place was largely due to the fact that Cocconi and Morrison and Drake were all three there. They obviously talked to each other.

And your connection with Cornell was before their time—is that correct?

Yes, I left Cornell in 1953, but I've always kept in touch with them.

Any final thoughts?

My role has been pretty modest. The most important fact about SETI is that so much can be done with so little outlay of funds. This is something the public ought to be more aware of. What it takes is not really much money, but just a certain amount of brains.

Kunitomo Sakurai

PHYSICIST

Born May 27, 1933

Kodama, Saitama Prefecture, Japan

Kunitomo Sakurai is a professor in the Institute of Physics, at Kanagawa University, Yokohama, Japan. One of the first scientists in Japan to seriously consider the possibility of ETI, he has been a leader in encouraging other scientists and the public to do likewise. Several of his publications deal with the subject, including a book entitled *Search for Life in the Universe* (1975). He translated into Japanese *The Listeners* by James Gunn (1980), and *Search for Life in the Universe*, by D. Goldsmith and T. Owen (1983). He meets regularly with a small group to discuss SETI and related issues.

Educated at Kyoto University, as an undergraduate he emphasized cosmic-ray physics and geophysics. In graduate school he concentrated on problems related to solar flare particles. After receiving his Ph.D. in 1961, he continued his work at Kyoto University until 1968, first as research associate and later as associate professor.

From 1968 to 1974 he was a senior research associate and then a resident scientist at NASA's Goddard Space Flight Center. From 1975 to 1977 he was a professor at the University of Maryland's Institute of Fluid Dynamics and Applied Mathematics. Then he assumed his present post at Kanagawa University.

He has published over 150 technical papers in English on solar physics and high-energy astrophysics, in journals including *Solar Physics, Astrophysical Journal, Nature, Astrophysics and Space Science*, and *Journal of Geophysical Research*.

He has written eighteen books, three of which are in English: *Physics of Solar Cosmic Rays* (1974); *Cosmic Ray Astrophysics* (1988); and *Neutrinos in Cosmic Ray Physics and Astrophysics* (1989). He contributed a chapter on cosmic rays to *The Solar Wind and the Earth* (1987). His newest book is on SETI, written in Japanese for children.

He was the chairman of the Research Committee for New Communication Techniques, created by the Ministry of Posts and Telecommunication of Japan, from 1985 to 1988. He served on the Advisory Board of the Institute for Cosmic Ray Research, University of Tokyo, from 1977 to 1987.

Now he is a member of the Committee for Frontier Technology in Electrical Communication for the Ministry of Posts and Telecommunication, and of the HOPES Committee on the Observation of Earth Resources from Space.

His other memberships include the International Astronomical Union's commissions 10 and 51, International Association of Geomagnetism and Aeronomy, New York Academy of Sciences, American Astronomical Society, American Association for the Advancement of Science, Astronomical Society of Japan, Physical Society of Japan, and American Geophysical Union.

Interviewed February 1980 at Yokohama, Japan

Where did you live during childhood and when you were growing up?

I lived in a small village near Kodama, about a hundred kilometers northwest of Tokyo. Kodama is surrounded by mountain ranges. I went to elementary school there and then to junior high school in the town of Honjo and to high school in Kodama.

What were your childhood interests?

In my childhood, up to the age of eight, I was physically weak. I therefore made model planes, using paper and bamboo, and built other things. Using handmade electric motors, coils and other things, I did experiments on electromagnetic induction and other topics. And particularly, I loved drawing and painting at home.

In elementary school one of my classmates once lent me a book on astronomy for children, and I made notes from this book. I was fascinated by stories of stars and the universe. Later, I came to know this book was written by Professor Issei Yamamoto of Kyoto University.

Junior high school was nine miles from my home. I rode there and back on my bicycle every day. In the morning it was downhill, but in the afternoon it was harder to climb back up. I never missed one day. I liked to play baseball very much so I stayed after school to play for two hours. Then I had to go home to feed our animals. We had one bull, four sheep and a goat.

I studied biology, in particular botany. As school activities, I belonged to the biology and Alpine clubs and went to collect plants and did fieldwork on vegetation. Every summer I went mountain climbing.

In high school I was interested in biology and physics but at first was very bad in mathematics. Then I became interested in mathematics and studied hard many books on higher mathematics.

I grew up on a mountainside in a family of farmers. My environment was very helpful for the start of studying life science. At the front of my house was a small river that had a variety of fish. In my childhood I used to go there all the time, independent of the season. I caught some of the fish and brought them back to my home. I learned their living cycle, from egg to mature state. I could observe how they fertilized the egg and made small fish and how they grew up. I became very much interested in doing so in my childhood, maybe six or seven years old, then up to fifteen or sixteen years old.

Also when I went to the mountainside I might find many different insects: locusts and others. I took some of them to cultivate, breed them in my home. I also could observe their living cycle. I made some experiments on heredity by using morning glories. I took pollen from some specific colored flower to make hybrids. Such experiments were done for maybe two or three years, during my junior high school age.

My family job was farming rice and some other crops. I thought I would just be a farmer some years later, because I was the first son of my family. Traditionally, the first son must succeed the family's job. I never thought I would become a scientist.

But in high school during the second year, 1950, there was an act of god. My father attended a PTA meeting. My teacher, the advisor-counselor of the class, suggested to my father that I should be given some higher education by going to a university.

That night after Father came home I was asked to come to his study. He told me of my teacher's suggestion and asked me if I was interested in taking higher education. I was so pleased to hear of this advice from my teacher.

I still thought the most interesting subject was biology, because I had known for years about the life of fish, insects and plants. Many times I went up on the mountainside to find some specific plants. I was very much interested in the vegetation on the mountain because, in climbing up the mountain road, the vegetation gradually changed. We could find general trends in the vegetation dependent on the height, the elevation. That was quite interesting to me to study why.

Also I was interested in studying cultivated plants because my father was planting apple and orange trees on the farm, but they were not successful. For the apple tree the climate was so warm, but for the orange tree it was too cold in the wintertime. Father tried to find trees that were well adjusted to that climate there, but he didn't.

Though reluctant at first, my father finally allowed me to study to pass the entrance exam for university. Most people at that time wanted to go to the University of Tokyo, but I wanted to go to Kyoto because the faculty members at Kyoto's biology department seemed wonderful to me.

Had you met them?

No, but I knew about them by name. Around that time at Kyoto there were very famous scientists working in biology. Professor Kihara was world famous for his research on wheat. He made an expedition to the Near East, and in junior high school I bought a book written by him, *The Ancestry of Wheat*. I still have that book, and when I open the pages I recall how excited I was in my young age to study biological science. That book is worn out. It was originally published around 1946 or '47, just after the war. Good paper was not available, so very bad paper was used for the book.

Also professor Komai, a very famous geneticist, was at Kyoto. I wanted to study biology under the guidance of those famous scientists. That's why I selected to go to Kyoto. At that time I never thought I would be a physicist.

The first two years for general education all students had to take three courses of science. Of course, I selected biology as the top. The two other subjects I took were physics and chemistry. At that time physics was taught by Professor Tamura, who was a teacher of Professor Yukawa, the first Japanese to win a Nobel Prize. I took physics from Professor Tamura for two years, starting as a freshman. I didn't understand so well his course, but his teaching style and his attitude and research in physics were quite appealing to most of the students.

I didn't understand very much because I didn't take a physics course in high school, since the teacher had been sick the whole semester. For me the first meeting with physics was at college.

By taking Professor Tamura's courses I gradually became very much interested in physical science, but still I wasn't so sure that I would major in physics the last half of my university education. But in my sophomore year I read in the newspaper that a geophysicist at Kyoto University received a prize for research in space physics, studying cosmic rays. That article impressed me very much.

At the end of the sophomore year all the students had to decide which fields they would major in. Then I picked physics, especially oriented to geophysics. In my last two years I studied mostly physics and mathematics.

Around that time I forgot about studying biology, but still some interest remained somewhere in my brain. I heard something about new discoveries in biology. Mostly I read articles in two Japanese science journals, *Kagaku* and *Shizen*. In English "Kagaku" corresponds to "science"; "Shizen" to "nature." Around that time molecular biology was just in its initial stages of development, because of the 1953 famous report about DNA. That report was very interesting to me, even though I was working in physics.

After two years studying physics, I decided to get more specific study at graduate school. Then the most interesting subject to me was cosmic-ray physics, not biology.

When I finished graduate school in 1961 I moved to the faculty of engineering, because a leader of radiophysics, Professor Maeda, was there. He was one of the pioneers in Japan to study radio propagation in the upper ionosphere, and also shortwave communication on a worldwide basis. He was one of Japan's leading scientists. I was picked by him to be his research associate.

Then I began to study radio science and ionospheric physics by using techniques of radio engineering. I started to study radio communication methods and radio propagation in the upper atmosphere. Also I studied some problems of antennas. That was in 1961. I could use some of my knowledge about space physics to study those problems.

Then I became interested in radio astronomy, too, because radio technique is also applicable to radio astronomy. If we need to make some measurement about ionospheric problems, our technique can also be applied to study cosmic radio noise and other special problems.

I made some experiments to observe galactic radio noise, using antennas at three different radio frequencies: 15, 30 and 50 Megahertz. The 30-Megahertz frequency is very well fitted to study the ionosphere, but the two other frequencies can be used for observing galactic radio emission, especially background radio.

I was very excited because I could have experience very similar to that of Karl Jansky. By using these same antennas I could study solar radio emission, too, because those antennas were very sensitive to catching radio disturbance in the solar corona. I published papers about Type IV radio bursts in low frequency in association with solar flares. I also did a paper about galactic radiation.

At that time the expansion of my research field was inevitable, because I belonged to the engineering faculty. I worked there seven years, 1961–68. Not much biology in this, but even so it still comes back repeatedly to my heart.

When did you first think about the possibility that extraterrestrial intelligence might exist?

People usually hear about it in their childhood, and when I entered the university I heard some programs related to it. At that time in Japan there was a very popular book written by Professor Schrodinger, the man who developed the idea of quantum mechanics: *What is*

Life? Biophysical Processes of Evolution. He never mentioned anything about intelligent life in the universe, but we read the book for courses in biology and some other classes. People sometimes talked about the possibility that intelligent life existed elsewhere, and I began to study some problems related to it in my freshman year.

Schrodinger's book has an interesting story. It was translated into Japanese in the last year of my high school. At that time, just after the war, it was very difficult to get books which were published in foreign countries, because most people had no foreign currency. Original copy of Schrodinger's book was sent from the U.S. by Professor Yukawa to his colleague who was working at the Institute for Physical and Chemical Research in Japan.

When he read it, he decided to translate it into Japanese, but around that time people who wanted to translate any kind of foreign book had to get permission from a special cultural office of GHQ, the American occupation government in Japan. Schrodinger's book was the first one permitted to be translated into Japanese by that office.

In 1952 the peace treaty was made between Japan and other countries, but before that it was very difficult to publish a Japanese translation of a book published in another country.

What or who brought your attention to the possibility of intelligent life elsewhere?

Professor Ryozo Yoshii at Kyoto University might have talked to us about the possible existence of ETI, though I am not sure.

What year was this?

1952 or '53. Not quite so long ago, about 1965 or '66, I got an idea about the possibility of communication between terrestrial and extraterrestrial intelligence when I read the book by Walter Sullivan. It was a very popular book. It was also translated into Japanese, but I had an English edition because the translation was bad. I got a copy of the original in 1965, while I was research associate at Kyoto University.

My laboratory was doing research about extraterrestrial radio emission, mainly in the earth's magnetosphere. Some people at that time may have thought about the possibility of communication with ETI because they could read that book.

Did you discuss this possibility with other people?

I often discussed this possibility with my friends while I was a student at Kyoto University, since I took the course of biology given by Professor Yoshii.

What was their response?

It might be just nonsense, always so.

What about communicating *with, or at least* searching *for, ETI? When did you first think about this possibility?*

After I studied radio astronomy and its technique in my graduate school.

Who or what suggested this to you?

After I entered the graduate program in space physics my interest gradually expanded to biological problems, especially molecular biology and molecular genetics, because I read some articles published by the very famous physicist George Gamow. He wrote some paper about the genetic code. I also recalled reading the book by Schrodinger. It showed some of the possibilities of mutation and other problems related to molecular basis.

In 1964 Walter Sullivan's book was published. It left a big impression on me. He referred to many names, many scientists, like Frank Drake, Philip Morrison and Cocconi. At that time I already knew their names because of their research in other fields of physics.

For example, Philip Morrison is a very famous nuclear physicist. He did research on parity conservation, and he also made a very important contribution to the modulation effect of galactic cosmic rays by the action of the solar disturbance, and he also wrote some papers predicting the very high gamma-ray emission from the universe.

Frank Drake also had done work on radio astronomy. According to my recollection he wrote a paper on the molecular evolution of the Martian atmosphere, or the origin of life on Venus. So I was surprised to learn that in addition he was studying intelligent life in the universe and publishing about it. I was very surprised that he could make an experiment to look for communications from life in outer space.

Around that time, 1959 or 1960, no one in Japan had such an idea. Maybe people like Professors Morimoto or Jugaku had similar ideas, but I didn't know them; I only knew of foreign names like Sagan and Drake and Otto Struve, and Morrison and Cocconi and Bracewell.

Also about that time a collection of early papers on cosmic life, *Interstellar Communication*, was edited by Alistair Cameron. I bought this book and studied it, and my interest came up again to do some work on biology. In that case my subject was not life on the Earth but life in the universe.

I came to know the name of Carl Sagan by reading some articles on the atmosphere of Mars or Venus. Then around 1966 I found his very important book written with Shklovskii, *Intelligent Life in the Universe*.

I ordered this book from a company that imported books. The salesperson was very much wondering why I had made such an unusual order, in addition to regular physics books.

Usually in Japan we have to concentrate on a very small field of research. We call it Michi, "road," some specific road, very explicit research. But when I read those books I was so surprised to learn that some people in the U.S. have a very wide view of research. Sometimes they study intelligent life in the universe, sometimes they study gamma-ray astrophysics, cosmic-ray physics—that amazed me, because traditionally in Japan there is not such broad thinking among most scientists.

That very much impressed me. Around that time I was just a young associate professor, doing research on the radiophysics of the upper atmosphere. I knew Watson and Crick's model of DNA, and also the paper by Melvin Calvin on chemical evolution. Then I started to look for some of the research papers by Philip Morrison and Frank Drake and others studying intelligent life in the universe. My interest expanded to include the possibility of communication with such life. For me it was a natural trend.

You read the famous Morrison-Cocconi article after you read Sullivan's book?

Yes, I guess so. But in 1961 or '62 I wrote a very short article about life in the universe for my high-school newspaper. I might have already read that paper by Morrison and Cocconi, but I can't recall.

The high school I graduated from was a very small one, out in the country. I was their first one to go to higher education. That's why all the teachers at the school were very proud of having me. So I was asked to write on any subject to appeal to students. They needed to hear from anyone who had graduated from that school.

While I was working on galactic and solar radio emissions at the faculty of engineering, sometimes I felt those radio waves might have some information, some signals from life in the universe, because already I had read some of the articles by Philip Morrison and Cocconi and also Ronald Bracewell. Also around that time many people became interested in molecular biology. Books were coming out on chemical evolution; also biochemistry and biophysics. I read some of the articles and books. Again I became interested in studying biological subjects by observing the galactic radio and solar radio emissions.

In June 1968 I went to the U.S. to study at NASA's Goddard Space Flight Center. I belonged to the radio astronomy branch there, and then my interest came back again about life in the universe. Radio astronomy is quite a good subject for studying it.

So your interest in communication with outer space had begun before you came to the United States?

Yes. Before I went I had already read several books published in the U.S. and other countries about life in the universe. Then, around the time I arrived in the United States, it was a kind of gold rush for the discovery of interstellar molecules, such as ammonia, formalde-hyde, carbon monoxide, and many others discovered around that time.

At Goddard Space Flight Center the first speaker in the fall season colloquium series was Frank Drake. The topic of his talk was the pul-sar. He was already the director of the Arecibo telescope, and he ini-tiated the observation of pulsars by using the huge antenna. In his talk he presented the initial data from his study, but my English was not good enough to catch every word of his talk.

I was very interested because I thought he would only be interested in searching for life in the universe, but he was also an expert in pul-sar physics. After his talk I found his book about intelligent life in the universe in the Goddard library. Also I read much about inter-stellar molecules, and also the experiments by Stanley Miller, and the possiblity that some of the molecules basic to life could be synthesized in interstellar space. At Goddard there also was a small group study-ing interstellar molecules, especially astrochemistry. All of this inter-ested me.

Before going on to discuss your ideas about SETI, let's get some background information on your family.

My father was educated at the high school for agriculture and hor-ticulture. He had a job with civil service, for the agricultural coopera-tive in Kodama. Before and during World War II he had been a high-ranking army officer. One day in June 1941 a black limousine drove up to our house in the countryside. He got in and went away, without notice to his children, including myself. Some months later I went to visit him with my uncle, somewhere in Tokyo. After that, we did not see him for a long time, because he was sent to Manchuria, the north-ernmost part of China.

Once or twice a year he came back to Tokyo secretly, for some spe-cial meeting of the army command. On these trips he brought several books to me. One was about marine life. It was quite interesting.

At the end of 1943 he was moved to a position responsible for army weapons factories in Tokyo. He kept that position for the rest of the war. In January 1946 he was finally released from his duty as an army officer. He died twelve years ago.

My mother was educated only at elementary school. Her occupational skills and work experience were those of a typical housewife. Her view of our society was modern enough, though.

They were quiet and warm-hearted, though my father was a little bit quick-tempered, sometimes.

How many brothers and sisters did you have?

Two brothers and four sisters. I am the first son, the second of seven children. My brothers are both working for a fabric industry. My elder and two younger sisters are housewives. A younger sister is working for a company.

What were you like during childhood and adolescence?

Quiet and timid, since I was not so healthy up to the fourth grade in elementary school.

What about your friends when you were growing up? Did they share your interests in science and ETI?

Yes, in science at Kyoto University, but nothing about ETI.

What recreations do you enjoy?

Collecting records of classical music and listening to them. Sometimes going to concerts. Reading books on history.

What has been your greatest satisfaction in life?

Nothing has come as yet, but maybe the writing of the book entitled *Physics of Solar Cosmic Rays*.

What has been your greatest sorrow in life?

I lost my parents so early.

What was the most memorable moment or event in your life?

I lived in the U.S. for eight years and visited many national parks in the western U.S.

What effect, if any, has your interest in ETI had upon you as a scientist?

It made me critically evaluate my view of the world civilizations, and always made me think about whether this civilized world is the only possible selection for human beings.

How important is SETI to you at this time? What priority does it have?

Not too high, because I am doing research on cosmic-ray physics, mainly solar-flare particles and solar neutrinos; some of the problems that may be related to the origin of cosmic rays. But in order to study these problems we have to know what is going on in interstellar space and interplanetary space and on the sun.

Our understanding of the origin of the solar system suggests that the sun was initially produced by action from a supernova explosion. If so, we must study also the sun's origin and the evolution of the planetary system including the earth, which was influenced by the shock wave produced by the supernova.

If we study some very old meteorites, called carbonaceous chondrites, we may find out facts about the birth of our sun and its planets. We may also find very interesting chemical evidence about the original composition of the earth's atmosphere and also the atmosphere of the giant planets and the terrestrial-type planets.

This knowledge is highly related to the possibility of the origin of life on those planets. Then we have to learn all that happened in the evolution of the planetary system in relation to ordinary life and its evolution.

That's why I have become so interested in studying extraterrestrial life. The existence of life in the universe is very closely related to all of the physical and chemical processes of the stars and solar systems. The evolution of life is very important in understanding what went on in the history of our universe.

Also, by referring to the research results on ETIs and their cultural and political systems, we people would be able to learn how to live together on this small earth.

In order to think critically about the current situation of human beings socially, politically and ideologically, it seems necessary for everyone to have his own clear-cut view of human society in the near future, against the possible collapse of this society due to nuclear war or other human activity.

If everyone could have such a clear-cut view, based on his study of world affairs and everything else, he or she would never become aggressive politically or ideologically, and would always try to find some possibility of reaching negotiable agreement for any kind of problem, such as the maintenance of world peace.

Any information about extraterrestrial societies would give us a crucial insight into how human societies can survive and coexist, in spite of many possible troubles and obstacles among the nations of the world.

Egoism, being often seen in foreign affairs, would be washed out of every nation in this world. Everybody would then learn the importance of his coexistence with all other people.

Do most scientists in Japan agree with your view?

I don't think so. The few who agree with me are considered by most of the scientists in the physics and chemistry community as the un- usual guys. Most of those people don't believe in the existence of in- telligent life elsewhere in the universe. Some do, but most are very skeptical, expecially about the possibility of communication between us and ETI. Some people say to us that all who belong to the SETI association are crazy.

Has another scientist actually told you to your face that you're crazy for study- ing this?

Yes, that is true. Even now, I don't say that one of my main subjects is to search for ETI, because some people ask "what is ETI?" and you need to explain the details of ETI. But they don't think seriously about it. They see it as a kind of fancy story or space opera.

This is the way the other scientists think?

Yes, that is generally what the scientists think. They would say, "If you have time to study ETI, you should do some meaningful research on other subjects in physics." Maybe within ten years the situation will be different, I hope.

People are very much confused and mixed up between the scientific study of ETI and some fancy story related to the visit of ETI to earth. Some people think ETI is identified with UFOs. Then we, with diffi- culty, try to explain how ETI is different from UFO.

Do you think some scientists are also confused about it?

Yes. They never think very seriously about ETI. When you say something about the ETI problem, usually they just joke that it's a kind of science fiction. We don't think so, but people say "s.f."

When we say we are studying about ETI by using radio astronomy, they ask how come we are studying such a crazy idea, because they never learned about the present status of radio astronomical research. If they had studied the status of radio astronomy, they might have had some idea why we are trying so much to study ETI on a scientific basis. But they never study; just make an assumption based on pro- paganda. They think they know all about that kind of study, and usu- ally say so.

The public attitude may be much different. Now we have two or three different magazines about UFOs and space travel. These jour- nals are very useful. Many people are very interested in learning what is going on in the universe, especially on ETI, but their interests are a little different from ours. They don't care much about the scientific basis. We need to tell them what we are doing exactly on the scientific study of ETI. That's why I wrote that kind of article, to let them know exactly what we are doing.

In my book also I tried to explain to the public what we are doing. This book will be read mostly by laymen. I just explain the current status of the search for ETI. I never use difficult or technical terms.

How about formulas for radio transmission?

No. I never use those. I used only one very popular formula, the Drake Equation. They can understand this kind of formula. It's not very difficult.

Do you think the public is more open to the idea of extraterrestrial life than the majority of scientists?

Maybe so, but the public's belief has no scientific basis for the existence of ETI. They like to have any kind of ETI in the universe; that is more interesting to them. But scientists' reasoning is quite different from that of the public, because they want to know the scientific basis for the existence of ETI. We need to let them know about all the knowledge. But we have never published any kind of textbook about ETI. I guess my book may be the first one.

First in Japan?

Yes. Many books here have already been translated into Japanese, but all are very technical. We need to write some popular books so any layman can read them. We need to make propaganda by writing general articles. We need to write books for children about ETI.

Let's be sure I understand who would be more sympathetic to SETI: scientists or the public?

I guess the public would really support us more. We plan to use the radio telescope under construction to search for ETI, but some people working at the Tokyo Observatory say our project is nonsense, so we have trouble persuading directors of the observatory to give us time to study ETI. Top priority for using the telescope may be to study stellar objects, or some interstellar gas clouds or so. Search for ETI is second graded. That is the general attitude of the astronomical society members. Not very good for us.

So you think that it is important to work for public support?

Yes. Then the general public, in a couple of years, will understand what we'd like to do in research about ETI. Then we'd be given much time to study and make use of the new radio telescope for examining ETI.

We in Japan cover enough on analytical technology, but we don't have enough money to study. The government, especially the Ministry of Education, has no interest. If the ministry was very interested in supporting us to search for ETI, Tokyo Astronomical Observatory maybe would give us much time to study ETI using the new radio telescope, and also might give us some research funds, because the Tokyo Astronomical Observatory is established by the Ministry of Education.

In Japan there is a very strong desire of some scientists and the public in general to receive influence from foreign culture. Usually, if we are going to study some new subject in Japan, we refer to what's going on in a foreign country in the subject we would like to study here. Then, if you know what people in the foreign country are doing, it is easier to get acceptance or approval for starting the research in Japan.

Usually, research is not thought of as very important here, because many people think that Japan is still learning many things from foreign nations. But nowadays, in the physics community, we don't think so. We don't need to learn so much from foreign countries. We may start ourselves to study the new subject, but administrative people still have that kind of inferiority complex.

If we know that American scientists have started to study and search for ETI by getting NASA funds or other funds, if that news comes to Japan, it may be easier for *us* to search for ETI. Many people don't know that other countries *are* working on the search for ETI. In the United States some people are doing it, we know, but such news usually does not come to the people here in administration. Our administrators have to know first what is going on in foreign countries. Then it is easier for us to propose the kind of research we would like to do—in this case, to find out the possibility of ETI.

I guess American people like to compete with the Russian people. Here, the same way: we are competing with some of the people in foreign countries. We have some potentiality for studying ETI by ourselves, but still we don't have the influence of foreign studies on the Japanese public or Japanese physical science community. Most people in Japan don't think that studying ETI is very important.

How many scientists in Japan are interested in extraterrestrial intelligence?

I guess about thirty people. We have a very small association we call the Association of the SETI. We call it in Japanese "Seti no Kai." "Kai" means small group or association.

We have meetings every month or two, and many different topics come up. We discuss the possibilities of existence of ETI and how to communicate with those ETIs by using radio telescopes under construction. Sometimes people eagerly discuss the possibility of visits by ETI to the earth by UFOs. Some people discuss technical processes like how to make recordings of the radio signals, and sometimes they discuss the designing of their communications, like using the 45-meter telescope now being built.

About five people working at the Tokyo Astronomical Observatory are members of the association. Also some biologists, chemists, economists, radio engineers and mechanical engineers are very much interested, and they are members of the association, too. Professors M. Morimoto and Jugaku belong also.

"Seti no Kai" is still a very private community. We cannot say that it is very well received, like the Japan Astronomical Society. We don't say too much about the importance of our group or the search for intelligent life in the universe because many people say we are crazy.

What effect, if any, have your interests in ETI had on the way other people behave towards you?

Our friends who belong to the association, who are somewhat interested in ETI, have a very favorable attitude toward us. My family, my wife likes to read this kind of story; I have no trouble there. Superiors here never have heard anything about ETI.

They don't realize that you are interested in it?

No, I have never mentioned my interests, because if I say something to the effect that I'm doing research on ETI, some people would be upset. But in the Institute of Physics some people are somewhat interested because I talk so much with them about the importance of searching for intelligence in the universe.

I have never written any scientific paper about it in English, but once I wrote an article in Japanese about the current status of radio astronomical research in relation to the discovery of ETI. That was published about five years ago. That was my only paper.

Was that a scientific paper?

Yes, in a Japanese professional journal for cosmic-ray research, because my subject is related to radio astronomy.

Recently, two of my colleagues published an article in *Icarus*, discussing the possibility of sending information with a kind of virus called 5X174. A virus is a small living system. It could be reorganized to send some information through outer space.

Also my colleagues working at the Tokyo Astronomical Observatory published an article in *Nature* discussing the possibility of using molecular-line emissions of radio waves around eighteen centimeters to search for ETI.

That kind of work is being done but those papers cannot be understood by the public. We need to write more popular articles for Japanese popular journals.

What fraction of your working time do you spend on SETI?

About 25 percent, including the study of stellar and planetary formation. It depends on my research. When I translated this book, *Listeners* by James Gunn, I studied lots of research papers and some articles in English and other foreign languages.

After I completed my article, which was published as the appendix of the translation of the book, I didn't spend any time studying ETI because I am now proposing some new research programs on solar neutrinos in Japan. That is the most important topic to me now.

I would like to spend as much time as possible on ETI. I subscribed to the journal *Cosmic Search*, and I'm a member of the AAAS and I receive *Science*, and study any kind of article on ETI published in that journal. And then I know what's going on about ETI.

Have your ideas about ETI changed much since you first began thinking about it or have they remained pretty much the same?

Almost the same. Not much changed about the idea of ETI for the last twenty years.

What would you do differently if you were starting over again to study ETI?

I would start to study the origin and the evolution of life on the earth as referring to the evolution of the solar system, since the origin of this system seems to have been well understood. Then I would start the search for life in the universe, based on the current evidence on the life on the earth.

What has been the most difficult aspect of your ETI work?

There is just beginning to be a feeling that it is all right to look for ETI, but until now it has been considered crazy.

What effect do movies and television have on public attitudes toward ETI?

I guess a positive effect. In Japan we saw the movie *Close Encounters of the Third Kind*. Many, many people watched that movie. I, too, did. Many people said that the movie was very interesting.

I was asked sometimes by people who watched the movie whether ETI could travel through space to reach us. Then I would explain the possibility on a scientific basis and tell them what we are studying. The people who asked me questions now have a good idea about the scientific basis of ETI.

That movie was very popular here, and so was *Space Odyssey 2001*. That was a very good movie, and not about ETI but a kind of spaceology. All in all, movies are very helpful for the general understanding of ETI.

What effect do you think public interest in UFOs has had upon SETI?

I think that the effect is positive, because in our association about five to ten people became interested in studying the UFO problem, and some of them believe in the existence of UFOs. Most of these scientists are not astronomers. Some are economists and some are aeronautical engineers. These people believe that UFOs exist, and they are very sympathetic to us because they know what we are doing about scientific research on ETI, and they think they need to study the possibility of interstellar travel to understand UFOs.

In this case it is okay, but the general public has never heard about this kind of research on ETI. That may be the trouble. We need to tell them the connection between UFOs, ETI, and the possibilities of interstellar travel.

What is your own view about the relationship between UFOs and ETI?

If we look at the technological aspects, it would be almost impossible for UFOs, even if they exist, to travel between the stars; it is too far. Interstellar travel has many serious problems—speed, inertia, and other problems—that have to all be resolved. But we still don't have a good idea how to resolve them. Therefore I think that there is no possibility now, or until we learn how to reach the other stars.

What do you think about the UFO phenomenon?

I studied the book written by Hynek. Last year he came to Japan. I met him to discuss the UFO problem. He showed me many interesting pictures. Also Peter Sturrock gave me a tape which described some things we cannot explain and showed some pictures. They were very interesting because we don't know how the pictures were taken or sketches made. Before we understand how such a picture was taken, it is impossible to deny the existence of UFOs. Some are unidentified and some are puzzling.

When do you guess that we will actually receive some kind of signal from ETI?

I am still very skeptical about the possibility that some ETI is sending us radio signals which contain information about their existence.

You are skeptical?

Yes, very much. We can send radio waves into the universe, to try to find ETI somewhere within one hundred light years or so. If we are following some evolutionary trend of the stars and their planets, and most of the civilizations are evolving in the same steps as ours, then maybe some ETI has just reached the stage of civilization that we have. Then they are also trying to send radio waves to communicate with other species. But they're all in the same stages.

Therefore it is difficult for me to believe that maybe one hundred years or two hundred years ago some extraterrestrial intelligences were already sending us information of their existence.

You are assuming that any other civilization is progressing at about the same rate that we are?

Yes, within a very close area around the Solar System, because most of the stars within one hundred light years or maybe two hundred light years are almost the same age as our sun. That means that if planetary systems evolved around some stars within that area, they may evolve life at about the same rate as ours. Perhaps beyond our reach some more advanced civilization may exist, but it is almost impossible to communicate with them. Too far, you know.

What would be your guess as to when we might actually make contact with them?

Maybe within a couple of hundred years or so we may reach the stage of *communicating* with each other by sending radio signals, or maybe some other different signals, by using gamma rays or neutrinos or laser beams. It may be necessary to develop new techniques.

Then you do not expect it to happen tomorrow?

No, I don't think so. But we could *detect* such an existence within a couple of years or so because in Japan the 45-meter radio telescope is now under construction. But now we have no possibility of studying ETI by any means.

Now there is no one in Japan who is actually studying it, because there are no radio telescopes?

Nobody else.

What might ETI be like?

The shape may be different, but the biological processes must be similar. Mainly carbon, and some kinds of organic substance including DNA and proteins and other material must be used to make ETI. Then their metabolic system must be similar. If their evolution and atmosphere are similar then the process must be very similar.

Then our civilizations may be similar because if they could build their own civilization they have to develop some way to use many different kinds of instruments, suggesting that they have appendages close to our hands, and they have also developed a highly organized language system.

Those two things are very important factors for them to build their own civilization. If their metabolic processes and physical organization are very similar to ours, their civilization may be very similar to us. I think that even if they are very, very intelligent but they don't have tools, instruments and language, then they don't have a very good civilization.

What scientists or other people in the past do you admire?

Einstein, Boltzmann, Galois, Gauss and many others who opened up new fields of sciences. Thomas Hardy and R. Rolland in literature.

Who are the outstanding people in SETI today, and during the past twenty years or so?

Frank Drake and Carl Sagan, whose influence on me was very great.

What should a scientist do upon receiving convincing evidence of extraterrestrial intelligence?

He should publicize all the results of his findings on the existence of ETI, and, before taking any further action, he should try to listen to every idea or opinion about what he should do.

What would be the public's response to this information about the existence of ETI?

The public would have some anxiety and feel uneasy about some danger that may be introduced by contact with ETI. However, they would soon understand that the existence of ETI is not, in itself, dangerous to their daily life. Advice by scientists would be desirable.

What would be the long-term effect upon humanity of evidence that ETI exists?

Everyone on earth would learn that other intelligent life exists in the universe, and he is not alone.

Using the data available from contact with ETI, comparative study of living systems would inevitably start to reveal every detail of the biological and chemical basis of life on earth, by referring to the observed nature of ETIs.

Jill Cornell Tarter

ASTROPHYSICIST

Born January 16, 1944

Eastchester, New York

Jill Tarter is an associate research astronomer at the University of California, Berkeley, attached to NASA Ames Research Center, and senior scientist, SETI Institute. She is one of the few scientists devoting most of her time to the search for extraterrestrial intelligence. She is project scientist for NASA's SETI Project. She conducts an ongoing observational program at Nancay and other observatories, and develops plans for a long-term microwave search program using ten-million-channel spectrum analyzers and state-of-the-art cooled receivers.

As a graduate student she worked on a search for extraterrestrial radio signals, Project Serendip, at the University of California's Hat Creek Observatory 85-foot telescope, Project Serendip.

She received a bachelor's degree in engineering physics from Cornell University in 1965 and a master's (1971) and doctor's degree (1975) in theoretical astrophysics from U.C. Berkeley. Her thesis, on the interaction of gas and galaxies within galaxy clusters, discussed small brown dwarf stars that never successfully fuse hydrogen.

From 1975 to 1977 she held a postdoctoral National Research Council fellowship at NASA Ames Research Center. Since 1977 she has been on the U.C. Berkeley research staff, on contract to NASA Ames to work on SETI. In 1985 she reduced her involvement there by half in order to become principal investigator for the nonprofit SETI Institute. There she is pursuing SETI and initiating a project entitled "Observational Exobiology" to assess the potential for orbiting "Great Observatories" to study the origin and evolution of the biogenic elements and compounds.

She serves on the NASA Committee on Radio Frequencies and NASA's

Space Science and Applications Committee, as well as others, and since 1983 has been cochairman of the annual SETI review meetings of the International Academy of Astronautics.

Her reports on SETI have been presented at two dozen conferences of the IAA, American Astronomical Society, the International Council of Scientific Unions Committee on Space Research, the International Astronomical Union, NASA and other organizations.

She is the author of more than seventy technical reports, in *Icarus*, the *Astrophysical Journal*, *Acta Astronautica* and other journals. Most of these reports deal with SETI, but she is also interested in brown dwarfs (a term she coined), the search for extrasolar planets, and water masers.

She is a member of the American Astronomical Society, International Astronomical Union, International Radio Science Union, International Academy of Astronautics and the American Association for the Advancement of Science.

Interviewed June 1982 at Mountain View, California

Where did you live during childhood and when you were growing up?

I was born and raised in Eastchester, New York, a town whose main claim to fame is that it's next door to Scarsdale, New York. I didn't leave there until I went away to college. It was a commuting, bedroom suburb of New York City. Most of the people there were employed in the city, and the trains took them back and forth in the morning and evening.

What were your childhood interests?

Being a child around water I was a fish. I happened to be brought up in a complex around a lake, and my mother claims that she had to fish me out of the water to remind me to eat, because I really enjoyed the swimming and all kinds of outdoor activities. My father was very interested in sports—in hunting and fishing and camping and that sort of thing—so that's the way I spent my childhood. When I could spend time with my dad, it was usually in the mountains somewhere.

Tell me about your earliest activities in science.

In elementary and high school there was a system of tracking, so I was always in advanced classes with the same people, and did all of the math and biology and chemistry that was available in that school system. From a very early age I knew that this was something that I was interested in doing. I liked the logic of it. Also I don't have a very good long-term memory, so biology to me seemed out of the question;

I couldn't remember all of the phyla and species and that sort of thing. Chemistry and physics really did appeal to me as a child.

I think astronomy or any glimmerings of interest in that field came really later. Astronomy was for me, growing up, something that was a lot of fun but I didn't think of it professionally. My father had done some astronomy in his schooling, and we used to spend a lot of time in Florida on vacations. The southern sky, or what you can see of the southern sky from Florida walking along the beach at night, is just fantastic. When I thought about astronomy, those were the kinds of images that came to mind.

Do you remember any particular incidents involving science?

All I remember was something that started then and was to continue, that was being the only girl in the class. Somehow it wasn't considered by my peers strange when it was chemistry in the junior year, but when it was physics in the senior year, I think most people expected that I would have gotten tired of it by then.

I remember a marvelous old curmudgeon physics instructor in high school who didn't seem to think it was strange that I was interested in the topic. He spent a lot of time with all of the students, but he seemed always to have extra time to spend with me discussing things, and he was always very supportive and encouraging.

It was just those years, the early sixties, when the Gemini project was getting off the ground. And I can remember going out into the hall in the high school, to listen to the loudspeaker systems, to the launch and the reports.

At that time there was a big push in science as being an important field for young people to get interested in. So I certainly benefited from all of that and got caught up in it. There were special programs that were instituted under the idea, "We have to catch up with the Russians; we're not training our young people in science and math well enough." So there were special Saturday programs at Columbia University that came into being right around that time for special students. They dealt with physics and math, and I was able to partake. Again, I have a difficult time remembering a period when I didn't think that I would be doing something in science and math.

How was your elementary schooling?

I don't remember it being very exciting or stimulating, and I do remember having a lot of trouble with one particular teacher. I found his presentation so dull and boring that I would "space out," I guess would be the term; I would daydream. I had no trouble performing on tests, or achieving, but I just found the classwork so terribly dull that I had other things to think about, and that didn't sit so well with this instructor.

I always thought of myself as being a model pupil but my mother has saved some old report cards from this period, just to make sure that I remain humble. There were four report cards in a year, and in each of the report periods the teacher would write comments. In the fifth grade the comments, which I would vigorously deny, had to do with my excessive talking in class. The teacher wrote in each period, "Jill talks too much"; "Jill still talks too much"; "Jill is talking way too much." Finally, on the last report period he said, "Thank goodness, Jill has stopped talking so much." So I guess I wasn't the ideal student that I remember myself being.

Do you feel your elementary schooling gave you a helpful preparation for a scientific career?

It wasn't a superb education but at least it didn't depress enthusiasm. I certainly liked going to school. I liked the social aspects as well as the educational aspects.

What about secondary school?

High school was a lot more interesting in many aspects. There was driver-education training, which is a big point if one is growing up in a metropolitan area. And there were some excellent academic opportunities there. I was able to get into advanced programs in mathematics and finish the first year and a half of college calculus. The chemistry and physics courses were good, and there was this opportunity to participate in special Saturday classes at Columbia University.

So high school was excellent, and it was also very competitive. It was in the New York City area, and going to college was a big thing and there was a lot of emphasis on performing well on examinations for college entrance. So I had a lot of competition from my peers, and encouragement from them. It wasn't something like the Bronx High School of Science; it wasn't a special atmosphere like that, but it was good and it was rewarding, and it was okay to achieve scholastically—there was no negative connotation.

On the other hand, as an aside, a few years ago I was able to attend a conference in Washington focused on young women in science. We did a couple of sessions where we tried to find similarities in background: what motivations worked for women continuing in science. I was interested to find out that in this group, 80 percent of the women had not only excelled in math and science in their high schools, they had all overcome negative counseling: "You should not expect to go on." They had also, 80 percent of them, been cheerleaders or drum majorettes or something like that, so not only had they excelled in the scholastics, but they had excelled in the socially acceptable activities. They achieved in whatever arena was the available competitive form.

This one group of students that I used to track along with was, in fact, the most influential group of people for me. They were a very good counter to the counseling that I was receiving from the high school, which was saying, "Don't take so many credits, don't take science courses predominantly, don't expect to go on to the university; you're going to be a housewife and what do you need all this for?"

It was silly and foolish because my grades were always good, and I was very strong-willed and determined. I would have thought any counselor interacting with me would not have discouraged my ambition, but it was easier, I think, for them to deal with a model or a mold that they were more used to.

I was very noticeable as a student in our high school, being the only girl in the physics class and the only girl who was really interested in science and math, and therefore I achieved a lot of awards when I graduated from high school. I also received a lot of scholarship awards from various universities and colleges that I had applied to.

I think my singular status focused attention and helped to bring about that recognition. I knew that it wasn't really the standard thing for a woman to be so interested in math and science. But it didn't matter because that's what I was going to do. And it was fun. I was also a drum majorette.

When did you first think about the possibility that extraterrestrial intelligence might exist?

I grew up in a generation when I can't ever remember *not* assuming that, of course, there is extraterrestrial life, because we had Flash Gordon and we had secret decoder rings in cereal boxes and I had every one, just like all of my friends. So it's hard to think of a time when, at least in some fantastic way, I didn't believe in extraterrestrial intelligence.

Certainly the idea of ETI, at least in the guise of science fiction, was something that I accepted and was part of my life from a very early age. I loved science fiction and read enormous amounts of it. Robert Heinlein was a particular favorite science fiction writer. Towards the end of my school I got a copy of Fred Hoyle's book, *Frontiers of Astronomy*, when it first came out. The nice thing was that here the real world, astronomy and astrophysics, was as exciting as the Heinlein and other science fiction. I really thought it was fabulous.

And maybe at around that time I began to think about the fact that there really were rational reasons for believing in extraterrestrial intelligence. Just the vast number of stars. It was quite a bit later, though, with the Shklovskii and Sagan book, before I began to put thoughts together about the ubiquitousness of carbon-based organic chemistry in the universe and that sort of thing, to support what I had always more or less had as a background assumption.

Did you discuss this possibility with other people?

I don't remember in high school, particularly, other than in the vein of discussions of Heinlein's books. There were a number of us who'd read anything and everything we could get our hands on; anything that was written by Heinlein or a group of other writers. We liked science fiction and particularly what appeared to us to be technically accurate science fiction. I certainly don't ever remember having early *negative* responses to the concept of extraterrestrial life.

During your college years, was there any stimulus to think about ETI and the possibility of searching or communicating with it?

Getting involved or thinking of methods of search, for communicating—no. I remembered Fred Hoyle's *Frontiers of Astronomy* and harbored a belief that there were wild and wonderful things to be learned about the universe in the very near future. ETs were one of the many possibilities for discovery, but I never thought I'd be involved in that search or any other. I was too busy trying to succeed at the nontraditional tasks I had set for myself. I think I dream more now than I did then. I was very serious as a student, and overly concerned about performance.

I did an undergraduate degree in engineering physics at Cornell, and Frank Drake was there at the time, in the physics and astronomy department. I did take eventually some courses from Frank, in radio astronomy, but I don't remember if the subject actually came up.

Again, I had perhaps a strange and late introduction in life to the concept of actually making a search. I didn't get actively involved in thinking about that until I was a graduate student. In my background there were no large discussions or philosophical arguments about the question. It was so deeply ingrained that I just grew up with the idea of ETI.

After you completed your schooling and began working as a professional scientist, who or what has influenced your thinking about ETI and SETI?

It actually began before I finished my schooling as a graduate student at the University of California at Berkeley. The Cyclops report had just been published, and I read through that with enormous interest. It was brought to my attention by an X-ray astronomer at Berkeley, Stuart Bowyer.

He had been wondering how he could become actively involved in the search, and he came up with the idea of using the University of California's radio telescope in a parasitic search mode; to hang a little black box on the back end of the telescope. This would enable him to analyze the data that the radio astronomers were gathering in a way that would be more suitable to the detection of coherent artificial signals, which otherwise would go undetected by the kinds of analyzers that the astronomers were using.

He got interested in the search by reading the Cyclops report himself, and he passed that on to me and I got very interested in it. It was one of these very fortuitous things.

This was to be a very inexpensive project, to be done with student labor and leftover equipment and donated equipment. One of the pieces of donated equipment was an old PDP8S computer that had originally been used to run the telescope, and had then been replaced by another computer. Early in my graduate student years, as a graduate assistant, my job had been to program this PDP8S computer to run an automated photometry package on an optical telescope. And by the time the idea of a SETI device on the radio telescope came along, it turned out that I was the only one around who still knew how to program this machine.

That was the coincidence that got me involved with this project, which we called Serendip, and which was the first opportunity I had to actually think about what could be done in terms of communication and detection.

Subsequently, that piece of apparatus was completed and did live on the radio telescope at Hat Creek for a number of years. It never worked as well as we had expected. It was not a reliable device and it was in an environment which was remote. It would occasionally malfunction and turn itself off and sit in the corner, and no one else in the observatory would note it wasn't functioning, so it was only during a periodic check that we would realize that in the last two weeks it had not been listening to anything.

But nevertheless, that effort is still ongoing. The device has served as a mechanism for three master's theses in electrical engineering. And in the past few years, the device has gone from Hat Creek down to Jet Propulsion Laboratory to the 64-meter telescope, where it's not quite so remote and there is someone to give it a little tender loving care. It is functioning there on and off in the background in a parasitic mode.

Is it still the same basic instrument?

It's still based on the very old Princeton Applied Research 100-channel auto correlator. That's the spectrum analyzer. We'd dearly love to build another model, but there isn't the funding to do that, at least from the university at the moment.

The people involved with the Serendip system are trying to learn something about how one does a search in absentia. The device is set to trigger at a threshhold level if any signal is detected above a given threshhold in its 100 channels, with a few caveats that try to exclude natural signals, astrophysical signals.

But beyond that, one is faced, long after the fact, with a record of signals exceeding the threshhold, and the question is, "How do you develop software to search through all that?" What kind of pattern-recognition techniques can you bring to bear so that you can pick out local transmitters or spacecraft or the Air Force flying overhead? We must first exclude those and then try to pick out from the rest what is just statistical noise and what might actually be signals of some interest. So we—very laboriously, very slowly, and with what seems like not great progress at some times—have been trying to develop the software to do this.

The Ohio State people are involved in the same sort of business. They do a sky survey all the time, remotely automated, and have data to look at after the fact. Surprisingly, they have not yet been able to put very much effort into doing a concerted analysis, a pattern recognition analysis, of the many years of data. They're just starting that now.

So we're hoping that maybe, by this effort, we'll be able to aid them or anyone else in developing the capability to do a parasitic search, because it's quite possible that major funding for dedicated systematic searches will not materialize in the near future.

The involvement with the University of California, Berkeley, system was really like getting my feet wet in doing SETI practically. Later, when I finished my degree at Berkeley, which was in high-energy astrophysics, pure theoretical work, I came here to Ames Research Center as a National Research Council postdoctoral fellow, and continued studies looking for other kinds of objects which are almost as invisible as extraterrestrial intelligence. I was interested in missing matter in space, so-called "missing mass."

We have examples where we believe systems to be gravitationally balanced, but when we add up the masses of all the visible components we find that there isn't enough mass there to account for the gravitational configuration. One striking example is a cluster of galaxies where there may be 10,000 or 100,000 galaxies which appear to be very regular, gravitationally stable, isolated configurations, but we just can't account for the total mass that's necessary for the gravitational field to be binding these galaxies together.

I was interested in trying to find out what constraints could be placed on the properties of whatever is providing the missing mass. Could it be that it was neutral gas? Could it be that it was ionized gas? Could it be that it was black holes or small stars that never turned on to nuclear burning so they were never really visible—what we call brown dwarfs, things like that?

One secretary in the astronomy department observed that I had been involved not only with that research but had then gotten involved with SETI. She pointed out to me that I had a passion for looking for things that were difficult to find.

When I came to Ames and I continued that research for missing matter, I also knew of John Billingham and the SETI group. I said to John that if he didn't mention it to the National Research Council, I wouldn't tell them either, but was there anything that I could do to help out the group? Gradually that grew into a full-time association.

What effect, if any, has your interest in ETI had upon your career as a scientist?

At the moment it's sort of a determining factor. I did two years as a post-doc and wrote a number of papers, the standard kind of thing. I love problem solving, and astrophysics and astronomy offered to me what seemed to be fantastically interesting problems to solve. But I always felt a little guilty, in the sense that I enjoyed doing what I was doing so much. But I didn't see why the man on the street should support me to do that, which in essence is what is happening with government funding. I always felt a little funny about why should I be so lucky to be able to do this and get this marvelous gratification for myself, but I didn't necessarily see what the rest of the world gets out of it.

So, when I began to think about what I would do after the post-doc, I thought that the political situation was right and maybe I should put in, say, five years of really doing SETI and SETI politics and whatever is necessary to make it visible and try to push it. John Billingham has a very good organization going here. There was a lot of expertise to draw from. It excited me to think that here was something where I could continue doing the kinds of things I love to do but there might actually be a practical payoff for the larger community.

I think the existence of extraterrestrial intelligence, while I took it for granted growing up, is a truly fundamental question, and it's one that we might be able to answer in the near-term future, in the next few decades. And it was exciting to me to think about actually being part of an effort to get that work started, to really make a contribution to a fundamentally important problem which could have an effect on the lives of the general population.

So it was tremendously appealing to me to work on SETI problems in addition (now almost exclusively) to astrophysics, because I could see that it could make a difference, and that matters to me. It's a really worthwhile effort.

As I was trying to decide whether to continue in a more traditional astronomy or astrophysics career, or to move into SETI, Tom Clark from Goddard Spaceflight Center came and talked with Billingham's group. Tom was involved in the very beginnings of what is called VLBI, Very Long Baseline Interferometry, where radio telescopes widely separated observe a single source simultaneously, and the data from these telescopes are tape-recorded with extremely accurate time standards so that the tapes can actually be brought together and the data can be correlated at a later stage.

Certainly, we have interferometers now, and the best example is the Very Large Array in New Mexico, where the telescopes are located sufficiently close so that signals can be brought together in real time and correlated. But the VLBI effort involved telescopes separated by continental distances, like Massachusetts to California. Now we have VLBI experiments using telescopes located around the globe, so we simultaneously observe with telescopes in Russia, Sweden, the East Coast of the United States, and California.

This technique of using magnetic tape, as a medium for recording data at high rates for a long time, was something that Tom Clark pointed out to us, when we began to talk about building instruments for SETI which had high-frequency resolution. These instruments would be fundamentally different from the instrumentation normally built for radio astronomy, which is interested in looking at the world with much coarser resolution, because nature, as far as we know, is not a source of coherent emission.

We are sources of coherent emission. We do it all the time with our technology and our transmission, which is one of the reasons why we think it is such a good signal for an extraterrestrial to transmit.

Tom pointed out that we could use a one-bit fast sampling mode, where the data are recorded continuously on the tape; that if one were willing to take these tapes and use a computer and to do very long Fourier transforms, that one can turn these long-time recorded data streams into high spectral resolution data. Then we could look at the universe at very high spectral resolution; that is, we could sub-divide the spectrum into a very large number of individual frequency bins, and given all these narrow frequency bins we could begin to approach the problem of how we might recognize signals, how we could actually build algorithms to recognize signals amidst the noise in the background that we're going to be observing. We could conduct a modest observational program with this approach.

That appealed to me enormously. I teamed up with Tom Clark and Jeff Cuzzi and Dave Black and other people at the Center here, and we started an observational program of that nature which is still continuing.

Those printout sheets piled there are from 200 tapes I recorded at Arecibo observatory last November. It's not the way to do SETI. We can collect the data and get high spectral resolution with this technique but we can't look at it right away, we can't analyze it. There's too much of it. There's a cartoon on the wall which shows the data coming in the radio telescope and going via tape through the computer, and then some poor human being completely swamped under the printout because we can't do the signal recognition. We will have to do it in real time, eventually. But this is a very good tool for learning what the universe looks like in high spectral resolution.

So Tom Clark showed up at just about the time I was thinking that SETI is really an interesting possibility and might be something I could do. Indeed here *was* something I could do, and it turned out to be very interesting for me and I think good for the program that we can, at least with small resources, collect data that represents the kind of data that we would like to collect with special purpose receivers and begin to learn to analyze it. That has been the focus of my interaction with the group.

What do your colleagues think about your interest in SETI?

Well, SETI is a subject about which no one has *no* opinion. Everyone we've talked with seems to have a strong opinion one way or another. I think the vast majority of my colleagues feel that it's quite probable that extraterrestrial intelligence exists. Maybe I choose my colleagues in a fashion that makes them sympathetic, but most of them view SETI as a question whose answer is an observational one. It's an experimental question; it isn't one that can be answered with pure thought.

So, many of them are actually interested in what experiments can be constructed to test the hypothesis that extraterrestrial intelligence exists and that their technology is visible to us in some sense. Many of them actually work with us, as an advisory group, on this one particular microwave search that we're trying to get established at NASA.

It's almost hard to get a good argument on this subject, because most of the people I deal with, colleagues, agree that previous searches were individual efforts exploring only a tiny fraction of the sky and of likely wavelengths, so it's worth an investment to make a systematic search in the microwave region with the concepts that Morrison and Cocconi first put forth in their article twenty years ago. Those concepts have really become the bible, and they've withstood a lot of challenges from other approaches and other concepts.

So we agree in principle; we may argue about details. Do you want to multiplex in frequency or time, or do you want to take one telescope and one receiver and have many small channels simultaneously and move it along, or can you get by with coarser frequency resolution and greater frequency coverage? What's the best antenna to use? Do things like the parasitic Serendip system make any sense; is it sensitive enough to see anything of interest? Or should you do a sky survey or do a targeted search? Which approach makes more sense? With very few exceptions my colleagues are willing to say it's worth some effort—and everyone's level of effort seems to differ—to try and test the hypothesis.

I went to a meeting in Maryland called "Where are they?" organized by Michael Hart and Ben Zuckerman. This was an "anti-SETI" meeting. These people feel that interstellar travel is so absolutely inevitable that if any other intelligences had existed, they would have colonized the galaxy long before now; they would be here. But they're not here, so where are they?

They made long speeches about exactly what extraterrestrials would or would not do. These were very interesting ideas but, with one exception, the speakers were all willing to say at the end, "But we really don't know, and so it's worth looking."

The only counter example to that, and he isn't really a colleague of mine, is Frank Tipler. He so believes that one can determine in advance, by pure thought, exactly what anyone else will have done, that he is quite prepared to argue that we shouldn't search because they aren't there. He *knows* they aren't there. But he's an exception. In general my colleagues support the idea of a search, although some of them are so forthright as to say, "I think that extraterrestrials probably exist, but I would never spend *my* time trying to detect them. But it's fine if *you* want to."

A particular example is Professor Townes at the University of California at Berkeley. Recently we were discussing a particular graduate student who had come to join us as a post-doc, and Professor Townes was saying the student was a very good student. He said if it had been one of his own students, he would have had a very hard time counseling the young student about entering into a career associated with an idea which was so speculative, in the sense it wasn't established in the scientific community and its funding prospects were not very sound.

Basically, Townes was saying SETI is okay for people who are established, but he wouldn't counsel any of his students to go into it immediately. They should first go out and develop a fall-back position for themselves, in case SETI were to go away, so that they'd still have an academic or a scientific credential that they could revert to. In the same conversation he was very supportive about the work that we had been doing at Ames, and said that it was fine for *me* to do it.

What about your family? What do they think about your ETI interests?

My husband is the director of the radio astronomy lab at Berkeley, and he's a very strong supporter of the concept. And my daughter is fifteen now. When she was about eight, she used to fill in her school forms—where it says "mother's occupation," she used to put "looks for little green men."

She's another case of growing up in an atmosphere where it's hard for her to conceive that people could seriously argue that extraterrestrial intelligence does not exist. She's been far more exposed, having grown up in our household, to the arguments of the large number of stars and the prevalence of carbon chemistry and the rational basis behind it.

But I think just her exposure to the media, even if she had not been raised in our household, would have predisposed her to believe in extraterrestrial intelligence. She does have an appreciation for the very large numbers and the many possibilities for what happened here to have also happened elsewhere. I think that she's supportive of it as well.

What about your father?

My dad died when I was twelve. It was one of many questions I didn't get to ask him but I would predict that he would be quite supportive. My dad's philosophy was that you should do whatever it is you want to do: just make sure you do it well. Any approach, any enterprise that I wanted to undertake, and was willing to do it with excellence, was something that he was in favor of.

What about your mother?

My mom is alive, and living in Florida. She has a subscription to *Cosmic Search*, thanks to me, and she reads it, and she cops out more than she should. She says "It's over my head. I can't understand the arguments, but I think there are probably other intelligences."

She's certainly in favor of my pursuing the subject in a way that tries to demonstrate that there are at least observational paths open that might or might not verify the hypothesis. She had a very religious-structured upbringing, and there might be something closeted way back there that is at odds with the idea of intelligence elsewhere. I'm not sure; she certainly vocally says, "Well, of course."

What about your superiors, bosses, and so forth? What do they think about it?

When I was a graduate student I had one professor who was extremely supportive of the idea, but it was still the University of California at Berkeley, and they do not give Ph.D.'s for SETI activities, so it was clearly understood that this was fine, and something one could engage in, as long as it was in *addition* to the traditional or the more standard astronomy and astrophysics. Since I've stayed in the San Francisco area, I'm still in contact with all of the faculty at Berkeley and I think my thesis advisor scratches his head and wonders, "Hmmmmmm, I wonder why she didn't decide to go to a university and teach astronomy to astronomical researchers."

But here I am in a very privileged atmosphere. There is an Extraterrestrial Research Division at Ames Research Center. John Billingham is head of the SETI program and is highly respected, and there are many people in the NASA system who believe in the program. Therefore I find myself in a very positive atmosphere and people are saying, "Yeah, if you want to do it," or "How's it going?" or "Are you getting support?" or "What's the latest hurdle?" and "What can be done?" and "How can we help?" and that sort of thing.

So we fight political battles and negative impressions at a higher level, but here, in a working environment day-to-day, it is a very well thought of activity. We don't really run into dissenters or people who disagree until we get into the headquarters structure in Washington, in the political arena.

I'm officially employed by the University of California at Berkeley, and I'm on contract, funded by the SETI Program Office at Ames. So I spend four days a week at Ames, and one day a week in Berkeley, where I live. That's just so that I don't have to commute that one other day. And I spend almost all of my time on SETI and a little bit of extra time on doing some theoretical research into missing matter in the interstellar medium. Right now it's SETI; it's trying to get the program established. It's more politics and public relations than I had anticipated, but it's lots of fun.

What about your friends? What do they think about your ETI interests?

Oh, again my friends tend to come from a community of colleagues, and I haven't encountered any great disbelief or controversy from any of my friends.

Before going on to discuss your ideas about SETI, let's get a little more background information. What was your father's occupation?

My dad was a securities investigator for the Securities and Exchange Commission.

How much education did he have?

He had an undergraduate degree from Swarthmore College.

What about your mother?

My mother had a high school education and no formal college. She was the assistant manager for Bonwit Teller in New York City and Palm Beach, Florida, which is an exclusive women's clothing store.

How would you describe them?

My mom was very gregarious and so was my father. He was a sportsman.

How did they influence your eventual choice of science as a career?

Well, my dad always wanted a son and I was the only child, so I always got taken along to activities that were more conventionally reserved for fathers and sons: sports, fly-casting contests for fishing tackle, turkey shoots, and camping trips and construction camp sites or places that they were remodeling in the country—that sort of thing. It definitely had an influence. What I learned from my dad was that I could do whatever it was that I wanted to do, and I didn't have to be so influenced by more traditional stereotypes.

And what about ETI? How did they influence you in this respect?

My father had worked at Sproul Observatory at Swarthmore as an undergraduate. If it hadn't been for the war, he would probably have gone for some astronomical training in graduate school. So he developed a love of astronomy early on. And inasmuch as it ever led to formal connection for a career, it was that astronomy background.

The other thing that he taught or encouraged me to do as a child was to work with my hands in construction. From that came my interest in engineering. That, in fact, was what I started out to do; my first degree was in engineering. Astronomy came later, but the engineering background is particularly germane to the signal detection problem for extraterrestrial intelligence, the digital equipment that we are trying to construct.

Were there other people we haven't mentioned who influenced your ideas about science and ETI?

As an undergraduate and in the beginning part of my graduate career, I went through engineering to astrophysics, deciding that what I liked to do was solve problems. The problems that were most interesting to me were astrophysical ones rather than the engineering problems to which I was exposed, which seemed to be merely recreating a solution that was already known in a different form. I was more interested in new problems, to which solutions were unknown. I really didn't get specifically involved in a SETI type activity until after I was a graduate student and almost finished with my astrophysics degree. Then, as I mentioned, I got involved with the program here at Berkeley, to do a parasitic study.

What religious beliefs did your parents have?

I was raised as a Protestant, a Presbyterian, but there wasn't very much emphasis on religion at all when I was a child. It was something I got interested in for a period just after my father died, but that was the only time I was ever involved with any formalized religion.

What about your friends when you were growing up? Did they share your interests in science and ETI?

In the group of students with whom I went through high school—gifted children, they could be called—the emphasis was on science. It was the post-Sputnik era of competition with Russia. There was a lot of suggestion that one pursue a scientific career. We all were motivated. The majority of the people in that group were guys, and all of them went into science and engineering, at least at the undergraduate level. One other girl in that group became a chemist.

What were you like during childhood and adolescence?

I was rather outgoing, but I was also quite a loner. I was very large, physically. I grew up very quickly and so I was larger than the boys in my class and I wasn't the pretty, social-type female. I was very friendly, but mostly I did things on my own. I was very interested in group activities, but I ended up pursuing the things I wanted to do mostly on my own.

What other professional interests do you have, apart from SETI?

I continued an interest that I started as a graduate student in various forms of so-called missing matter in the universe: configurations where it appears that there is gravitational binding, but when we add up the visible forms of matter that we know about, we can't come up with the mass that's needed. Clusters of galaxies are one example. In the flattened plane or disk of our own Milky Way galaxy, the dynamics of the stars and the neutral hydrogen gas within the disc seem to imply that there might be a very large spherical distribution of matter outside the disc. That has a stabilizing effect.

One hypothesis for what this mass might be is a very large number of very small stars that were not sufficiently massive to ever stabilize into a hydrogen-burning main sequence. As these stars contracted, they developed a degenerate electron pressure which halted the collapse before the temperature of the core was sufficient for nuclear fusion to become established. They burn deuterium but they cannot burn hydrogen well, and so, if they exist at all, they slowly cool off into what must be something like rocks in the interstellar medium.

Now we don't know that such things form; we've never seen one. We in fact have made proposals to look for these objects with infrared, with the IRAS satellite, for example. We don't know any theoretical reasons why the fragments of this mass should *not* have formed in the general process of star formation. It's hard to understand how a collapsing cloud will know that if it forms a fragment of such small mass that it will never make a stable star. And we haven't found any break, or any reason to terminate the fragmentation process at a mass above that which will stably burn hydrogen, so these are theoretical possibilities. These objects would be quite invisible after a billion or two billion years because they just do not have a sufficient source of internal energy to shine for very long.

I've continued an interest in those exotic objects. And I have more recently been looking at interstellar masers, particularly water masers in molecular clouds. And I'm working on a theoretical model for a means of pumping this maser action without requiring an internal source of photons, infrared photons; some way of transferring energy into a small, dense region without requiring that in the center of that region there be an internal source of energy which might eventually become a star.

The SETI activity has led to an enormous fascination with paleontology and biology and botany. Before you came in I was studying about the particular properties of dust and molecules in the interstellar medium, trying to get more familiar with exactly the types of organic chemistry that we seem to be finding everywhere in space.

My main interest is what we do in terms of flight experiments that NASA is planning in the future, with large telescopes in space or perhaps with ground-based telescopes. What are the experiments that we can perform to learn more about the composition of the dust grains in the interstellar medium, how they are formed, how they're incorporated into pre-stellar nebula, and whether in fact this long chain of events has a direct bearing on the chemical evolution that we think took place on the surface of the primitive earth? Did it have a head start? Did it really start with only the organics in a reducing atmosphere, and concentrate those, or were there some seed materials on the surface of the earth, brought here by meteoritic or other bombardment of particles from pre-solar nebula, which might have had some complex organic form?

Fred Hoyle carries this to an extreme and says all the dust in the interstellar medium is living bacteria. I think that's rather absurd, but there may in fact be a connection between the organic chemistry that we find preserved in meteorites and the actual organic chemistry, or the chemical evolution that led to biological evolution. How do we find out what molecules nature likes to make and how they survive and how they are augmented? It's SETI related but it's a closer tie to traditional astronomy, and it's fascinating.

There are all kinds of opportunities in the next few decades with high resolution spectroscopy in the infrared, the submillimeter and the millimeter, to get above the earth's atmosphere and to do some rather good experiments on the chemical composition of the interstellar medium.

What recreations do you enjoy?

I love flying. I love piloting. I'll take airplane rides in any form, but I do like to *fly* airplanes. I have a license and I belong to a flying club, and my husband actually owns a lovely airplane with one other person. It's a pressurized Cessna 210. We use it a great deal. He has a rationale for a great deal of flying, by virtue of going back and forth between here and the radio observatory at Hat Creek, and we manage to be able to use the airplane to go to a lot of meetings and to transport a lot of astronomers from here to there, as well as for strictly recreational trips.

My flying is a bit more limited because it's mostly recreational, which means I have to pay for it out of my own pocket. It's a lot easier when it's a business expense. I love flying, I like making things with my hands, I like crafts and sewing, and I love dancing. Dancing is another passion.

Folk dancing, social dancing?

Anything; all of that. I do a lot of jazz in a class situation rather than performing. And I have done, over the years, a lot of folk dancing. Whenever I can convince anyone else that what we should do tonight is to go dancing socially, we do that. Actually, it's one small benefit of being the minority sex in a profession. When meetings occur, there are often many occasions for social dancing, and very few women.

What has been your greatest satisfaction in life?

I don't think I have any single greatest satisfaction. I'm constantly excited and satisfied to be able to pursue a career where I'm doing something that I find so fascinating and interesting. It's been really fun watching my daughter grow up and watching Jack's children, who were fairly grown by the time I got to know them. Each day is satisfying, because I'm doing what interests me, what I wanted to do.

What has been your greatest sorrow in life?

My father died when I was twelve and I think there was a great deal that I could have learned from him. He was a rather wise gentleman, but I was busy being a child and very concerned with myself, so I missed a lot of opportunities to interact with him and learn from him. I am very sad about it.

I learned a lesson from this, which is the fact that opportunities exist for only a finite time and one should take advantage of them.

What was the most memorable moment or event in your life?

I remember the birth of my daughter, a moment where I suddenly realized that I was to be completely responsible for another individual for many years to come. The intellectual anticipation of her birth had focused on more pragmatic questions and I was totally unprepared for the emotional experience, so that was very memorable. Flying an airplane for the first time was a very memorable instance.

I also remember, as an engineering student, trying to understand some of the basic properties of the transmission of light, and realizing that certain numbers, which to me had been just numbers in a book, had a very fundamental relationship. I can remember being very startled when I understood that the speed of light, the dielectric constant, and the magnetic permeability of free space really were connected.

I can also remember a very special moment on the top of the enormous feed structure at Arecibo, Puerto Rico. It's a magnificent sight, and even when you sit in the control room and you look at this device day in and day out, you still are not prepared for what it looks like when you actually go out on it and climb up to the other end, the working end of the telescope, and look at it, and the island surrounding it. It's very impressive that our technology was able to conceive of that and keep it going. It is not even very sophisticated technology. That telescope has always overachieved what its initial conception was.

It was particularly impressive for me because the very first course that I attended as a freshman engineer at Cornell University was a class called Engineering Problems and Methods, in which examples of achievements in each of the different engineering fields were presented to engineering students who might not have decided yet what they wanted to do professionally. At the very first lecture, someone got up and talked about these engineers from Cornell who went down to Puerto Rico and lined this spherical bowl with chicken wire, and made a telescope out of it to listen to cosmic noise.

At that time, as a student, I was not impressed with the feat or the wisdom of it. But many years later, on top of the feed structure at Arecibo, I thought back to that original lecture and looked down at the telescope below and thought that perhaps this instrument will someday detect the signals that we're searching for.

It's really magnificent. It's almost unbelievable. In some senses, it's terribly crude. The feed support is like walking along a bridge. It's big, big heavy girders bolted together, and it doesn't at all coincide with the idea you have when you're in the control room some distance away, looking at this instrument as a functioning receiver, and you see these feed arms suspended on thin cables and it looks very graceful and delicate. But when you get up there it's just massive: real nuts and bolts technology, and it doesn't seem elegant and graceful, but it certainly is powerful.

The telescope has aged well and has continued to grow in its ability to do not only the atmospheric and ionospheric work it was originally conceived for, but also radio astronomy, which it was never intended at its inception to deal with. It remains an excellent, fine instrument in spite of its increasing age. Now there are designs and plans to improve its capability in the years to come. It's an impressive piece of engineering.

Turning now to SETI, about what fraction or part of your working time do you spend on it?

I spend probably 80 percent on SETI. It's difficult to draw a line; so much is related to it. Perhaps the water maser study that I'm working on now could be taken out and put in a separate category, but SETI has so many tentacles in all these different fields I was describing—questions about the composition of dust, which is really a SETI related or motivated question, but yet it's also a different activity, a different endeavor in astronomy. I guess you could say that 80 to 90 percent of my time is spent on SETI-related activities and the other 10 is theoretical: the water maser study. I'm one of the few people who does have the opportunity to spend that much time on SETI.

What stages has your thinking about ETI and SETI gone through?

First, when I really began to think about it seriously, was as a result of the Cyclops report, and the Cocconi and Morrison article, which I read as soon as I was aware of it. Perhaps the salesmen were so persuasive that I never looked back, but I think that those arguments have really withstood the test of time. I do believe that it is a reasonable hypothesis that electromagnetic signals are being used by extraterrestrial intelligence, for their own purposes, but also for the purposes of attracting the attention of emerging civilizations such as ourselves.

I think that the rationale for the frequencies that are currently being investigated still holds. I haven't come up against any very compelling arguments for other spectral regions, although I don't have a closed mind to them. It is just that, when we look at the universe, the microwave region is really the quietest place.

I haven't been any more impressed, as the years have gone by, with people who argue that they *know* exactly what extraterrestrial intelligences or our own technology in the future must do, or with the idea that space travel is inevitable, that interstellar distances are crossable and *must* be crossed by intelligences.

As I've looked at it over the years, I've begun to apply various weights and measures and to actually put numbers down for what we've done vs. what we're talking about doing, and what we can do and what we might have to do. I now have a better idea than when I first started thinking about it, of the enormousness of the search task—even the particular, small, well-defined microwave search for signals that we're talking about—and how little we've done toward it. But by the same token, I feel it's even more important that we get started on it than I did originally.

What would you do differently if you were starting over again to study ETI?

I don't think that anyone should start out to study ETI, nor did I. I think that one should start out to be educated in whatever excites them and to gain the greatest possible breadth in an educational sense. The thing that I would change from my own education is to allow myself a little more scope while still keeping the science, mathematics, physics, and engineering focus.

I was very linear in my education. I did many more courses than were required for my degree, but they were all of the same type. They were additional mathematics courses, or other physics courses, or electrical engineering courses that I wasn't required to take. But I took them.

I neglected economics and I neglected biology, and civilizations, and paleontology. I'm delighted that I'm having the opportunity to think more about these subjects now, but in hindsight I would counsel myself that I should have been less focused as a student, as an under-graduate. But I still think that I did as a student what was exciting to me, what I wanted to do, and I think that's what everyone should do.

Why is ETI important to you?

That's very easy for me to answer. The detection of an extraterres-trial signal—whether or not there is information encoded on it, whether or not it takes us a hundred or thousand years to decipher what the information is—the simple detection of the signal will tell us one thing, and that is that it is possible to survive the technological adolescence that we seem to be going through.

If we, with our relatively primitive technology, do detect a signal in the very near future, the implication is that there are many more sig-nals to be detected, and there are many civilizations. And *that* can only be if the civilizations are relatively stable and long-lived. We have no other indication there is any possibility that we can somehow get through this stage of potential for nuclear destruction.

We've been accused, by some people who are not pro-SETI, of cop-ping out and seeking extraterrestrial salvation. I don't think that's it at all. I think it's a fundamental question: are we alone in the uni-verse? Most people at one time or another must get close to asking that question. It's exciting to think about being able to answer it, and it's exciting to think that an answer, a positive answer, would imply that it is possible that we won't be foolish, and we could manage to have our species survive.

And what if we get a long-term *negative* answer, and it really ap-pears that there isn't anyone out there? Some people say that, in that case, we will be chastened, we'll be more prone to protect the only form of intelligence that we know, which is our own species.

But to answer why I work on it, for me it is a fundamental question and the potential benefits are enormous. And it's fun!

What has been the most difficult aspect of your ETI work?

I have no particular skill at political maneuvering, and this program is at a stage where that skill is a fundamental requirement. Trying to master that, trying to not get upset with people who are actively ig-norant, willfully dumb, has been very difficult for me. I'm much more prone to get very upset rather than to continue a rational conversa-tion or to deal with someone whose intellect or whose intellectual con-cepts I doubt.

To have to continue dealing with such a person because politically it's to my advantage to do so, has been very difficult. I'm not a statesman. To come by that skill, to whatever degree I possess it, has been hard work. I'd much rather tell somebody, "I think you're a jerk. That's ridiculous," but that's not a way to promote a program.

How do you think most scientists today view SETI?

"Most" is the operative word in that question. I'll answer it this way. We spent two years making presentations before the Astronomy and Astrophysics Survey Committee for the 1980s, which was chaired by George Field and was established by the National Academy of Sciences. As a result of those efforts, SETI was endorsed by this rather conservative group of the scientific establishment as being a very worthwhile program that should be pursued at a modest level.

We got the clear indication from many people that, because funds were tight and because this was by no means a sure thing, that it shouldn't be an enormous program, that it would be wrong to sink very large amounts of funding into it. But we did get agreement to the concept that it was quite a legitimate long-term research effort, to which a modest amount of money might legitimately be channeled.

We ran into opposition. There are a lot of people who stand up and say, "No, I know there's no one out there, because they're not here," or "No, we should wait a billion years until the national debt is settled before we pursue this program." But in the scientific community I think most people would agree that it's worth trying, it's worth at least some expenditure.

What about public attitudes toward SETI?

We, the people that I'm associated with at the SETI program office at Ames, do an enormous amount of speaking to public groups: a) because we get asked a lot, and b) because the public that we speak to are almost uniformly enthusiastic. Whether it's just the current generation of Star Trek and Star Wars fans, whether it's just an environment in which people are more accepting of the concept, or whether they're desperately looking for something upbeat, I'm not sure, but we do have large turnouts. It's a subject on which almost no one has no opinion.

Occasionally, there are people that are disruptive at such a talk. They are of two types. There are some people who are absolutely convinced that they have already contacted extraterrestrial intelligence, and we are wasting our time in the approach that we're taking because they can tell us all the answers. There are other people who think it's all nonsense, there's no one out there. They usually don't choose to come to such lectures, but when they do, they occasionally stand up and say it is foolish and why should we devote any money to it. But they're certainly a minority.

By disruptive, I meant that they often throw the schedule out of kilter. They either interrupt the information flow because they are sure that you are wrong and they have the right thing to say, or the question period gets channeled into not answering questions that most people have, but into a polemic on one view or another. I don't mean they throw rocks; at least they haven't yet.

Realizing that it is only a guess, when do you think contact with ETI will be made?

I don't think it will be in my lifetime. There are two reasons why not: one, an appreciation of the magnitude of the search that might have to be done; and two, pessimism about the political climate for actively supporting and funding a program whose extent is likely to be far longer than the politician's term in office.

What might ETI be like?

A number of studies indicate that function determines form, and that we can expect bilateral symmetry with a head end and manipulative digits and some sort of relatively efficient metabolizing process that doesn't radiate a lot of heat—they're not breathing smoke or fire. This is likely to be the norm, but size and details will depend on the local environment in which they evolve. Certainly, the forms of life on this planet are enormously diversified, but all intelligent species are recognizable as life forms.

A qualification to this is that we are talking about a search for extra-terrestrial intelligence; but we are really looking for extraterrestrial *technology*, because that's how we are searching; "technelligence," as someone called it. Therefore, those forms that we do find in this manner will be more similar to life as we understand it than other forms that may exist. We put a filter on the problem. It's probably carbon-based life because the chemistry that we see everywhere includes carbon.

SETI is becoming so established on this point that it has its own counter-culture, subdisciplines of people who are critical for specific reasons. One of their suggestions is that we should keep our eyes, or our instruments, open for anomalous phenomena within our solar system, in particular for any evidences that asteroids have small satellites, which may possibly be an extraterrestrial probe or colony or robot in a parasitic relation with the raw materials being furnished by the asteroid.

I am still very impressed by the difficulties of sending physical matter through space, as opposed to sending photons, because of the energy requirement. But of all suggestions I have heard about how to look for evidence that interstellar travel or colonization is in fact a reality, the most persuasive is that we might find such interstellar colonies among the asteroids in our own planetary system, simply because such colonies are nothing more than a small version of our own planet going through interstellar space. They are self-contained worlds in themselves where generations are born and live and die. They have no desire to interact with any other, except for their requirements for raw materials, and asteroids may be the best source of these materials.

I would certainly go along with suggestions that say, "As long as you're looking at the nearest stars for things that are anomalous, why don't you include some of the nearby bodies in the solar system and see if in any way, shape, or form, they appear abnormal?"

There are also other suggestions that are tenable or interesting because they probably will pay off in terms of understanding astrophysics and astronomy, and these are the suggestions of exactly how one might look for a Dyson Sphere: looking at what otherwise appear to be normal, O-type stars, but which have anomalous infrared excesses.

It's worthwhile to go through catalogs of infrared sources, for example, the kind of catalog that the Infrared Astronomical Satellite will provide us with in a few years. Then we can look for sources in the infrared that correspond to otherwise normal, solar-type stars, for objects that don't exist, that can't be explained, that don't correspond to any of the standard sorts of infrared sources that we know about.

Suggestions of that nature are really worthwhile pursuing, because we're going to learn something, one way or another. I hope that's also true with the microwave program for SETI. It's the first time that we will look at the universe systematically with very high spectral resolution. Everyone bets that nature is not a producer of coherent radiation. We've not yet seen an example of it, but this is the first time where we're setting out to systematically search for such. And we're exploring a volume, a parameter of search space in the physical universe that hasn't been tapped yet. So we might serendipitously find something of astrophysical interest that we weren't expecting in our search.

It's really a different approach to looking at the universe. It's quite different from the often-used approach where one extends techniques that one's now using just a little farther, a little bigger, or a little more sensitive. This is really a different look; it's a different set of eyes that are being focused on the universe in terms of spectral resolution.

What, in your opinion, are UFOs?

Unexplained phenomena. Whether they are physically real phenomena, or unexplained and manufactured in the minds of people for whatever reason, I keep an open mind. I think that the universe is grand enough and large enough and mysterious enough so that there are probably phenomena that we are not even aware of. Some of these UFO sightings might, in fact, turn out to be valid reports witnessing such a phenomenon that doesn't yet have an explanation. I say, "might be." Certainly the vast majority of them are not anything even that credible. I see no evidence at all that these sightings have anything to do with little green men in flying saucers or extraterrestrial interstellar travelers.

I have an open mind on whether anyone has seen anything that has a reality to it; a closed mind on the idea that what has been seen is a manifestation of extraterrestrial visitation, and an absolute demand that people who are pursuing this subject attempt to treat this the way we would any other scientific investigation, and that is to arrange for hard, reproducible, believable evidence.

Assuming that we do get evidence of the existence of extraterrestrial life, what then would you do with that information?

I would make that information public, without question, because it is not information that belongs to any one person or one country or any group. The information is intrinsically important to all people. It is something that mankind has been searching for or asking about throughout recorded history, and mankind deserves to know that at least we think we have found one instance of an answer. They deserve to know it, even if my belief that the signal is extraterrestrial turns out to be wrong.

In fact, one of the best reasons for making the information public, as soon as we have exhausted all the tests we can think of, is because there are a lot of other people who have the ability to attempt to verify our interpretation of whatever it is we have observed. And if we are wrong, especially if we are being hoaxed, which is a possibility that we can't escape, then having a lot of people trying their hand at it is the quickest way to find out whether it is real, if it is what we think it is, or whether someone has successfully managed to fool us.

If I am wrong, then that's just the way it is, and the next time I am as equally convinced that I have evidence, I would do the same thing.

You have to be prepared in this business, because there are so many things that are unknown, to cry "wolf," and cry "wolf" again, and perhaps be wrong, but at least not be secretive.

At what stage of your own certainty that you have evidence would you announce it to the public?

There are lots of different levels to that question. One is "what's required to make me certain," and that is something we can debate and do debate. The other thing is I am a member of a team that is being funded by NASA, and the form of the announcement therefore will follow the protocol that is adopted by the NASA program for such an announcement. At this point in time that protocol has *not* been formalized.

It's not that we haven't been working on it; it's just that lots of different people have lots of different ideas as to what it should be, and who should make the announcement. NASA is a bureaucratic agency, and therefore it is necessary to percolate this kind of plan up through the agency, and it never goes through in one straight shot. Someone always sends it back because they have a different view of it.

It is a unique problem we're posing to the agency, one that they have not had to face exactly in this way before. So at the moment I can't tell you whether it will be the head of the observing team or the program manager at NASA headquarters, or perhaps the director of the agency, or perhaps the president of the United States. There will be some protocol established, as we go along, that will provide a mechanism for making the announcement, and the only thing I can say is that I would abide by that mechanism and certainly would make use of it.

I would resist any attempt to supersede that procedure and enforce secrecy. If I began to see the established protocol being bypassed and secrecy being imposed, I would probably cease to be a member of the team and would do something as a private individual, because I think the information should be made public. I have never been provided with any overwhelmingly convincing evidence that the announcement of the detection of something that I believe to be there would pose a threat to my countrymen or the people of the world. In the absence of that, I think the information should be made public.

What is the present status of that protocol? When will it be ready?

We're working on it. There have been a number of drafts and a number of discussions. It's been discussed briefly at headquarters; it's been discussed among our group, and members of the science working group from the university community. The easy parts, that we really know how to grapple with, are the steps in verification: what do you think is really going to convince you that you really have something that's coming from the stars?

We tend to allow ourselves to get caught up in those arguments because, as I said, those are the technical things, that we know how to deal with. The political process is a more difficult one. So I would say the protocol is probably not going to be forthcoming for several years yet.

On the other hand, NASA's SETI program does not intend to be ready for awhile. Even if everything goes swimmingly well, we probably won't be engaged in a systematic, large-scale observing program until the beginning of the next decade, the early 1990s, so we do have some time to get this protocol in place.

We'll be doing some small searches along the way, but since we have argued that a systematic approach is necessary, then we don't expect that the sporadic searches, needed in order to learn how to do the job correctly, will have success. If they do, we might be in the embarrassing position of wanting to announce but not having the protocol in place, and then we'd just have to use our own best judgment, as scientists and individuals and people with a sense of responsibility.

But I think that won't happen. I think we have time to consider what protocol is appropriate and to get it in place, and then the opposite might transpire, which is that the search will go on into the future, and it may have to be extraordinarily long before success comes, and then you wonder whether the archaic, ancient, and dusty document which is the protocol you had originally envisioned will still be the correct one, given whatever the future political and social milieu is.

If evidence is found and announced, what do you suppose public response to this might be?

My guess would be influenced by some studies that Mary Connors did a few years ago here at Ames. I'm impressed by her results. I believe it would be big and splashy. Everyone would be interested and it would be instantly known around the world, and there would be headlines, and there would be good reporting and irresponsible reporting, and different members of the news and broadcast media would deal with it more or less hysterically or responsibly.

But the excitement will die down, very quickly, because the mere detection of a signal answers the question about the existence of other intelligent life in the universe. It doesn't necessarily follow that the detection of a signal brings you additional information right away, and for me it's enough to answer that first question, that life does exist elsewhere. I think that's the exciting part.

Any further information that might be associated with the signal would probably take a while to extract. The most detectable signals are ones that, by design, have a very low information rate. Let's put it this way. Those signals which we will detect best, as being distinct from other natural radiation in the universe, will *not* have a very high information content, because to put a lot of modulation and information on the signal makes it look more like noise.

So it will be an exciting problem for a relatively large group of scientists and technologists, and perhaps philosophers and code breakers, around the world, to try and find how or where the information associated with this detected signal can be extracted. Then, understanding that, once you know the method for encoding the information, then trying to *de*code it, then trying to learn to speak Galactic or whatever, in order to make sense of the information that is presented, will take time.

We think of ourselves as very intelligent, but we are probably very technologically backward with respect to a civilization that is capable of sending a signal that we, with our primitive technology, can detect. And so it will probably take us a while to figure it out, even though I expect that if they have gone to the trouble of sending it, they will have tried to make it as easy as possible.

Meanwhile, there will be a tidal wave or an earthquake or a man-made disaster that will happen somewhere around the world that will divert attention away from the signal and onto things that have a larger impact on our daily life. The extracting of information will be a problem, which will excite and keep a whole bunch of people busy, but the world at large will know we are not alone and will only slowly become aware of the additional information, if any, that has been provided by the signal.

It will be really exciting at first, and then will just become a part of the consciousness of the species, that there are other beings out there, and what they have to tell us will come out piecemeal and slowly. I think we can accommodate to that, and we'll probably have some pretty lively debates about what we should do in terms of a response, and there may be a lot of energy expended by other than the technologists trying to decode the signal.

There probably will be a lot of worldwide debate about who should do what, because there is absolutely no way to enforce against individual actions. I'm sure there would be a lot of people who would make an attempt to respond in some manner. Whether they are able to organize sufficient technical capacity to make a response that is detectable or meaningful or successful is something I can't predict, but I'm sure people will try, and people will debate about whether we *should* try.

What about the concern sometimes expressed that news of this sort might touch off panic?

That is what Mary Connors was looking into, and it's very hard to come up with a sociological study that will give a good answer to that question. But from the clues she was able to piece together, her conclusion was "No, the information network that currently exists globally is too secure and too pervasive to allow the sort of War of the Worlds panic that was reported following the H. G. Wells broadcast."

Information about exactly what the evidence is and what it is not would be available to everyone around the world very quickly. Unless the journalistic handling of the announcement is totally out of hand, the information that a signal has been detected can be and will be conveyed in a way that will *not* cause panic, because it is not "The Russians have landed" kind of thing.

What might be the long-term impact upon civilization of verified contact with extraterrestrials?

I hope that it would be very stabilizing. If we, with our relatively primitive technology, detect evidence of an extraterrestrial civilization, then it has to mean that those civilizations live for a very long time. The reason being that if we can find a signal easily when we first try, they have to be, on the average, fairly close to us.

That means that there have to be a lot of civilizations, and this just can't occur if the average civilization only lives for a hundred years and then blows itself up. It's totally improbable that you could have civilizations that are detectable and close to you and easy to find if they, as a rule, only live the life span that we've had on this planet, and then do themselves in.

So the detection of a signal, even without any information content, at least lets us know that it is possible to stabilize a society and have it live for a long time. That is the really major impact it would have on our civilization.

It must make us take a look at ourselves from somebody else's point of view. It must make us look at ourselves as one small planet in a big universe, and the geographical boundaries and the political and religious differences we have must become, even to our own eyes, minuscule and somewhat trivialized by this experience. That's all very positive as far as I'm concerned.

So hopefully, it would be one way to convince us that there is a solution, that there really is some way to survive, and then what we have to do is look for that solution. Knowing that there is a solution to an apparently bleak, hopeless problem gives one a lot more incentive to try and find what that solution is.

I am not saying that they will give us the answer, but the detection of the signal would motivate us enormously to solve our own problems, and give us a perspective that we don't now have, that might allow us to actually make progress.

What scientists or other people in the past do you admire?

Historically, I'm fond of Gauss. He's a very, very inspiring figure. I was also impressed by Madame Curie, who was a two-time Nobel winner.

Who are the outstanding people in SETI today?

I keep saying that my work is enjoyable. It has to do with being able to interact with such people: a Barney Oliver and a Frank Drake—really, really inspiring individuals. They constantly astound me with their ability to conceive new ways to do things. Phil Morrison is delightful and outstanding in every respect. I'm sorry that I don't get more chance to interact with him. He's the father of this field of intellectual pursuit. Carl Sagan, for a number of reasons, is outstanding in the field because he has popularized it to such an extent, and he has brought money into astronomy and planetary science—and SETI. Although we don't interact with him on specific details on the SETI program, Carl is a very large figure in the world today in terms of public enthusiasm for the field.

In the Soviet Union I've always enjoyed Dr. Kardashev. I'm intrigued by his ideas about the positronium line providing the frequency standard for interstellar communication and his suggestion that some sort of large-scale astroengineering is responsible for the observed activity at our own galactic center. The classification of civilizations, whether it has any basis in fact, at least is a correct intellectual theoretical way to go. There are certain energy levels that we might have at our disposal, and Kardashev has correctly pointed out that is a way one might legitimately characterize the efforts of a civilization.

A very important person in SETI today who is perhaps not as visible as some of the others is Dr. Billingham. He's largely responsible for the fact that, despite many setbacks and political changes, SETI has continued at all as a NASA endeavor at *any* level. He's certainly a key figure in extraterrestrial biology now (a science without a subject, it is often called), the endeavor to try and learn from other planets in our own solar system, or from the conditions present in the early solar system, and to infer from our own evolution what might have happened elsewhere.

John is so strong, personally, in terms of just keeping this effort going, in a politically not very favorable climate. He's been really quite inspirational. If things go wrong, it doesn't matter; he's got another plan, another scheme. And he deserves an enormous amount of credit for this. Should we be successful in establishing a NASA-based research effort of some scope, then John will be primarily responsible for that.

Finally, what else should future historians know about you that would not otherwise be preserved?

I have already mentioned that I thoroughly enjoy what I'm doing; and it's important to me to do something that is enjoyable and worthwhile. I like interacting with people, and I like learning new things, and all of those are a part of SETI. I feel very fortunate to be able to have a family and close emotional ties with people; that's very important to me. I often run when most people would walk and I tend to be very enthusiastic, more enthusiastic than a lot of the people that I deal with daily.

When I was a new undergraduate engineering student I had things all charted out. I figured that about the time that I would be getting a master's degree in engineering, NASA would be looking for some women astronauts for the moon program. A huge disappointment in my life was that NASA turned away from that, and it has only been lately that women have been accepted into the space program. I hope that at some time there will be an opportunity to fly with an experiment, with a payload that would be SETI-related, or astronomically related.

If there's an instrument, I always have a bias for putting it in space, because maybe I can go along with it. I hope sometime I will.

Conclusions

SETI progressed in its first quarter-century from the lunatic fringe to scientific respectability. When Morrison and Cocconi published their landmark report in 1959, extraterrestrial intelligence was widely regarded as fantasy, beyond the borders of legitimate science. It was good entertainment but not to be taken seriously by reputable researchers.

By the mid-1980s it had become acceptable, endorsed by prestigious scientific organizations and funded by the governments of the United States and the Soviet Union. The factors in this transformation are illumined by these interviews with the first scientists involved. This final chapter examines two questions: Why did SETI emerge at this point in history, and why did these particular scientists become involved? The answers are not simple.

Timing

Why did scientists begin at this particular time to examine an issue that had been around for thousands of years?

TECHNOLOGY

The most obvious difference between SETI and earlier work on extraterrestrial intelligence was technology. By the latter part of the twentieth century, technology had matured to the point at which tools for examining these ancient questions were finally becoming available—the technological base had developed sufficiently that long existent ideas could at last be empirically tested. And motivation to apply these tools came from various sources, ranging from opportunities to test theories in several scientific disciplines to a general awareness of space that was activated by Sputnik.

The first artificial satellite was launched on October 4, 1957. The tiny Soviet sphere, orbiting the Earth every ninety-six minutes, had a profound effect upon the world's population. Many scientists were as startled as laymen. The United States responded by forming a new agency, with a multi-billion dollar budget and a mandate to put men on the moon within ten years. Virtually overnight Sputnik created a supportive environment for space activities, including what was later to become SETI.

Technology aided SETI in several ways. Most apparent was the actual equipment used. Drake could not have searched for messages, nor sent any, without a radio telescope, nor could Kraus, Troitskii and others who followed him. The first major radio telescopes became available in the late 1950s. England's Jodrell Bank 250-foot radio telescope was completed in 1957. Green Bank's 85-foot radio telescope was finished in 1959, and the 300-foot dish in 1962. Parkes Observatory in Australia, with a 210-foot dish, opened in 1961.

A second consequence of technology was that it stimulated thinking. The availability of radio telescopes encouraged people to consider ways they might be used, by others as well as by us. If we on Earth have this capacity, might not intelligent beings elsewhere have similar or even better technology? The rapid development of our own capabilities in a few decades suggested that older civilizations might have advanced far beyond us. Could their signals already be reaching Earth?

Thinking was also stimulated by other kinds of technology. For example, Morrison and Cocconi realized that they knew how to make gamma rays. "We were making lots of them downstairs at the Cornell synchrotron. So Cocconi asked whether they could be used for communication between the stars." The possibility of using gamma rays prompted them to examine the whole electromagnetic spectrum in order to find the best wavelength for communication.

Third, technology enhanced communication, enabling people to discuss ideas regardless of time or distance. Scientists anywhere in the world could exchange information, using copy machines, computers, or low-cost telephone service. And high-speed travel facilitated meeting in person at Green Bank in 1961. A decade later, interaction had reached the international level with the Byurakan conference in Armenia, sponsored jointly by the academies of science of the United States and the Soviet Union.

These developments transformed concern about extraterrestrials from an individual to a group activity, providing stimulation, criticism and reinforcement through opportunities for interaction. A scientist was no longer on his own. He now had the benefits of collective effort. He could utilize resources previously unavailable because of time or distance or cost.

Thus Drake conducted his search on complex equipment designed, built and paid for by a federal organization, the National Radio Astronomy Observatory. Oliver in California, reading of Drake's project, "dropped in" by airplane to see him in West Virginia. Cocconi and Morrison, based in New York, finished their report in Switzerland and published it in England. It was soon read by scientists in the

Soviet Union and Japan as well in the United States, and news of it followed Morrison as he traveled around the world.

A critical mass had been achieved through technology, enabling and even stimulating scientists to do things they could not have done by themselves. Thoughts that previously would have lain dormant were now activated by interested colleagues, who might live and work thousands of miles away.

Technology lowered the threshold of participation in SETI to the level at which it became practical for busy scientists to take time off from their mainstream activities to work on this unconventional, low-priority topic.

OTTO STRUVE

Technology alone, crucial though it was, does not fully explain the rise of SETI. Although SETI started in the United States, other countries also had radio telescopes. England, for example, had pioneered in radio astronomy, and its Jodrell Bank dish was for many years the world's largest fully steerable radio telescope, until Germany's Effelsberg telescope opened in 1971. Yet neither nation was the birthplace of SETI, nor subsequently very active in it.

We must therefore look beyond technology to other factors. These are illumined by SETI's most prominent precursor, Otto Struve. He enables us to clarify the genesis of SETI. In particular, he helps us to distinguish between two distinct though related elements: one, evidence that extraterrestrials may exist; and two, the idea that we can search for them, probably in the microwave region.

What he did and what he did not do are both illuminating. He ardently expounded the first, but did not engage in the second, though he certainly encouraged others to do so.

Otto Struve was one of the most eminent astronomers of this century. Born in Russia in 1897, he was the fourth generation of his family to achieve distinction in that occupation. Friedrich Georg Wilhelm von Struve (1793–1864) had been director of the Dorpat Observatory in Estonia, and was invited by Tsar Nicholas I to supervise the building of the Pulkovo Observatory near St. Petersburg. In 1864 Struve was succeeded by his son, Otto Wilhelm Struve (1819–1905). One of O.W. Struve's sons, Hermann (1854–1920) became director of the Berlin Observatory, while Ludwig (1858–1920) was appointed to the chair of astronomy at Kharkov. Ludwig's son was Otto.

Otto was an artillery officer in World War I. He fought on the losing side during the Russian Revolution and fled to Turkey. Emigrating to the United States in 1921, he earned a Ph.D. in astrophysics at the

University of Chicago. He eventually became head of the astronomy department and in 1932 was appointed joint director of the university's Yerkes and McDonald observatories (Ridpath, pp. 201–2).

He moved to the University of California at Berkeley in 1950, and in 1959 became the first regular director of the new National Radio Astronomy Observatory at Green Bank, West Virginia.

He was an influential champion of the view that many stars like the Sun may have planetary systems. In his youth, he and his father had read Percival Lowell's speculations about life on Mars, and Otto's subsequent studies of slow-rotating stars had persuaded him that planets are plentiful in the universe. In 1960 he estimated that there are fifty billion solar systems in the Milky Way Galaxy alone.

On the question of how many have produced intelligent life, he stated: "An intrinsically improbable single event may become highly probable if the number of events is very great." If the probability of finding intelligent life on a planet at one given time is substantially more than one in ten billion, "then it is probable that a good many of the billions of planets in the Milky Way support intelligent forms of life. To me this conclusion is of great philosophical interest. I believe that science has reached the point where it is necessary to take into account the action of intelligent beings, in addition to the classical laws of physics" (Sullivan p. 200).

Struve had stimulating ideas, and they carried weight because he was so distinguished. Concepts that would have been dismissed if voiced by other scientists were likely to be taken more seriously because of his eminence. In addition to planting ideas in some people's minds, he also reinforced scientists who were already thinking along such lines.

Struve had been in all the major SETI locations. He was director of observatories or astronomy departments in Chicago, and on the east and west coasts. He published in popular as well as professional journals and lectured in institutions such as Cornell, where Drake heard him in 1951. Drake remembers the excitement of hearing the distinguished astronomer discuss issues that he, too, was thinking about.

> In 1951 Struve was invited to Cornell to give a set of lectures in an eminent lecture series, given once a year there, called the Messenger Lectures. He lectured primarily on the topic of stellar evolution, and in the course of the lectures he did discuss observational data on the rotation of stars, and the fact that the data showed very massive stars rotating very quickly, whereas the less massive ones, the solar type stars, were rotating much more slowly.

And then he played out a very provocative fact, that the total angular momentum of the solar system is about equal to the angular momentum in the rapidly rotating massive stars, and he suggested that this meant that the rapidly rotating stars did not have planets but the slow rotating ones did, because the angular momentum was consistent, and because of that he proposed that essentially *all* of the solar types—small, less massive stars—possessed planetary systems. To me, that was very exciting.

Q: You remember being excited by what he said?
A: Yes, I was excited because he had some real *evidence* for the existence of extrasolar planets. At that time it was *not* at all accepted. And this was the world's most eminent astronomer, favoring the existence of the great abundance of planetary systems.
Q: And was this in the back of your mind from then on, during the rest of the fifties?
A: Yes, it still is.

Within the decade, Drake was working under Struve at the Green Bank National Radio Astronomy Observatory (NRAO), conducting Project OZMA, the world's first modern search for extraterrestrial signals.

Seeger's comments are helpful here; in addition to providing background on that historic activity, they also clarify Struve's role in SETI, and the process by which the pioneers, particularly the early ones, became interested in the search.

I met Struve many times. He was a very good friend of Oort's, and whenever he was abroad he would stop in Leiden and we'd talk about many things. Struve told me that his reason for pushing the Ozma thing was to help make a better image of NRAO, during the period when it was absorbing money and producing very little because it wasn't built yet.

NRAO was founded essentially by a scientist politician, Lloyd Berkner, a pioneer in wave propagation, and one of the early explorers of the ionosphere. He was a compatriot of Vannevar Bush's at MIT (who first proposed the International Geophysical Year of 1957–58), and headed the team of nine universities which had established the Brookhaven National Laboratory. He was now supervising Kitt Peak and NRAO for NSF. He saw what was happening in Europe, Australia, England in radio astronomy, and said the United States must have a radio observatory, too.

NRAO was run from a long distance by Berkner and his engineering associates—mainly his associates. NRAO was not in good repute with the astronomical community because of the huge funding that had been spent on it; a 140-foot telescope was enormously expensive for those times, and largely couldn't show that it was any good for anything.

At any case, NRAO was without a scientific head, or at least an engineering head. Struve was brought in because he was a name scientist who wanted to get the hell out of Berkeley at that point (this was in 1959, with all the student unrest) and he was persuaded to be the head.

His meeting up with Drake at NRAO was the reason Drake did his OZMA experiments. I'll never be able to separate in my mind who was more responsible, Drake or Struve, but without Struve it would not have happened, certainly not then. Drake was very new, very young, relatively inexperienced at that time, and it would have taken someone of Struve's drive and reputation. It was important at that time to have visible projects, because NRAO was under great crititicism for being so slow. There was risk in OZMA, but it brought a lot of attention to NRAO.

Struve was really the driving force that put OZMA on the boards. Drake would certainly have done it at one time or another. He would have persuaded someone eventually, but Struve pushed it because of his own interest in the subject, which was quite independent.

Thus Struve was vitally important in approving Ozma, encouraging it—in fact, pushing Drake to do it. But the actual *idea* of using radio telescopes for such a search was already in Drake's mind. He did not get it from Struve. We may recall that Drake, when asked who or what first suggested to him the possibility of searching for ETI, replied: "Nobody. I've always been interested in life and space, and had thought about it over the years."

To be certain of Struve's role in Drake's thinking, I asked Drake again about the 1951 Cornell lecture:

Q: Did he actually give you any new ideas or was it that he was simply encouraging you to proceed along what you were already thinking?

A: The latter. He gave *no* new ideas as to how you *search*. He never even *talked* about searching. He discussed evidence—the first anybody really had—that there were large numbers of planetary systems.

Q: And this was something you were already thinking when you heard him?

A: Well, I was *hoping*. You couldn't really *believe* such an idea, because there was zero data at that time. This was the first data to support the idea. But in his talk he said nothing whatsoever about searching for planets, or anything of that sort.

Struve had a different impact on Bracewell. Struve had brought Bracewell to the UC Berkeley astronomy department in 1954. Bracewell recalls that for many years he thought the 1959 *Nature* report had started his thoughts on ETI. But then he "suddenly realized, with a bit of a shock really," that his association with Struve "whom I came to know quite well, must have prepared my mind for the subject of life in space. . . . I realized that that must've been lying dormant in my mind, and that when I read Cocconi and Morrison's article, it struck more sparks off me than it did off some of my friends. I had been prepared for that sort of concept."

Several of the pioneers crossed Struve's path, yet none named him as the source of their ideas on searching. His statements were intriguing, reassuring, exciting, but not sufficiently specific. He didn't supply the final ingredient: the means for searching. He, and others, provided stimulation but not the method.

The SETI pioneers were not given to impulsive leaps into the unknown. They would wait until they saw a realistic way to conduct a search.

This way had to be practical for them, personally. Even the trailblazing example of OZMA was not enough to get all the pioneers involved.

Seeger and Kraus didn't take up SETI until a decade after OZMA. This lag was certainly not due to lack of expertise; Seeger and Kraus had been active in radio since childhood, and when radio astronomy emerged, they were among the first people in it, and contributed much to its development.

Seeger had known Struve since 1946 and had discussed OZMA with him in the early 1960s, " but my first real thought of doing something about it came up with Billingham and Oliver. It didn't cross my consciousness until Project Cyclops crossed my desk in 1971—and I jumped at that, because there was something I could really get my hands into."

For Kraus, there was a lag of half a century between the idea of radio communication with Martians, suggested in the amateur radio journal *QST* during the 1920s, and his actual involvement, which began in 1973 at the suggestion of his coworker and former student, Robert Dixon.

The fact that Struve did not actually engage in SETI illumines the thought processes of the SETI pioneers. They were ready to start, but needed a specific way to go about it. Whether theorists or experimenters, they were practical, tough-minded scientists. No matter how fascinated they might be with the possibility of ETI, they would proceed methodically, step by step, toward answering the age old question.

The Process of Involvement

Neither technology nor the presence of a very distinguished astronomer fully explain why these particular scientists got involved in SETI while others did not.

In earlier times, the natural ability of a Leonardo, Newton or Galileo might have been a reasonable answer, but today it is difficult to tell what sets one scientist apart from the rest. Education? There are multitudes of Ph.D.s, not to mention masters of science and baccalaureates. General levels of training and competence have risen enormously in many fields of human endeavor, not only in science but in such diverse areas as arts and athletics.

In this respect, science today resembles a professional sports league, with dozens of teams, all staffed with highly competent players. Differences between teams and between individuals are very slight, making it hard to explain why one is able to do better than the rest.

Yet this analogy is faulty in one major aspect: competition. The SETI pioneers were not competing with other scientists, at least not with respect to finding ETI. Of course, scientists do compete for funds from the same limited sources; the more scientist X gets for his project, the less will remain for his colleagues. But as for the topic itself, there was no race to see who would detect ETI first. Unlike the search for the secret of DNA, there were no Linus Paulings breathing down the necks of Drake or Morrison or Cocconi, trying to beat them to the discovery of extraterrestrial signals.

No one else was in the game. The pioneers were the only ones willing to play. They didn't know where the goals were and had no assurance that they even existed. The playing field was unmarked and literally limitless, comprising the entire universe. There was no detailed set of rules, only rudimentary assumptions, such as the expectation that the laws of nature are everywhere the same. And when would the game be over? No one knew; it could last a lifetime, or ten thousand years.

CONVENTIONAL APPROACH

Willingness to venture into fringe areas may occur in two kinds of situations, very different from each other. On one hand, people may deviate from conventional, standard positions if they are not strongly attached to an occupation, society, or a relevant system of ideas. Unfettered by knowledge or by social restraints that might curb their thoughts or their behavior, they are free to let their imaginations influence their perceptions.

On the other hand, freedom to explore new areas can also come from the opposite extreme, from very firm integration within a system. The person who has mastered standard techniques and ideas can combine them in new ways, not perceived or attainable by his colleagues. So, deviation from the norm can come from knowledge as well as from ignorance.

These pioneers were solidly grounded in science. Thus they differed markedly from laymen who are interested in ETI but are unfamiliar with basic scientific principles. The SETI pioneers also differed from other scientists, who, though competent in their particular fields, were not sufficiently versed in the specific areas relevant to SETI.

A notable point about these scientists was their general manner during the interviews. They were serious, calm and thoughtful, and what was not said was almost as revealing as what they did say. They did not rhapsodize about the importance of the search or the thrill of communication with alien species. Their discussions were low-key, both in content and delivery, in contrast to the stereotype of a zealot. An observer watching the interviews, without knowing what was being discussed, would have no clue that it was anything unusual.

These pioneers were, in many ways, conventional people. There was nothing in their origins, appearance or lifestyle to distinguish them from typical members of the middle class. Their parents were of modest means, earning their livelihood as merchants, musicians, salesmen or teachers. Some parents had technical training, but few had much formal education beyond high school. The scientists themselves all adhered to the traditional pattern of getting married and raising children. Their recreational pursuits, also, were like those of other professional groups. It was their interest in extraterrestrial intelligence that set them apart from other scientists.

Skeptical of claims that we are being visited by aliens in flying saucers, these scientists embarked on a long-term project to detect extraterrestrial civilizations. Their approach was conservative, suggesting that they got involved *because* of, rather than in spite of, their position in the scientific mainstream. They moved toward SETI because they were firmly grounded in the fundamentals of their field, not because they were ignorant of them. They were so well versed in relevant technology or theory that they could see where these might lead.

But there were thousands of other traditional scientists who did not get involved in SETI, so we must look further to learn why these did.

ACCUMULATION OF ADVANTAGES

The pioneers gained an edge over their colleagues through a process of "accumulation of advantages." This process, a "composite of self selection and social selection, operates to enlarge opportunities for further effective work and role performance which, in turn, open up further opportunities" (Zuckerman, p. 95). In less formal terms, it was a "snowball" effect. A small step in the right direction, whether taken deliberately or by chance, could have significant consequences later.

First, some possible influences toward SETI can be dismissed. We can rule out direct, deliberate influences from their families. None of the pioneers felt that their parents steered them to the study of ETI or, for that matter, towards science of any sort, either as an activity or a career. This was true even for the one person, Kraus, whose father was a professor in the sciences.

Siblings, too, were reported as not being an influence. Extending the inquiry further, to teachers and friends, still did not bring reports of influence toward ETI. In sum, there did not seem to be any particular type of person in the scientists' first twenty years or so who gave them the idea. It was not typically their parents, or a high school science teacher, or the occupant of any standard role who inclined them toward SETI.

Instead, interest in SETI came from other sources, outside of family and school. It was endemic in popular culture, with which the pioneers had been familiar since childhood. And, as a serious endeavor, it was suggested to them by various elements in their adult, professional environment, including people, ideas and technology.

The pioneers had been avid readers in their youth. They repeatedly mentioned publications about astronomy, radio, science fiction and fantasy concerning life on other worlds. Jules Verne, H. G. Wells and Edgar Rice Burroughs were often cited. The book, *Splendour of the Heavens*, was mentioned by both Freeman Dyson and Bernard Oliver. Other influential publications were *Radio News*, *Amazing Stories*, *Popular Science*, *Frontiers of Astronomy*, and *Fantastic Adventures*. Troitskii recalled the book, *The Pressure of Light*, and Shklovskii reported the impact of *Novi Mir* (New World).

Another aspect of this accumulation of advantages can be seen in their childhood activities: Morrison working with a chemistry set; Drake's fascination with the Museum of Science and Industry; Troitskii's interest in airplanes; Dyson's involvement with calculations and star-gazing; Shklovskii's artistic talent; Billingham's chess; Kardashev's visits to the planetarium; Bracewell's puzzles and chemistry experiments; Sakurai's fascination with the natural world around his mountain farm; and Calvin's investigations of grasshoppers.

Several pioneers grew up with a technology that was to become so vital to SETI, namely radio. Kraus, Seeger, Troitskii and Morrison were avid radio "nuts" in their youth.

World War II gave the young pioneers experience with the most advanced technology of that period. Drake's extensive electronic experience in the Navy led him into radio astronomy in graduate school. Dyson was involved with operations research for the Royal Air Force. While Oliver was at Bell Laboratories working on radar he calculated what the range would be if the energy sent out from the microwave transmitter was not reflected back to the transmitting antenna. He realized that this technology could be used to communicate with Mars or more distant planets.

Kraus worked at the Naval Ordnance Laboratory, Washington, D.C., on a project to protect ships against magnetic mines, and developed a radio telephone. Later, he moved to Harvard's Radio Research Laboratory to develop radar counter-measures. Bracewell designed radar equipment at the Commonwealth Scientific and Industrial Research Organization in Sydney.

Seeger was an instructor in the Army pre-radar training program. He observed that "the war provided me with opportunities and experiences that would have been totally beyond the pale in peacetime. Above all, it got me into research."

LOCATION

Geographical location was another factor. As we see where the pioneers were born, where they went to college, and where they were and are employed, we notice a pattern of clustering which becomes increasingly concentrated as we approach the present. Furthermore, the location of the pioneers follows closely the pattern described by Maurice Richter for science generally, even to the scientist who became the least involved:

Centers of science are unusual places. Science has flourished only in small segments of a small number of societies during a comparatively brief period of human history. Since the decline of Italian science in the early seventeenth century, Britain, France,

Germany, the United States, Russia, and Japan have accounted for an overwhelmingly large portion of world scientific achievements. Large areas of the world have remained comparatively barren from the standpoint of modern science. . . . Within scientifically important countries, scientific activity has been strongly localized: thus large parts of the United States and large segments of its population remain scientifically quite barren." (Richter, p. 39).

Richter observes that science has become strongly international, yet "there has nevertheless usually been a single geographic area or locality which has constituted an unofficial center of world scientific activity." He mentions that, since the Scientific Revolution, such centers have included northern Italy, England, France, Germany, and now the United States (Richter, pp. 38–39).

This is certainly the case with SETI. While scientists from several nations are interested in it, SETI activity is concentrated in the United States and, on a smaller scale, in the Soviet Union.

The SETI pioneers came only from countries on Richter's list: three from the British Commonwealth, three from Russia, eight from the United States, one from Japan and one from Italy. There is even a parallel between one of the nations and its SETI scientist son. Northern Italy, which at one time led the scientific world, is represented by the individual who likewise relinquished his leading position in SETI.

The pioneers were born and raised in or near cities—New York, Moscow, London, Chicago, Sydney, San Francisco, Tokyo, Milan. These locales, including some of the world's major cities, were more than just urban areas; they were, and are, centers of government, commerce, art and, most relevant, science and technology. They are stimulating places, attracting creative people and supporting new ideas.

The pioneers' pattern of birthplace and childhood was unusual for those times. It was not typical of the general population. Urbanization then had not proceeded as far as it has today. Statistics from the U.S. censuses of 1910 and 1920 show that half the population was still rural. In contrast, all of the eight American pioneers lived in or near cities. Even Oliver, who grew up on a ranch, was only seventy miles from San Francisco. Shklovskii was the only one who lived farther than 80 miles from one of these major cities. He was 150 miles from Kiev and Kharkov, and 300 miles from Moscow.

Among the advantages of growing up in these places was proximity to major universities. The pioneers in California attended Stanford,

Cal Tech and U.C. Berkeley. The universities of Michigan, Minnesota and Chicago were available in the Chicago area, and Cornell and Harvard were in the vicinity of New York. The British pioneers had Oxford and Cambridge within fifty miles of London, and Moscow, Milan, Sydney and Tokyo each harbored major universities.

After graduation the location of the SETI pioneers became even more concentrated, and remains so to the present. Eleven of the sixteen are in the United States; three in the Soviet Union; and one each in Switzerland and Japan. Most of the pioneers are in one of three clusters. In the Soviet Union, two are in Moscow, and the third is in Gorkii, 240 miles to the east. The Americans are clustered into two regions: Cornell, MIT, and Princeton on the East Coast, and the smaller San Francisco Bay area in the West. Kraus is the only SETI pioneer in between.

The West Coast contingent is extremely concentrated. Five of the pioneers work in Palo Alto or at Ames Research Center five miles away. Two others are within thirty miles: Drake in Santa Cruz and Calvin in Berkeley. Thus, in the 1980s, almost half of the the fifteen living SETI pioneers were within an hour's drive of Stanford University.

Seeger and Oliver were born in the San Francisco region and have worked there for many years. Calvin has taught at Berkeley since 1937. Bracewell, after a year at Berkeley, has been at Stanford since 1955. In 1965 Billingham moved to Ames Research Center, and Tarter came to Berkeley in 1968. Drake joined them in 1984, moving from Cornell to Santa Cruz.

The only geographically isolated pioneers are Kraus in Ohio, Sakurai near Tokyo, and Cocconi in Switzerland. Kraus and Sakurai remain active in SETI and related activities, while Cocconi has, since the early 1960s, been the least active of the pioneers. He was also the most reluctant to be interviewed.

Kraus, the most isolated American pioneer, was one of the last pioneers to become involved in SETI; he did so in 1973. He was also among the most conservative during the interview, most likely to say, "I don't know," least willing to hazard guesses on such questions as when ETI might be detected, what would they look like, how most scientists view SETI or what he would do if he were starting over again.

Apparently, modern technology, despite its tremendous contributions to communication, does not completely overcome the influence of geography. What particular aspect of location might have encouraged SETI on the East and West coasts of North America, and discouraged it elsewhere? Seeger suggests an answer:

The East and West coasts were the strong scientific centers in the country before, during and after the war. After the war they were the centers of modern technology. Look at the people who came back to California from working on military technological development. Terman came to Stanford. Whinnery was head of electrical engineering at Berkeley for a long time. During the war, Ramo and Whinnery produced one of the major teaching texts for introduction to microwaves. I taught out of their book at Cornell. Oliver knew both of them, I think.

The return from the East Coast of people who had been in the MIT or Harvard radiation laboratories gave California a whale of a boost because, with the federal funds they brought in, they were able to support a lot of developments and graduate students. Sagan, who was always interested in ETI, went out to Berkeley for his post graduate work.

SETI was a manifestation of that combination of science and engineering on the East and West coasts. In contrast, the Midwest is more provincial. Its influence on Kraus, the one pioneer who remained there, was noted previously. Seeger adds that

Kraus is a rather remarkable person, and he has never had the credit due him nationally. He got local recognition, but being in a conservative part of the world he was discriminated against by the East and West coasts, which had developed the ability to raise federal funds. It was very hard for Kraus to get any. The eastern and western clubs in science give each other mutual support. It's hard for someone else to break in sometimes. One deals with one's acquaintances and friends, the people you know. The bulk of science was weighted to the East and West coasts.

A significant factor underlying science is support for intellectual activity and openness to new approaches. While this support is important for science in general, it was essential for SETI. SETI emerged in nations, regions, cities and organizations receptive to new ideas. Such locations provide a critical mass, in which people of similar interests can meet and perhaps develop something new.

The San Francisco area has been a birthplace for innovation of many sorts, ranging from semiconductors to student movements—and even sexual revolution. It also has been hospitable to SETI. While the three main events launching SETI occurred on the East Coast—the Morrison-Cocconi report, OZMA, and the Green Bank meeting—by 1970 the center of SETI activity had shifted to the San Francisco Bay area.

By that time Oliver, Calvin and Bracewell had been there for many years. The return of Seeger and the arrival of Billingham provided the critical mass, which in 1971 produced Cyclops, and subsequently NASA's SETI program—minuscule, but nevertheless an official government activity.

CORNELL

Analysis of a possible "Cornell connection" provides further perspective on the origins of SETI. This university has had the largest contingent of SETI pioneers of any single institution. Seven have been affiliated with Cornell at one time or another, six as members of the faculty. Their presence began in the late 1930s and continues to this day as the base for Carl Sagan. Cornell had already attracted other distinguished scientists, including future Nobel laureates Hans Bethe who arrived in 1935, and Fritz Lipmann in 1939 (Zuckerman, p. 155).

The long association of SETI pioneers with Cornell began in 1939 when Seeger, already an experienced radio technician, enrolled in the School of Electrical Engineering. During World War II, while still a student, he taught courses in radar, antennas and networks. He founded the Cornell Radio Astronomy Project in 1946 and served four years as its chief scientist.

Morrison joined the physics faculty in 1946, as did Cocconi the following year. Both remained at Cornell until well into the 1960s. Dyson also arrived at Cornell in 1947 as a postgraduate scholar, becoming a professor of physics in 1951, and leaving two years later for Princeton. Drake was there as an undergraduate, receiving his bachelor's degree in 1952, and returning to join the Cornell faculty in 1964 as professor of astronomy. In 1971 he became director of the National Astronomy and Ionosphere Center, with headquarters at Cornell.

Thus, at mid-century, five of the SETI pioneers were at the same university, as were other scientists who had distinguished themselves in related fields.

While there are no indications of a specific concern for SETI at Cornell in the 1950s—no sign of a formal group or a cell or a course specifically concerned with seeking alien life—the evidence strongly suggests that Cornell was a fertile location for SETI. Publication of the historic Morrison-Cocconi report by two members of the Cornell faculty would, in itself, ensure the university's place in the annals of SETI.

In addition, there were other indications that Cornell was an unusually hospitable place for SETI, or at least for its precursors. Prominent among these indicators is the Arecibo telescope. It was not accidental that the world's largest single radio astronomy dish was constructed by Cornell rather than by some other institution. It was completed in 1963, but planning for it had started years earlier and had required the sustained, cooperative effort of many people at Cornell.

What was behind such activity? One ingredient was radio astronomy, started at Cornell in the 1940s by Seeger. Another was engineering physics, which also goes back several decades. Engineering physics departments are not common now and were even less so back in the early days. Jill Tarter, who graduated from Cornell's engineering physics department, observed that despite being a small department, it produced at least two dozen graduates who subsequently went into radio astronomy—including Drake. She notes that these people, like herself, did not want to build mundane engineering projects such as bridges or dams, but on the other hand they were too pragmatic to be completely satisfied with physics alone.

If SETI itself was not discussed at Cornell in the 1950s (or anywhere else, for that matter), engineering physics and radio astronomy were very close to it, providing skills and concepts which later became an integral part of SETI. Theory and technology both were encouraged at Cornell. Consequently it could attract, support and train the broad array of scientists who would later launch the unusual new field.

Under such conditions it is not surprising that the report that inaugurated SETI was prepared by scientists at Cornell.

Drake summarizes the situation:

> You certainly can't draw the conclusion that Cornell was an institution trying to support SETI. There was no dean or department chairman who ever made any appointments to assist SETI or to bring the SETI presence to Cornell.
>
> But Cornell was, and still is, a place which is very cordial to creative and imaginative people, so a lot of people of that type went to Cornell or passed through it. It was a very friendly place for creative, unusual ideas. If there was a reason so many SETI people were there, that was it.
>
> Q: No discussions of SETI before Cocconi and Morrison got together in 1959?
> A: No, there was none. It started with them.

BIRTH ORDER

We turn now to an unexpected finding: birth order. When we examine the pioneers' position within their families, a striking pattern

Table 1. Birth Order of SETI Pioneers

Pioneer	Order	Siblings
Billingham	First-born son	brother
Bracewell	First-born son	brother
Calvin	First-born son	sister younger
Cocconi	First-born son	sister older
Drake	First-born son	brother, sister younger
Dyson	First-born son	sister older
Kardashev	First-born son	sister younger
Kraus	First-born son	sister older
Morrison	First-born son	sister younger
Oliver	Only child	
Sagan	First-born son	sister younger
Sakurai	First-born son	2 brothers, 1 sister older and 3 younger
Seeger	First-born son	3 brothers, 3 sisters younger
Shklovskii	First-born son	sister
Tarter	Only child	
Troitskii	First-born son	brother, sister younger

emerges. *All* the scientists were oldest sons or only children. Ten were the oldest children, two were only children, and the remaining four boys had an older sister but no older brothers. This favored position of first sons held true even in the two cases in which there were seven children in the family. In other words, not one of the scientists had an older brother.

The privileged position of first-born sons has long been noted. In many times and many cultures they have inherited the best their families offered: land, wealth, titles, desirable occupations, leadership positions. There is even a word for it—primogeniture—but more significant is the fact that it denotes a widely recognized principle. In modern societies emphasis has shifted from inheritance to competence, from ascription to achievement, yet the advantages of being the first-born male or an only child persist. A review by Schachter observes that "the numerous studies of the correlates of fame or genius have, with astonishing consistency, reported a marked association between birth order and eminence. Repeatedly it has been demonstrated that those who are more productive or creative or eminent tend to be first-born or only children" (Schachter, p. 757).

This association has been studied for more than a century. Francis Galton noted it among British scientists in 1874, and Yoder observed it twenty years later in an international group of fifty eminent men.

Subsequent studies continue to report the pattern among American scientists, Italian professors, ex-Rhodes scholars, Nobel laureates and many other groups (Schachter, p. 757).

A comparison is provided by Nobel prize winners. Data provided by Roger Clark on seventy-seven Nobel laureates in physics and chemistry reveals that twenty-five of them did *not* follow the pioneers' pattern. Fully a third of these eminent scientists were not in the privileged position of being first-born males or only children. The SETI pioneers, therefore, are a very unusual group (Clark, unpublished data).

"The dice are loaded in favor of the first-born" (Altus, p. 48). Two specific advantages are relevant here. Dean found that the first-born show more curiosity. They ask more questions (p. 21). It is therefore not surprising that the SETI pioneers might be somewhat more likely than their colleagues to wonder about such topics as alien life and how to find it.

A second benefit associated with birth order is college attendance. The first-born are overrepresented among college populations, and there is some evidence that the more selective the college, the greater the overrepresentation (Altus, p. 48). This pattern certainly held for the SETI group. The institutions they attended are among the world's finest.

In the United States these included Cornell, Harvard, Michigan, Berkeley, Cal Tech, Chicago, Stanford and Minnesota. All but the last two were among the thirteen colleges judged by Berelson in 1959 as "elite"; Minnesota's ranking is nevertheless very high, and it has produced as many Nobel laureates as Cornell and Wisconsin. Calvin is one of them. Stanford is recognized today as being at the highest level (Zuckerman, p. 280n).

The three pioneers from the British commonwealth also came from top universities: Bracewell and Dyson from Cambridge, and Billingham from Oxford. The remaining foreign pioneers were educated at eminent institutions in their respective countries.

PRIVILEGED POSITIONS

The accumulation of such advantages led the pioneers to places conducive to SETI. Working in the most prestigious universities and research organizations gave them access to resources not readily available to other scientists.

This was true for both the first and second waves of pioneers. The pioneers can be divided into two groups, depending upon when they started thinking seriously about searching for ETI. The time of

OZMA and the Morrison-Cocconi paper, 1959–60, forms the dividing line between the two groups. The first wave created these historic events that stimulated the second wave.

The early group presents the biggest challenge in explaining the birth of SETI. In addition to Cocconi, Morrison and Drake, this first wave also included Oliver, Calvin and Sagan. None of them named a specific person as giving them the idea of SETI, of actually searching, although they did mention people who stimulated or encouraged them.

For the first wave of pioneers, ideas about searching came out of their own work. In Drake's case, it was radio astronomy. For example, when he was doing the observations for his Ph.D. thesis he detected a strong signal that "was clearly an intelligent signal, and that just rang a bell. Could it really be an intelligent signal from the Pleiades? It was in the right frequency. It was very exciting."

For Calvin, the idea emerged from his research in chemical evolution. For Morrison and Cocconi, it was gamma wave theory; for Sagan, the origins of life; and for Oliver, experiments with radar.

They were stimulated by the most advanced technology in existence at that time, available to the very few scientists fortunate enough to be at the world's leading research facilities. Large reductions in cost, which might later make it possible for amateurs to engage in "backyard SETI," were still some years in the future.

Morrison and Cocconi had a linear accelerator in their basement, capable of producing gamma rays. The cyclotron in Calvin's basement provided the energy for his experiments. Drake had the radio telescope at Green Bank. Bracewell was designing and building radio telescopes at Stanford. Oliver had worked with radar at Bell Labs and later with other state-of-the-art equipment at Hewlett Packard. Sagan had worked in laboratories and other research facilities, ranging from astronomy under Kuiper to zoology under Nobel laureate Muller.

For the pioneers, technology wasn't something in a distant lab. Being at top scientific institutions, the technology was right there, perhaps in the same building.

Of course, scientists employed elsewhere could have used it, too, but they were likely to meet some obstacles. At the very least, it was less convenient, and more difficult to squeeze in a little time between formally scheduled uses. Not being employees of the organization that owned the equipment, they would probably have to submit a formal proposal, requiring them to plan weeks, months or years ahead, describing a specific use conventional enough to be approved by the funding agency, a review panel of professional peers, and the director of the research facility.

Such requirements discourage thinking about novel uses of expensive equipment, but these obstacles did not apply to the fortunate few—including the SETI pioneers—who worked nearby. For them it was simply a part of the environment, a potential resource available to them because they were part of the scientific elite.

Summary

The pioneers became involved in SETI through an accumulation of advantages, each in itself hardly perceptible but, in combination, sufficient to launch them in this new direction. Like much other human behavior, the process was subtle and complex. No single factor explains it all.

The pioneers had a solid scientific background, expertise in a particular area of technology or theory, and a high level of performance enabling them to win a place in a leading research institution. This placement, in turn, gave them access to ideas or equipment, which they utilized in their SETI work.

Conspicuously absent were reports of encouragement from family, friends or teachers to become scientists, let alone to seek extraterrestrial intelligence. Conspicuously present was their birth order as eldest sons or only children. Was their interest in ETI somehow related to their position in the family?

Being first-born sons or only children, they probably had more interaction with adults than their siblings did. This interaction stimulated their intellectual and emotional growth. While their brothers and sisters were still playing with other children, these pioneers were being exposed to the more complex issues characteristic of the adult world. They developed the mental capacity, curiosity and self-confidence to tackle challenging problems, of which SETI is a supreme example.

SETI began in essentially the same locations as American centers of high technology: Silicon Valley and the Boston-New York area. It was another manifestation of the same basic forces that produced transistors, microcomputers, lasers and other electronic marvels. These locations attracted and supported a critical mass of talented, trained scientists and engineers who were at the leading edge of several fields.

There is no indication that any of these scientists, when they first became involved in SETI, intended to form a group, or that they foresaw that a collective effort would someday emerge. Nor was their own entry into SETI a deliberate career decision, planned years in advance, even though some, like Drake, had wondered since childhood about the existence of extraterrestrials.

Instead, it is more accurate to say that they were drawn into SETI, carried along on the currents of their more traditional scientific interests. Searching for alien civilization was a logical step in the work in which they were already engaged. Chemistry, biology, astronomy, physics and engineering were reaching the point at which the question of ETI was materializing on the horizon. Because of a complex accumulation of factors, these pioneers became the first scientists to embark upon this fascinating quest.

Epilogue: An Interview

Paul Horowitz

PHYSICIST

Born December 28, 1942

Perspective on the first generation of SETI pioneers is provided by a leading member of the second generation. Harvard physicist Paul Horowitz conducted his first search for extraterrestrial signals in 1977 at Arecibo. In the early 1980s he developed the portable equipment known as "Suitcase SETI." After a brief search at Arecibo, the equipment was installed permanently at the Harvard radio telescope, where it was renamed Project Sentinel. In 1985 it was expanded to 8.4 million channels as Project Meta, engaged in a continuous all-sky survey. Not content to be an imaginative designer, Horowitz personally soldered over 100,000 joints.

Born in New York City, he grew up in Summit, New Jersey. He resembles his SETI predecessors in lifestyle, educational attainment and professional eminence. His three degrees are all from Harvard, in physics: an A.B. summa cum laude in 1965, followed by an A.M., and in 1970 the Ph.D. He then joined the Harvard faculty, and four years later was a full professor.

His interests range from the very small to the very large: from tunneling microscopy to searching for new pulsars. He is the coauthor of *The Art of Electronics* (1980) and has written numerous articles, book chapters and reports on diverse topics in experimental physics and astrophysics, electronics and national security.

Photograph by Mike Blake.

While generally following the pattern of early SETI scientists, Horowitz differs significantly from them with respect to family influences during his formative years. He is not a first-born son, and his family strongly affected his choice of science as a career. In fact, his older brother taught him electronics, so essential to SETI.

Interviewed October 1988 at Cambridge, Massachusetts, by telephone

Let's get some information about your family background.

My father was a businessman. He owned a textile mill. My mother is a marriage counselor. She got a degree in that kind of work. They always encouraged me to do science and to become an academic, if that's what I wanted to do, and do it well. It's good to have encouragement of that sort because it makes you not feel guilty about doing what you really want to do, particularly if it's something you *should* be doing. I think this is the right stuff for me to be doing.

Do you have any brothers or sisters?

Yes, I've got one of each. I'm the squirt. My brother is six years older than I. I got a lot from him. He went to MIT. He's an electrical engineer. He taught me everything I knew for a long time about electronics, and got me to really love electronics. Being able to do electronics is what has made most of these projects possible. SETI is hopeless without electronics.

Were there any other influences in your family that might have oriented you toward looking for ETI?

I don't think anything specific. Certainly they didn't train me in the prevalence of civilizations. If we discussed it, and I can't remember any specific instance but it probably did happen—I'm sure they were of the sensible point of view that there is probably other life up there, but without Drake's Equation to prove it. "Look at all those stars up there. Can you possibly believe we're the only ones?" is what my mother and father would have said to me.

What percent of your time do you spend on SETI?

It depends on the year. When I was building the META Project, it was close to 100 percent. It was an enormous job. Half a million solder joints done by hand, as well as the design for the thing. Right now it's down to a fairly small percent; the experiment runs itself rather nicely. So it's not more than 10 percent of my time now. However, we are planning this year to build a second system to go to Argentina to do the southern sky, which is building up my effort again.

We don't want to be too static, just doing the same old experiment, so we'll expect to see every few years some new kind of gadget, or some new strategy which will occupy some significant amount of time to get it launched. You let it run for a year or two, and see what you get. You don't want to be dogmatic, to get stuck in a rut. You want to use the results of new scientific findings such as the findings about the planets, new findings about optimum strategies for signaling. So you can see SETI as a rather dynamic activity, not just building a receiver and letting it run. So the time scales up and down, depending on what new ideas we have, what we're building.

Why is SETI important?

There are a lot of things we don't know about the universe, for which we have only one example, such as the form that life can take, or the way that living forms evolve from simple, single-celled organisms to the complexity we see today. If there were other examples of life in the universe it would certainly give us needed points as to which things we see here are general and which are peculiar. So just from a scientific point of view, it would be invaluable to have some other examples of life.

That probably is too low-key a way to put it, because, if there was other life in the universe, it would have a big impact on the way people on earth think. We're not aware of other life out there; we tend to go our way whether there is or whether there isn't. We just do our thing.

There either is or there isn't other life in the universe, and either way it would have a profound effect. If we know there is other life, and particularly if we know that there is other highly evolved life, it would mean that it is possible to have a technological civilization survive over significant periods of time.

I don't think it would have an immediate and direct effect on the way we do business on earth in terms of, say, peace overtures to be generated in international bodies and so on. It's something that would have to soak into our consciousness over a period of time, but I think it would have a profound effect in the long run.

Perhaps even more profound would be the implication, if it turns out after much searching, that life is very scarce or that we are the only advanced civilization in the galaxy or in the universe. It would mean that life is something very precious, very rare. We better take care of it—it just doesn't occur very often.

From a scientific point of view, we would want to understand why life had not formed elsewhere. I would be very uncomfortable with that, because there have been enough hints in our scientific knowledge at this point that there ought to be a lot of life: hints of the widespread availability of the chemical constituents of life in meteoric bodies (carbonaceous chondrites); in interstellar space in the form of molecules; in experiments like those of Urey and Miller that showed that it is rather easy to form these essential biological building blocks.

If we found there wasn't other life, we'd have to understand that from a scientific point of view. If there are materials out there, why don't they form life? What are the conditions under which life does occur?

The life scientists and chemists have a real puzzle there, something to keep them busy. For the rest of us, it just would be an astounding fact that we have 10^{21} stars in the universe, and life formed only around one of them. I personally find that very hard to swallow; I just don't think it is that way. But if it is, a lot of people like me would have a lot of thinking to do, to try to understand how that could be. What is so special about Earth and the Sun that caused life to happen here and not elsewhere?

For the philosophers among us, serious thoughts would be raised as to what you do, if you are the sole steward of life in the universe, to make sure that it doesn't go away; and that would lead to a pretty serious examination of the way we do business on earth, and what our chances are of maintaining it over some significant amount of time. There are a lot of reasons to be pessimistic about that if you just look at the way things are happening. I guess there are other reasons to be optimistic as well, but perhaps we can play more fast and free with life if we don't think we are the sole example.

What about the other case? What would be the implications if we find there is life out there?

There are all kinds of implications. First of all, there's the timing of the way it happens. Somebody announces, or the word leaks out—I think it's more likely to leak out, because we'd have to go through a process of confirmation involving several radio telescope observatories. The word's gotten out, let's say, that from a star a signal has been detected that is clearly of unnatural, artificial origin.

It hits the front page. It lasts longer than things usually do, perhaps a whole week. Astounding. All kinds of people scrambling for position. The philosophies, the religions of the world, the financial institutions, the "I told you so's"—there's going to be pandemonium for a while. The experts on the subject will be on TV and there will be articles on the front of every magazine in the world. You'll have to look hard to find an article on another subject. The world will stand still for a few days, just thinking about this event.

After a while there will be other things that may displace it. Sooner or later, it might take a few weeks, it will drift to the back pages. But the fact won't go away. People will start to make jokes about it. It will get into the language of science and diplomacy and everything else. Little phrases will slip in, becoming part of our consciousness, to affect everything we do. It might take a whole generation before the idea is really firmly entrenched that there is other life out there that is surely more advanced.

By the way, that is almost an axiom in this business: that if you have a civilization you can communicate with, it will be more advanced. I can demonstrate that fact if you don't already know it, but let's assume it's true, that we've got some more advanced creatures out there. They know how to do things better than we do in every way. They're like the Japanese, only even better.

There's all kinds of interesting questions we'd like to ask them. We'd like to understand how they do what they do, what the questions are they're interested in. What are the things they do that *aren't* science? What is their music like, what is their art like, if they have such things? Maybe those categories don't exist for them, just because of their nature.

Anyway, this leads to the question of dialogs and discourses. If the spacing from our civilization to theirs is close, say ten, twenty or thirty light years, we can talk about dialogs, each dialog taking a full human lifetime. But never mind; our ancestors asked questions and we find the answers, and we get to ask questions for the next generation, and so on.

If it's a great distance, a thousand years, we'd better resign ourselves to exchanging monologues, long monologues in which we grow up as we are asking the questions, and by the time the answers come, two thousand years later, they might seem rather foolish to us, but never mind, they've been sending us a rather steady stream of stuff in the meantime that we've been pecking away at, trying to understand.

Some of it we'll probably understand with a fairly small amount of effort because it's meant to be simple and clean, and it is objective kinds of realities or propositions. Some of it probably will be very difficult to understand; it's culturally relative, it's political—maybe a lot of this material never gets picked apart at all.

We have enough trouble on earth just talking with civilizations made out of the same stuff, living on the same world but existing just a few hundreds of years apart, or a few continents apart, so we can count on some extraterrestrial material never making sense to us. What would Chaucer have thought if he was faced with particle accelerators and pocket calculators, although he spoke nominally the same language as we do?

The other impact of finding a signal, and therefore knowing there is life up there, is that life scientists have a lot to learn about how life starts, and physical scientists have a lot to learn about the other planets and the possible precursors to life on them, and so on. For these scientific practitioners there's going to be a lot of interesting stuff coming down in the form of signals or discourse, and it will change forever the way we do science on earth, particularly when you consider that there's a civilization up there that probably has asked all the questions we are interested in now, and answered them thousands of years ago.

It's like having the answers in the back of the textbook, and wondering whether you should take all the steps through the book or whether you should take a peek. I don't know how people will cope with that. I don't think we'll stop asking questions or trying to answer them ourselves, particularly if the word from the extraterrestrial is awkward or involves long time delays, but it is sure to have some profound effect on science.

I'm no more qualified to talk about this than any man in the street, but all of us in this business have thought about this a bit, and have these random thoughts.

Of all the possible topics you might have examined, why did you focus on SETI?

That's pretty easy to answer. If you take a global view, the Earth is about four and a half billion years old. Life has existed on Earth about three and a half billion years. It spent most of that time evolving, and only in the last eye blink of that time have we talked about humans and recorded history and science and technology. We've emerged, but a generation or two behind us you could really not talk about even looking for signals from space in any realistic way.

So here we are. We've just reached the moment when we can talk seriously about contact. We're really the first generation on Earth that could establish contact. There's only a few generations back that people had enough information about astronomy to even make realistic guesses as to whether there might be other life.

The technology is not that difficult, once you know what you're doing, yet there are amazingly few people doing this. There are now two, maybe three searches going on in the whole world. In fact, until the last few months, there was only one.

People mount a little experiment; they look a little bit and they give up. They're terribly underfunded. The total expenditure on actual searches is a few hundred thousand dollars, probably, over the whole history of the search, going back to 1960. Maybe half a million, if you really stretch things. That's not terribly much effort to devote toward something that could yield an incredible discovery. Contact with another civilization that spent its three or four billion years evolving would be a bridge across four billion years of evolution to a completely separate life start. It would be the end of the Earth's cultural isolation in a very deep sense.

It would be, without exaggeration, the greatest event in the history of mankind. I can think of nothing, no other single event, that could match it in terms of broad implications, in terms of an astounding discovery, or anything else.

So here it is. We have this chance to do this incredible bridge across this period of time to a separate life start, to begin learning incredible things that we have no way of knowing right now, to answer this age-old question of whether we are alone, and there's only one or two people in the whole world doing it. I know how to do this kind of work, and I enjoy doing it, and I'd have to be damned crazy not to. That's the answer.

Who are the other two searchers? I assume the Ohio group is one of them?

I get the feeling that Ohio isn't quite on the air again. They've been off for a couple of years while they've sorted things out there and changed their system. Bob Dixon and his colleagues in Ohio is the main one, and they have been on for a long time. They started back in the early seventies.

The other group that I believe is running right now is the parasitic or serendipitous experiment at the Green Bank radio telescope being run by the folks from Berkeley: Dan Wertheimer and Stu Bowyer. I don't know whether that's on the air right now, but it's been on enough the last few years that I count it as a running experiment.

Apart from that there's just nothing happening that I know of. There are little fits and starts of experiments that get done here and there, but they tend to be short efforts, abandoned after some period of time; the people doing it go off and do something else.

Is Troitskii doing anything these days?

I don't think so. We've heard about some plans the Soviets have in mind to do some all-sky searches but it sounded as if they're not doing it right now. I get the feeling that there's not a lot going on now.

Let's talk about the roles of the first pioneers.

Who knows what we'd be doing without Cocconi and Morrison, for instance. They really came out with a pretty amazing paper for 1959. I gather from talking to Phil that they hadn't really thought they were going to look into *radio* communication. They were playing around with gamma rays and particle accelerators, and they got to thinking, "Could you communicate with another star with gamma rays? They are, after all, high energy electromagnetic radiation, with the speed of light."

They did some calculations and decided that it would take a lot of gamma rays, but there are better ways. And as they looked into the problem of what is the best way, they wound up with radio waves. That's nice, because radio waves are a heck of a lot easier to make than gamma rays, and we now know in retrospect that radio waves may well be the optimum way, or if not the optimum, they're not far from it, of all the methods we can understand right now. We owe them a lot for having figured that all out, and having figured it out thirty years ago, and giving us all a jump on the thing.

Of course, if Cocconi and Morrison hadn't done it, Frank Drake had already more or less figured it out and was quietly doing his experiment, not even knowing of the Cocconi and Morrison paper, so perhaps the world was ready in 1960 to start searching for radio signals from extraterrestrials.

Actually, if you look at history you'll discover that people had tried earlier. Marconi had tried, and Tesla, in ways that had no realistic chance of hearing signals. They were done at frequencies that were below the plasma frequency of the ionosphere, so there was no chance of detecting signals, but they didn't know that. So the idea of listening for radio signals from extraterrestrials was not new. What was new in 1960 was a method that had a chance of succeeding, as opposed to a method that was fanciful.

Sagan certainly did a lot to put the subject on a firm footing at a time when it really needed to be, with his book with Shklovskii, examining many of the aspects we now think of as the Drake Equation: the issues of the origin of life, the evolution of life, what it takes to have a habitable planet, calculations of signal margins, and so on, following up on the Cocconi-Morrison type calculation, and I think that was a pretty wonderful book for its time, 1966.

I remember that date because I had a graduate student roommate who was taking Sagan's course, and Sagan was using the manuscript as notes for the course. Pretty good stuff. Until then a lot of people had consigned the subject to science fiction. They said it's all speculation and we really don't know much about it. We don't know how life started, and it's all just a big guessing game.

What Shklovskii and Sagan showed pretty well in that book was that there are aspects of the subject that are extremely speculative, but there are other aspects that are anything but speculative. For instance, it's possible to calculate with exquisite accuracy how much signal will be detected over some distance—pick a distance, pick a kind of radio telescope, pick a wavelength, and you can do the communication calculations. We know how to do those things. We've proved it over and over with our space probes that we know how to communicate over long distances, and they showed that this could all be done quite cleanly.

There's still lots of speculation here, mostly because science doesn't allow us to fill in the gaps. We don't know enough yet about the origins of life and the distribution of planets. We're working hard on that, of course, and we've heard interesting reports on the prevalence of planetary systems, of brown dwarfs, of heavy planets, and all the things we'd like to know, but it's going to take a while to figure this all out.

Science is progressing nicely, but we still don't know nearly enough to answer the question: "How much life is out there?" And certain things we'll never know, unless we contact other life and they tell us. For instance, the average lifetime of civilizations in their technological state.

Anyway, to get back to the question, I think Sagan and Shklovskii really did a service in pointing out to many people that the subject wasn't humbug, that you could calculate plenty of things, that you could guess about other things, you could make plausibility arguments, you could make all the arguments you need to suggest that the search for extraterrestrial intelligence, with radio telescopes in particular, was an entirely legitimate thing to do, with a plausible chance of success, and with no chance if you did not do it.

I think it inspired a lot of people. I was certainly inspired by Sagan's course, even though I didn't take it. I took it vicariously through my roommate. That was one of several inspirations.

Barney I think of mainly in terms of the Cyclops work in '71. One of those summers at NASA Ames they carried on a study to see what it would take to engineer a real system for detecting extraterrestrial signals, and Barney led that study. Billingham was part of that effort in a major way, too, but I understand that it was really Barney who was responsible from the scientific point of view.

The Cyclops report is really an amazing thing. It addresses not just the system for detecting signals but all kinds of motivational issues: Why should we search for signals? What are the dangers that might happen? and so on. The first forty to fifty pages of the Cyclops report are mandatory reading for anyone in this field. They demonstrated that it was possible, albeit at a cost of ten to twenty billion dollars— really an Apollo Project scale of money, but still less than SDI, however—to engineer a system that could detect leakage radiation from nearby civilizations. It wasn't even necessary for these folks to be transmitting to us purposely.

Cyclops used techniques that in 1971 seemed to be the right way to do it: holographic signal processing schemes; you probably wouldn't use that anymore, but never mind. What they designed in that report would do the job and would have a reasonable chance of detecting leakage radiation from radio civilizations up to fifty to a hundred light years away, and purposely transmitting civilizations out to the edge of the galaxy.

So it was "put up or shut up." They said, "Here's the system. All you have to do is build this thing, and we can do SETI today." In retrospect their proposal was too grand. The world wasn't ready for a twenty-billion-dollar project, and still isn't, to detect extraterrestrials. But it was nice of them to demonstrate that it could be done, and I think of Oliver as the primary architect of that particular demonstration.

Ronald Bracewell?

I associate his name with *The Galactic Club* and with some interesting calculations to demonstrate that there's a critical density of civilizations, such that if you have more than that, they're close enough together so that they can have round-trip travel of radio signals, and if there's fewer than that it probably means their lifetimes are short, and also the distances apart are great so that there's no round trips possible. So the whole thing leads to the concept that there's either almost no communication at all in the galaxy or there's so much of it that it all could form one big club.

Bracewell did the math for that and made us all realize that if we communicate with some other civilization it is not going to be the first time it's happened, nor is it going to be the only place it's happening in the galaxy. So we might as well be ready for a whole network in the galaxy that's already out there talking to itself, and just waiting for us to merge into it. I'm sure Ron's been responsible for a lot of other calculations, but they haven't had any particular effect on the way I've been thinking about the problem.

How about Shklovskii, apart from his book with Sagan?

I'm not sure who did what. They used typographical techniques to try to keep straight who did what in that book, but it all blends together in my mind. I really think of them in the same thought, so I have nothing separate for Shklovskii.

Nikolai Kardashev?

Kardashev did Russian SETI of one sort or another. I think of him in terms of some rather bold thinking about what an extraterrestrial civilization might really be able to do if we get out of our mind-set that it has to be like us, that it lives on a small planet and is able to burn fossil fuels and perhaps burn nuclear fuels and do things on the scale we do on Earth.

Kardashev pointed out that you could have grander civilizations, that are able to capture most of the energy coming from their sun or perhaps the energy coming from many suns, and he gave these things Type One, Two and Three designations. At first it sounds like somebody rambling about science fiction that shouldn't be taken seriously, but I think the right way to look at it is that we've progressed pretty far in the last couple of hundred years, from riding around on horses to launching rockets to the moon and the outer solar system, having pretty good control over lots of technical processes—microelectronics, nuclear fission and fusion, and that kind of stuff.

It's probably way too timid to think that we're going to stay the way we are now, in the search for energy, particularly if we can't get what we need out of fossil fuels, and it is clear we cannot, on the scale of these Kardashev civilizations. We probably will start fiddling with astroengineering and large-scale projects to accumulate large amounts of free energy.

So we probably should give advanced civilizations credit for being able to do those things. And if we do, it changes the kinds of things we can expect to learn from them, but also it changes the kinds of signals we might expect to hear.

I don't think it's outrageous to expect an advanced civilization to have a beacon somewhere in the center of the galaxy that could be heard anywhere in the galaxy, transmitting omnidirectionally for periods of hundreds of millions of years, for the sole purpose of attracting new members into the 'galactic club', to us Bracewell's term. Kardashev's role was in pointing out to us that that is entirely plausible and really not in the realm of science fiction at all.

I tend to be less good at going to libraries and reading about what these folks have said, and more interested in getting into the laboratory and building experiments, and as a result I'm probably rather uneducated about the real contributions of some of these folks. I've been too busy soldering joints to do my homework.

How about Billingham?

JB had a big effect on me because I spent a year at the Ames Research Center. He's head of life sciences there, or at least close enough to head that it was indistinguishable to me, and therefore was my host. I was aware that he oversaw not only the technical SETI programs but also the life science areas that border on SETI; for instance, experiments and research directed at the way life may have started, the clays that may have catalyzed early life, and so on.

I don't have enough information about what Billingham did in the Cyclops project to be able to say anything intelligent about that, but he is an awfully nice guy, he's got the long-term point of view, and he keeps NASA on track. And for many years, when there were enemies of SETI about, who wanted to cut it off because they claimed it's a waste of money and time, Billingham doggedly kept making reasonable arguments and rebutting the arguments that were anti-SETI. Therefore he's responsible in large part for the fact that NASA is where it is today in SETI. Of course the technical program is headed by Oliver.

And Charles Seeger?

I knew Charley from the Ames days. I'm not sure what he did ten or fifteen years ago. Charley is a wonderful guy with an unconventional point of view; very knowledgeable about SETI, a real curmudgeon. I love him. I'm not sure what his deep role is in SETI because I don't know enough about it.

John Kraus?

Kraus is the designer of the Kraus telescope, the Ohio State telescope. If I remember right, Kraus is really the one who got that thing started on SETI, not Dixon, since that was 1973, when Dixon was a babe in arms. I've met Kraus a couple of times but I haven't had enough discussion with him to get a feeling about what he believes today or the major things that he accomplished.

I guess getting Ohio State into the continuing SETI business is what he did, and given that that was the *only* telescope for a decade doing sustained SETI, he's got to be counted as one of the obvious pioneers in real observational SETI. Drake went for two months with his Project Ozma, but that was it.

Freeman Dyson?

Oh, Freeman, yes, he's wonderful. Freeman is an unconventional thinker in every way, and a brilliant man. He's best known in SETI circles for the two short papers he wrote on the so-called "Dyson spheres." He didn't call them that, but they are basically a variation on a theme of Kardashev, that one could have a civilization that would construct a shell around its star in order to use much more energy than just hits the planet, and he showed a nice example of what Kardashev was illustrating with his Type II civilization, namely, that you could imagine a sphere of orbiting particles to collect the energy.

Although there are technical problems in building such a thing, you really could have it, and its signature would be a star that is very bright in the far infrared, where this warm mass would radiate, but essentially invisible in the visible. It wouldn't show as a normal star. That's clever thinking. Some people look for these Dyson spheres. No one has found one but maybe we will.

The other idea was that a civilization could achieve a reasonable supply of energy, at least for a finite time, if it happened to live around a double star and get the right kind of acrobatics, with particles and masses launched into its double-star system, to extract energy from the double star.

The signature of such a civilization would be that, after a period of such greedy energy acquisition had subsided, you'd be left with two stars just about touching and you wouldn't dare extract anything more from them. We should be looking, therefore, for very close double stars. They may be what's left over when a civilization like that has gotten all the energy from its system.

Clever thinking. I don't know what else Freeman may have done but I'm sure he's part of the gang of eight or ten that thought about these problems back in the sixties and seventies. I talked to him somewhat recently about SETI. He's still a daring and clever thinker on that subject.

Jill Tarter?

Jill Tarter has been plugging away at SETI for at least a decade, rebutting bad ideas, going around actually *doing* SETI on the world's telescopes instead of *talking* about it, basically calling everyone's bluff and keeping the subject from becoming too theoretical. The amount of energy she's put into this thing is incredible, and she deserves a major award for almost singlehandedly keeping SETI very much alive in bureaucratic circles of NASA and elsewhere.

Let's turn now to you. How did you get interested in SETI?

I suppose we all go back to childhood for thinking about the existence of extraterrestrials. I'm sure I gave it some thought then, as did everyone, but at that point it wasn't clear to me that I could do anything about it, except to think about it in the way you think about other hard questions, like what is beyond the end of the universe, and what's beyond the end of time?

In college was the first time I really remember thinking about this as a subject you could actually *do* something about and not just think about. It occurred when I was an undergraduate at Harvard. I had a professor named Ed Purcell, famous among other things for discovering the 21-cm line in space. He wrote a very nice piece called "Radio Astronomy and Communication Through Space."

He pointed out that communication through space by radio astronomy is extremely efficient, for extremely small amounts of energy, and you can get all the information you probably want, whereas space travel to distant star systems deserves to be back where it came from, on cereal boxes. That got me thinking about the problems of communicating through space.

The next thing that got me going, as a graduate student, was Sagan's course, taken by my roommate. A little later, Frank Drake gave a series of talks in the Loeb Lecture series. Frank is a wonderful speaker, and attacked all aspects of the problem. He sort of did a Shklovskii and Sagan book for us in four or five lectures, and really got me quite excited about the subject. So, when I had a sabbatical a few years after I got my degree, I asked Frank if I could "look for life in Puerto Rico," as I put it.

He thought that was a great idea, since he was at the time head of the Arecibo telescope, and had long been interested in signals from extraterrestrials, so we talked about what kind of experiment might be interesting and different from what had been done. I wound up going down there in 1977 for three or four months and doing a search for ultra narrowband carriers, which had not been done before, and is the kind of thing I've been doing since.

Then I went back to ordinary physics for a while. In 1981 the NASA Ames center had a fellowship available. Jill Tarter, aware of the SETI I had done at the Arecibo telescope, called me and asked if I'd like to apply for one of these fellowships, and I went there in 1981–82.

The original idea was to work on the NASA multichannel analyzer, but when I got there it was behind schedule by enough so that it was really still in the design and planning stage. It would seem frustrating to put in a year on that and leave with it still at the planning stage, so instead I cooked up an idea for a little independent SETI machine that could be carried around to various radio telescopes. This was the thing that got the name "Suitcase SETI." I built it at Stanford with several colleagues from Ames, notably Ivan Linscott. It was a 128,000 channel portable analyzer, fitted into three boxes that really were suitcase size. We took it down to Arecibo and did a search similar to the earlier effort in 1977.

So that was a very nice proof that you could build hardware to do rather powerful SETI if you wanted to. When we were done there, that apparatus was left over and we set it up at Harvard and that became "Sentinel," and you know the rest. Sentinel became "META," and that's where we stand now, with the 8.4-million-channel capability.

Is there anything else we ought to cover about the influence of the older pioneers upon you?

Yes, I think so. The thing about these older pioneers is they know what they're talking about and they're clever, and they come up with new ideas that blow you away. Phil Morrison is the one who astounds me more often than anyone else with his insights, but both Drake and Morrison have made suggestions that have profoundly affected what I've done.

Drake pointed out what a signal would look like when it has come across the galaxy, what a pure carrier wave looks like a thousand light years later, spread out to the order of a tenth of a hertz. It was that calculation he did with his student Helou that motivated the SETI I did at Arecibo in 1977, the ultra-narrowband SETI, matched to the physical properties of the galaxy, which Drake pointed out to the world. All our later searches, including META now, are still based on the Drake-Helou bandwidth.

Phil Morrison made a comparable contribution to deciding what we do in these experiments. He recommended that not only should you look for a special frequency, such as the hydrogen line that he suggested with Cocconi in 1959, but that one should look also in a guessable reference frame to solve the problem of rather large motions of astronomical objects. Things really zip around out there, and lead to very large Doppler shifts that are difficult to handle otherwise.

Phil wrote me a letter back around the early eighties pointing out that there are preferred reference frames, and probably the most preferred of all is the remnant of the Big Bang radiation with which the universe started. There is a preferred frame in which that radiation looks the same in all directions. Extraterrestrials in our galaxy or nearby would surely be aware of that, and it would seem an obvious procedure to transmit a signal at a special frequency (such as the Cocconi-Morrison hydrogen frequency), corrected to that reference frame which we all can observe, in order to solve once and for all the problems of relative motions between source and observer.

Our META project now is based on that idea as well as Frank Drake's calculation of the spreading, and Cocconi and Morrison's original suggestion of the hydrogen hyperfine frequency.

Appendix A. Research Methods

The process of selecting the specific individuals to interview for this study differed from that of many research projects because of the unusual nature of this group. The number of SETI pioneers was small enough (sixteen) so that all of them could be interviewed. It was, therefore, unnecessary to use standard sampling procedures; I was not working with a sample, but rather with the entire "population," to use a statistical term.

This not only simplified the process of obtaining the data, it also permitted greater confidence in interpreting the results. There is no question about how well these scientists represent the entire group of SETI pioneers; they *are* the group. Since we are not generalizing from a sample, we do not have to get involved with "confidence levels" or "tests of significance" that encumber much research today.

Several of the pioneers were obvious, and they nominated others. Giuseppe Cocconi and Philip Morrison wrote the landmark 1959 paper which started the scientific community thinking about SETI. At the same time, Drake was preparing the first actual search with a radio telescope.

Drake, whom I interviewed first, recommended Morrison and three others: Bernard Oliver, director of the Cyclops project; John Billingham, Oliver's colleague who worked within NASA to establish the SETI program, and Jill Tarter, the first scientist to pursue SETI full-time, or nearly so. They in turn called my attention to team member Charles Seeger, a veteran radio astronomer doing important groundwork.

Other obvious candidates were two who had attended the first scientific meeting devoted to the subject, at Green Bank radio observatory in 1961: Melvin Calvin, whose work on chemical evolution won a Nobel Prize; and Carl Sagan, a researcher and publicist for SETI.

The NASA group suggested that Ronald Bracewell, author of *The Galactic Club*, could provide a somewhat different perspective, and he, for similar reasons, recommended Princeton's Freeman Dyson. John Kraus was conducting a SETI search at Ohio State, and published the first journal devoted specifically to SETI. A Stanford physicist told me about Kunitomo Sakurai, who was working to establish SETI in Japan.

Iosef Shklovskii and his pupil Nikolai Kardashev had pioneered the study of SETI in the Soviet Union: Shklovskii and Sagan coauthored the first serious book on the subject, and Kardashev conducted the first Russian search for extraterrestrial signals. Vasevolod Troitskii, vice-director of the Gorkii Radio Physics Institute, was conducting a major search.

This list includes all of the most active or prominent scientists in SETI during its first fifteen years. Others have become involved more recently, but the pioneers themselves confirmed that all of the leading figures of the early days are included in this book, with the possible exception of Otto Struve, who died in 1963. Although not actually concerned with searching for extraterrestrial intelligence itself, he had contributed to the knowledge of stellar evolution, suggested that many stars may have planetary systems and, as director of the National Radio Astronomy Observatory, encouraged Drake to conduct Project Ozma.

I wrote each person a one-page letter introducing myself, explaining the project, requesting their help, and stating that I would telephone to make an appointment for an interview. They were cooperative, both in scheduling interviews at times convenient for me and in answering my questions.

I prepared a list of questions and set out to interview each scientist, tape recording the responses. Fortunately, six scientists were near San Francisco and two more were in the Boston area. I talked to them in person, as I also did a ninth, on a trip to Japan. I interviewed four more by telephone: three in the United States and one in Switzerland. The final three, Russians, were interviewed in Estonia by Bernard Oliver, whom I had interviewed earlier.

Arranging these interviews with the American scientists was simple enough, considering how busy they are. Arranging interviews in other countries was more difficult, due principally to problems with mail and telephone service. For example, I arrived in Yokohama before my letter to Kunitomo Sakurai, so he had no idea who I was or what I wanted. Nevertheless, he graciously consented to an interview the next day. The three Soviet scientists all knew Bernard Oliver and responded cordially to the questions when he talked to them in person. Afterwards, however, clarifying details through the mail was a long process; Vasevolod Troitskii's correspondence eventually arrived, but I never did receive more information from Iosef Shklovskii.

Negative answers or non-reponses were usually excluded, but in a few instances they have been included in order to highlight a specific aspect of the person's thought or perspective.

This procedure was also followed for other items, particularly about the person's background. For example, all of the scientists were married and all had children. Most reported no influences toward the sciences or SETI from parents, teachers or friends.

Similarly, the questions were usually asked in full, but often abbreviated for presentation in this book. In the case of the three Russian interviews, there was some variation in wording the questions, but two major benefits far outweighed any hypothetical disadvantages of deviation from standardized wording: I could not otherwise have obtained these interviews, not only because of economic limitations on travel, but also because it was a dark period in the Cold War, these scientists were not often allowed to visit the United States, and extended telephone interviews were not only expensive but difficult to arrange. The second major benefit of the Russian interviews was capturing conversations between SETI pioneers.

After the main interviews, some additional material was obtained and added, chiefly for clarification. Nothing was added that would have been out of the time frame of the initial interviews. For example, Morrison's suggestion about preparing for a supernova was made in 1981, six years before the spectacular 1987 occurrence.

Because I was financing this entirely on my own, I could not afford to travel to the Soviet Union or to buy release time from my regular duties to finish this project more promptly. A consequence was the death of Iosef Shklovskii before I could obtain all the desired information about him.

My inability to obtain research funds was due, at least in part, to the uniqueness of this project. Although it involves natural science, history, social science and philosophy, it does not rest firmly in any one discipline, but instead falls into the cracks between them. While this research was generally acknowledged to be interesting and worthwhile, it lost out to higher priority activities. Each discipline, each agency preferred to grant support to other, more conventional projects, closer to its established mainstream.

Will this process be repeated for SETI itself? Will more immediate concerns delay funding for SETI until the rising level of radio noise from our strident earthly activities drowns out possible signals from other worlds? Eventually, telescopes in space or on the moon could continue the search. In the meantime decades will have passed, and we will be deprived of the knowledge of whether we are alone.

Appendix B. Interview Questionnaire

Introduction

In this interview I'd like you to tell me about your life. What were the events, the people, the situations that led to your present ideas about SETI?

I'm particularly interested in the things that are *not* in published articles, books or other standard scientific sources.

In future centuries, historians may look back to this time, when scientists first began to think seriously about extraterrestrial intelligence, and they will want to know about you as a person, a human being, and they will want to know how you came to your present ideas about SETI.

Thus we have two related topics to talk about today: 1) You as a person—what were your hopes, your fears, your joys and frustrations? What recreations did you like? 2) The development of your ideas—how did your SETI ideas evolve? What stages did they go through? What alternative paths or blind alleys did you explore?

If we have time at the end we might summarize your present ideas about SETI, but let's begin now by looking at you chronologically. Let's go back to your childhood.

1. When were you born?
2. Where were you born?
3. Where did you live during childhood and when you were growing up?
4. Was that in a city or a town or a rural area?
5. What were your childhood interests?
6. Tell me about your earliest activities in science. What incidents involving science do you remember from those early years?
7. Experiments, ideas, anecdotes?
8. What about during adolescence? What activities do you recall from those years?
9. How was your elementary schooling—exciting and stimulating, dull and boring, or just neutral?
10. Did it give you helpful preparation for a scientific career?
11. What about secondary school—was it stimulating or dull or neutral?

12. How would you assess its helpfulness to your scientific career?
13. When did you first think about the possibility that extraterrestrial intelligence might exist?
14. Who or what suggested this possibility to you?
15. Did you discuss this possibility with other people?
16. Who?
17. What was their response?
18. What about communicating with, or at least searching for, ETI? When did you first think about this possibility?
19. Who or what suggested this to you?
20. During your college years were there any ideas, books, teachers, classes, students or events which stimulated you to think about ETI and the possibility of searching for or communicating with it? If "Yes," explain.
21. After you completed your schooling and began working as a professional scientist, who or what influenced your thinking about ETI and SETI?
22. What effect, if any, has your interest in ETI had upon your career as a scientist?
23. What do your colleagues think about your interest in SETI?
24. Do they ever discuss it with you or ask you about it?
25. What about your family? What do they think about your ETI interests?
26. What about your friends? What do they think about your ETI interests?
27. What about your superiors, bosses, and so forth? What do they think about it?
28. Before going on to discuss your ideas about SETI, let's get a little more background information. What was your father's occupation?
29. How much education did he have?
30. What about your mother; what formal education did she have?
31. Did she have occupational skills or work experience other than that of the typical wife and mother of that time?
32. How would you describe them? Quiet, outgoing, reserved, serious, gregarious?
33. How did they influence your eventual choice of science as a career?
34. And what about ETI; how did they influence you in this respect?
35. Were there other people we haven't mentioned who influenced your ideas about science and ETI? Teachers, relatives and so forth?
36. What religious beliefs did your parents have?

37. How many brothers and sisters did you have?
38. Where were you in the birth order: oldest, youngest, second of four children, and so on?
39. What occupations did they enter?
40. Did any of them influence your thinking about science and ETI?
41. Who and in what way?
42. What about your friends when you were growing up; did they share your interests in science and ETI?
43. In what way?
44. What were you like during childhood and adolescence: outgoing, gregarious, with many friends; or timid, introspective with few friends; or average?
45. How many children do you have?
46. What is your present marital status: single, married, divorced, and so forth?
47. What professional interests do you have, apart from SETI?
48. What hobbies or recreations do you enjoy?
49. What has been your greatest satisfaction in life?
50. What has been your greatest sorrow in life?
51. What was the most memorable moment or event in your life?
52. Turning now to SETI, about what fraction or part of your working time do you spend on it?
53. What stages has your thinking about ETI and SETI gone through? Have your ideas about it changed much since you first began thinking about it or have they remained pretty much the same? Please elaborate.
54. What alternatives, what other possibilities about ETI have you considered?
55. What would you do differently if you were starting over again to study ETI?
56. Why is ETI important to you? Of all the thousands of topics a scientist might investigate, why did you choose ETI?
57. What has been the most difficult aspect of your ETI work?
58. How do you think most scientists today view SETI: as a legitimate activity for scientists, as legitimate but not important, as not legitimate?
59. What about public attitudes toward SETI? Do you think the public is more enthusiastic about SETI than scientists are, about the same, or less enthusiastic than scientists are?
60. Realizing that it is only a guess, when do you think contact with ETI will be made?
61. What might ETI be like; what form might it take?
62. What scientists or other people in the past do you admire?

63. Who are the outstanding people in SETI today, and during the past twenty years or so?
64. What role has each played in SETI?
65. How would you describe, characterize your own role in SETI?
66. What else should future historians know about you that would not otherwise be preserved in regular archives, journals, books, and so forth?
67. What are UFOs?
68. What should a scientist do upon receiving convincing evidence of extraterrestrial intelligence?
69. What would be the public's response to this information about the existence of ETI?
70. What would be the long-term effect upon humanity of evidence that ETI exists?
71. How well do you know the other SETI pioneers? (See list.)

Glossary

AAUP   American Association of University Professors

angular momentum   A momentum given by the product of a mass, its velocity and its distance from a reference point. Its special significance is that it is conserved, or unchanging unless forces are present.

ASEE   American Society for Engineering Education

astroengineering   Construction of very large artificial objects in space, presumably built by highly advanced extraterrestrial civilizations

AU   Astronomical unit: the mean Earth to Sun distance (1.5×10^{11} meters)

bandpass   The range of frequencies which can pass through a filter such as one in an electrical circuit

bandwidth   The range of frequencies in a signal; the difference in cycles per second, or hertz, between the low and high frequencies

Cas A   The celestial source of strongest radio emission at most wavelengths

CCD   Charge coupled device, capable of detecting very faint images

CERN   Centre Européen Recherche Nucléaire (European Center for Nuclear Research), Geneva, Switzerland

CETI   Acronym for Communication with Extraterrestrial Intelligence. Now obsolete, it implied sending messages as well as detecting them, a more complicated process than SETI: simply searching

coherent   A wave form which is smooth and continuous, without noise-like properties

commensal   Two entities existing together, but not at the expense of one another

COSPAR   Committee On Space Research, International Council of Scientific Unions

CTA 102   One of the earliest quasars discovered

CW   Continuous wave: an essentially single-frequency electromagnetic wave

Cyclops   Project Cyclops (1971). The first major study devoted to the design of powerful search systems for SETI. It proposed an array of many antennas linked together.

dielectric constant   Parameter defining the permeability of a medium by an electric field

Drake Equation $N = N_*f_pn_ef_lf_if_cL$, where N^* is the average rate of star formation in the galaxy since its origin, f_p is the fraction of stars with planets, n_e is the average number of planets at a distance from the central star suitable for life, f_l is the fraction of these on which life appears, f_i is the fraction of these which develop intelligent life, f_c is the fraction of these that develop a technological civilization, and L is their average lifetime in a signal transmitting mode that might be observed at interstellar distances

Dyson Sphere   A shell constructed around a star in order to capture the energy radiated from that star

ecliptic   The plane traced by the Sun in the course of a year, and near which most solar-system objects are found

ephemerides   Tables of the predicted positions of planets, asteroids and other celestial objects

flux   The amount of energy passing through a unit surface area per unit time

Fourier analysis   Breaking a complex signal into an ensemble of sinusoidal components

f_p In the Drake Equation, the fraction of stars possessing planets

G_2V A category of stars classified according to their spectra

gain   A power ratio: P^{out}/P^{in} describing an amplifier circuit

gaussian   Distribution of probabilities; the traditional bell curve

Gemini Project   American manned-space project of the early 1960s

GHz   A unit of frequency equal to one billion hertz

hertz   A unit of frequency equal to one cycle per second

heterotrophic   Incapable of synthesizing proteins and carbohydrates

high-resolution spectroscopy   The study of the spectrum of objects in great detail

high-spectral resolution   The ability to distinguish fine detail in spectra

hyperfine structure   The detailed rearrangement of energy levels within an atom that results when interactions between the spin and angular momentum of the electrons and the spin of the nucleus are important. The very slight difference in energy when the spin of the electron and the proton of a hydrogen atom are parallel or anti-parallel is the origin of the 21-cm wavelength radiation from neutral hydrogen gas in the universe

IAF   International Astronautical Federation

IAU   International Astronomical Union

ICSG   Interstellar Communication Study Group

IEEE   Institute of Electrical and Electronic Engineers

interferometry   Linking a pair of receiving devices, such as radio telescopes, to achieve, with limitations, improved resolving power

IRAS   Infrared Astronomy Satellite

isotropic   Uniform in all directions

jansky   Unit of strength of cosmic radio waves (10^{-26} watts per square meter per hertz)

lambda doublet   Each of two energy levels of the OH molecule responsible for the 18-cm wavelength radiation is actually double, consisting of two energy levels very close together. In this case the small energy difference in the double levels depends upon which direction the electron is orbiting. These two pairs of double levels permit radiation at four discrete radio frequencies (1612, 1665, 1667 and 1720 MHz) characteristic of OH.

LSI   Large-scale integration of solid state technology. Describes miniature computer circuits

magneto-hydrodynamics   The study of the interaction of plasma and a magnetic field

maser   A source of very intense, narrow-band, coherent microwave radiation

meson   A type of fundamental particle

meta galaxy   A large group of galaxies

micron   A very small unit of measurement: 10^{-6} meters

monochromatic   Of one wavelength or color

narrowband   A limited range of frequencies

nonlinearities   An often undesirable situation in which the output of a system is not proportional to the input, and/or in which waveforms are distorted

O Type stars   The hottest category of stars classified according to their spectra

parametric amplifier   A sensitive amplifying device which employs the properties of nonlinear elements

parasitic   Making use of something created for another purpose

parity conservation   The conservation of net "spin" of all electrons in a system

perihelion   Closest point to the Sun of a body orbiting around it

positronium line   The frequency associated with the simplest transition in bound configuration of an electron in orbit about a positron

pulsar   A pulsating radio source caused by a rapidly rotating neutron star

quantum electrodynamics   Quantifying Maxwell's equations regarding electricity and magnetism according to the principles of quantum mechanics

quasar   A starlike object believed to be the active nucleus of a very distant galaxy

radio telescope   An antenna or set of antennas, with supporting equipment, for detecting radio radiation from space *recombination line*   Radiation resulting from the addition of an electron to an ion when the electron subsequently jumps to a lower energy state

secular   A process of change which is not cyclical

SETI   Search for Extraterrestrial Intelligence

synchrotron radiation   A type of nonthermal radiation emitted by electrons moving at relativistic velocities in a magnetic field

VLBI   Very Long Baseline Interferometry: a procedure in which widely separated radio telescopes observe a single source simultaneously, tape recording the data with extreme accuracy. The data are carefully combined later to simulate the performance of an interferometer

UHF   Ultra high frequency (300–3000 MHz)

waterhole   A band of frequencies in a region of relatively low noise in the radio spectrum, considered a likely place to find signals from extraterrestrials. This band lies between the frequencies of hydrogen (1420 MHz) and the hydroxyl radical (1662 MHz). These two components of water "beckon all water-based life to search for its kind at the age-old meeting place of all species: the waterhole" (Project Cyclops).

Zeeman effect   Broadening or splitting of radiation from a substance into several wavelengths because of the presence of a magnetic field

References

Altus, William D.
"Birth Order and Its Sequelae," *Science* 151, 7 Jan. (1966) 44–49.
Billingham, John, ed.
Life in the Universe. Cambridge, Massachusetts: The MIT Press, 1981.
Bracewell, Ronald N.
The Galactic Club. San Francisco: W.H. Freeman and Company, 1975.
Clark, Roger D., and Glenn Rice.
"Family Constellations and Eminence: The Birth Order of Nobel Prize Winners," *Journal of Psychology*, 110 (1982) 281–87.
Dean, D.A.
Thesis [no title listed], State University of Iowa, 1947, p.21, quoted in Altus, p.48.
Drake, Frank D.
Intelligent Life in Space. New York: The Macmillan Co., 1967.
Goldsmith, Donald.
The Quest for Extraterrestrial Life: A Book of Readings. Mill Valley, California: University Science Books, 1980.
Humphries, Rolfe, trans.
The Way Things Are: De Rerum Natura of Titus Lucretius Carus. Bloomington, Indiana: Indiana University Press, 1968, p. 82.
Maxim, Hiram Percy.
"What Is It All About?" *Scientific American* (April 1932): 201.
McDonough, Thomas R.
The Search for Extraterrestrial Intelligence. New York: John Wiley & Sons, 1987.
Morrison, Philip, John Billingham, and John Wolfe, eds.
The Search for Extraterrestrial Intelligence. NASA Special Publication SP-419, 1977. Reprinted, with minor deletions, by Dover Publications, New York, 1979.
Oliver, Bernard M., and John Billingham, eds.
Project Cyclops: A Design Study of a System for Detecting Extraterrestrial Intelligent Life. NASA Contract Report CR-114445, 1972 (revised 1973).
Richter, Maurice N., Jr.
The Autonomy of Science. Cambridge, Massachusetts: Schenkman Publishing Co., 1980.

Ridpath, Ian.
　　The Illustrated Encyclopedia of Astronomy and Space. Revised edition.
　　New York: Thomas Y. Crowell, 1979.
Schachter, Stanley.
　　"Birth Order, Eminence and Higher Education," *American Socio-
　　logical Review* 28 (1963): 757–768.
Shklovskii, I. S. and Carl Sagan.
　　Intelligent Life in the Universe. New York: Dell Publishing Co.,
　　1968.
Sullivan, Walter.
　　We Are Not Alone. New York: The New American Library, 1966.
Tipler, Frank J.
　　"A Brief History of the Extraterrestrial Intelligence Concept,"
　　Quarterly Journal of the Royal Astronomical Society 22 (1981):
　　133–45.
Zuckerman, Harriet.
　　Scientific Elite: Nobel Laureates in the United States. New York: The
　　Free Press, 1977.

Index